Advanced GNVQ Engineering

Gwyn Davies
John Eames
Martin Froud
Gordon Hicks
Ian Spencer
Roger Timings

 LONGMAN

Addison Wesley Longman Limited
Edinburgh Gate, Harlow
Essex CM20 2JE
England
and Associated Companies throughout the world

© Addison Wesley Longman Limited 1997

The right of the contributing authors to be identified as authors of
this work has been asserted by them in accordance with the
Copyright, Design and Patents Act 1988.

All rights reserved; no part of this publication may be reproduced,
stored in any retrieval system, or transmitted in any form or by any
means, electronic, mechanical, photocopying, recording, or otherwise
without either the prior written permission of the Publishers or a
licence permitting restricted copying in the United Kingdom issued by
the Copyright Licensing Agency Ltd., 90 Tottenham Court Road,
London W1P 9HE.

First published 1997

British Library Cataloguing-in-Publication Data
A catalogue entry for this title is available from the British Library.

ISBN 0-582-29090-2

Set by 32 in Sabon $9\frac{1}{2}/11$
Printed in Great Britain by Henry Ling Ltd, at the Dorset Press,
Dorchester, Dorset.

LLYSFASI COLLEGE	
C00011927	
Morley Books	9.3.98
620	£19.99
2574	5991

Contents

Preface **vii**
Acknowledgements **viii**

Unit 1 Engineering and Commercial Functions in Business 1
   1.1 Investigate business functions which involve engineers **2**
   1.2 Investigate engineering and commercial functions in business **11**
   1.3 Calculate the cost of engineered products and engineering services **16**
   1.4 Unit test **20**

Unit 2 Engineering Systems 22
   2.1 Investigate engineering systems in terms of their inputs and outputs **23**
   2.2 Investigate the operation of engineering systems **28**
   2.3 Unit test **48**

Unit 3 Engineering Processes 50
   3.1 Select processes to make electromechanical engineered products **51**
   3.2 Make an electromechanical engineering product to specification **80**
   3.3 Perform engineering services to specification **85**
   3.4 Unit test **90**

Unit 4 Engineering Materials 92
   4.1 Characterise materials in terms of their properties **93**
   4.2 Relate materials' characteristics to processing methods **114**
   4.3 Select materials for engineered products **123**
   4.4 Unit test **129**

Unit 5 Design Development 131
   5.1 Produce design briefs for an engineered product or an engineering service **132**
   5.2 Produce and evaluate design solutions for an electromechanical engineered product and an engineering service **141**
   5.3 Use technical drawings to communicate designs for engineered products and engineering services **149**
   5.4 Unit test **161**

Unit 6 Engineering in Society and the Environment 163
   6.1 Investigate the effects of engineering on society **164**
   6.2 Investigate the effects of engineering on the working environment **179**
   6.3 Investigate the effects of engineering activities on the physical environment **195**
   6.4 Unit test **206**

Unit 7 Science for Engineering 210
- 7.1 Static systems **211**
- 7.2 Dynamic systems **216**
- 7.3 Thermal systems **221**
- 7.4 Electrical systems **226**
- 7.5 Electromagnetic systems **233**
- 7.6 Unit test **238**

Unit 8 Mathematics for Engineering 240
- 8.1 Use number and algebra to solve engineering problems **241**
- 8.2 Use trigonometry to solve engineering problems **257**
- 8.3 Use functions and graphs to model engineering situations and solve engineering problems **267**
- 8.4 Unit test **287**

Answers to numerical self-assessment tasks **289**
Answers to unit tests **291**
Index **292**

Preface

Qualifications for technician occupations associated with engineering are undergoing development by the National Council for Vocational Qualifications. This has resulted in a qualification known as the General National Vocational Qualification: Engineering. The General National Vocational Qualification (GNVQ) replaces the former BTEC First Diploma and National Diploma awards, and is available at Foundation, Intermediate and Advanced levels.

The GNVQ qualifications are awarded by the City & Guilds of London Institute, the Royal Society of Arts (RSA) and EdExcel (formerly the Business and Technology Education Council). GNVQs consist of mandatory, key skills, optional and additional units; the syllabi for the mandatory and key skill units are common to all three awarding bodies. This textbook has been designed to meet the revised requirements of the three bodies for the eight mandatory units of the Advanced award.

While the text has been primarily designed to satisfy the requirements of the Advanced GNVQ, it will also be useful reference for the Foundation and Intermediate awards, plus the relevant optional and additional units, and National Vocational Qualifications (NVQs).

The book has been written in units which follow the order and titles of the eight mandatory units of the Advanced award. These are:

- Engineering and Commercial Functions in Business
- Engineering Systems
- Engineering Processes
- Engineering Materials
- Design Development
- Engineering in Society and the Environment
- Science for Engineering
- Mathematics for Engineering

Each **unit** is divided into sections which generally reflect the various **elements** of the units. These are then further sub-divided into more specific topics which address the requirements of the **performance criteria** and their associated **range statements**. Key words and phrases related to these are highlighted throughout the book.

Spread throughout each unit are **self-assessment tasks** designed to encourage the reader to reinforce their learning, some of which integrate the key skills of Communication, Application of Number and Information Technology. At the end of each unit are sample **multiple-choice questions** which provide examples of the questions which the student might meet in the **external mandatory tests** for the units.

The presentation of the information in the individual units has not always slavishly followed the exact order of the unit specifications. This has been done to allow the text to be presented within a more logical structure and allow relevant and complementing issues to be placed together for the reader's benefit. It is hoped, however, that the careful organization and sign-posting of information, together with the comprehensive index, will allow students to easily satisfy the requirements of each unit.

The book is not meant to be an exemplar of engineering and associated information, but rather to indicate the basic approaches which may be adopted to fulfil the various requirements. Where dimensions are stated, they are intended to give an idea of scale rather than be prescriptive in meeting particular requirements.

The authors, in writing this book, have been conscious of the need to reflect the philosophy of the GNVQ awards and realise that there will be important omissions apparent to the informed reader. These omissions have been made in order to reduce any possible confusion in the student due to the inclusion of material not required by the various syllabi. The authors and publishers, however, would be pleased to receive any constructive comments or suggestions that may be incorporated into future revisions.

Acknowledgements

We are grateful to the following organisations for permission to reproduce copyright material:

EnTra Publications Ltd for our Figs 6.15, 6.16, 6.18, 6.22 & 6.23; the Health and Safety Executive for our Figs 6.9 & 6.10 (Crown copyright is reproduced with the permission of the Controller of Her Majesty's Stationery Office); Myford Ltd for our Fig. 6.24(a); Silvaflame Ltd for our Figs 6.25(b), (c) & (d); PowerGen plc for our Fig. 6.27 (from *Environmental Performance Report 1995*); ICI plc for our Figs 6.29 & 6.30 (from *Environmental Performance Report 1990/1995*); Biffa Waste Services Ltd for our Figs 6.31 & 6.32 (from *Waste: Somebody Else's Problem? Our Choices and Responsibilities*).

While every effort has been made to trace the owners of copyright material, in some cases this has proved impossible and we take this opportunity to offer our apologies to any copyright holders whose rights we have unwittingly infringed.

Engineering and Commercial Functions in Business

The demands that business and commerce place on engineers have never been higher. The time when Britain's developed engineering industry had access to most countries of the world is gone. Nineteenth-century Britain manufactured a range of products whose uniqueness and invention were the envy of the world. Today many of those countries who so assiduously bought our products are our major competitors.

Uniqueness and invention are rare; look at cars, washing machines, aeroplanes, pens – the list is too long, we make them but so do our international competitors. To be successful in this marketplace we must have an 'edge' – to make things cheaper to run, easier to service, or more attractive. Salespeople know that without this 'edge' products are unlikely to succeed. But while advice from the salesforce is invaluable in the fight for the markets, it is engineers who have to meet their requirements.

This unit describes the way in which the engineering functions of a company, such as design, R&D and manufacturing, relate to the commercial functions such as sales and marketing, finance and purchasing. For a company to be successful it is vital that these functions are integrated effectively. There are many products that have generally been recognised as wonderful examples of engineering excellence, but have been commercial failures – think of Concorde, De Lorean cars, or even the Sinclair C5.

In order to complete the unit you will need to prepare reports on various job functions and organisational structures within companies. This may be done by contacting companies within your own area, and asking them for details of their operations. Ideally, you should visit a local company and talk to engineers and other employees about their jobs.

In this unit the student will learn how to:

- Investigate business functions which involve engineers.
- Investigate engineering and commercial functions in business.
- Calculate the cost of engineered products and engineering services.

After reading this unit the student will be able to:

- Specify the job roles of engineers working in different sectors.
- Identify and describe functions and ownership patterns within business in general terms, giving examples.
- Explain the interface between engineering and the commercial function.
- Describe the influence of financial factors on engineering and commercial functions.
- Explain the costing procedures involved in production and servicing.
- Use calculations involving direct and indirect costs, when pricing an engineered product or service.

1.1 Investigate business functions which involve engineers

Topics covered in this element are:

- economic sectors
- patterns of ownership
- organisational structures
- engineering and commercial functions

The headings above move from industry in its widest sense down to activities which clearly involve specific people. The word industry needs some explanation. *The Longman Dictionary of the English Language* defines it as:

(a) systematic work, especially for the creation of value
(b) a department or branch of craft, art, business or manufacturing concern with large personnel and substantial capital
(c) usually a specified group of productive or profit-making enterprises.

If that sounds like just about everything that gives people employment, that is because it is. In the next section we will break down industry into logical and manageable parts. The first step in this process is to divide industry into three sectors: primary, secondary and tertiary.

Economic sectors

The primary sector

This is the sector concerned with the basic or raw materials of manufacture. These are derived from the Earth, either from below the surface or from above it. From above there is fishing, agriculture and forestry; from beneath, gas and oil extraction, mining and quarrying.

Contained within the primary sector is primary manufacture, i.e. industrial processes that allow primary materials to be changed into more transportable products. Examples of this are: steel mills, creameries and wood sawing and pulping mills.

> **Self-assessment task**
>
> The smooth running of the primary sector depends upon the efforts of engineers and their activities, which cover among other things: transportation, production, installation and maintenance. Taking one industry within this sector, select two engineering occupations and compile a report describing the detailed functions involved and the training or education necessary to perform these functions.

The secondary sector

This sector is the main manufacturing part of industry, turning the raw materials produced by the primary sector into finished goods or structures.

This is the sector that immediately comes to mind when we think of engineering. It is here that manufacturing, processing and construction take place, and it involves a huge range of engineering occupations.

With the exception of the narrowly restricted type of manufacture that occurs in the primary sector, all manufacturing occurs in the secondary sector – everything from silicon chips to oil tankers, or personal computers to furniture. Also in the sector are two forms of the construction industry, general building and civil engineering.

In summary: the secondary sector is where things are made, refined or shaped from transported primary products.

> **Self-assessment task**
>
> Identify two *engineering* job functions within a secondary sector business, and investigate the detailed tasks involved and the skills and qualifications necessary. If there are any large-scale manufacturing companies in your area, see if they can help you.

Departments in which you are most likely to find engineers are:

- *Production* – This department is responsible for the manufacture and assembly of finished products, including the setting up of new machinery and the maintenance of existing machines.
- *Research and development (R&D)* – The department in which new products are developed and tested before going into mass production.
- *Product support* – When a product has been sold to a customer, it may be necessary to assist the customer in setting up and using the product in the most efficient manner, and engineers may be employed in this way.

More detailed descriptions of these and other engineering functions will be given at the end of this element and in Element 1.2.

The tertiary sector

Often called by the more explanatory name the service sector, this sector covers a wide area from the straightforward retailing and transporting of products from the primary and secondary sectors, to education, health provision, tourism and insurance. As a definition it is perhaps easier to say what the tertiary sector does not do: it does not make things. It services things, it entertains, it transports, it administers but it does not make tangible things. The profits, the wages, the employment that arise from this sector are the result of services rendered.

Table 1.1 illustrates the relationship of the sectors to each other.

Table 1.1 Economic sectors

Primary	Secondary	Tertiary
Barley	Brewing	Public house
Wood pulp	Print	Newsagent
Steel mills	Car plant	Show room
Petroleum	Refinery	Garage forecourt

This may seem straightforward, but consider these questions. In what sectors would you place the following?

(a) vehicle mechanic in a mining company
(b) brickmaker
(c) building surveyor

The mechanic in (a) is in the primary sector although performing a servicing role. (It is where the activity takes place that determines the sector.) A brickmaker (b) is in the primary sector (the sheer volume of material needed for brick manufacture precludes transportation in any other form than bricks). The answer to (c) depends on the sphere of activity. If the surveyor is employed by a construction company then the answer is secondary, but if employed by an estate agent or housebuyer then the answer is tertiary.

The key point here is that the terms primary, secondary and tertiary refer to the *company* and not to individual employees within it.

Many of the occupations found in the tertiary sector do not follow the logical progression shown in Table 1.1. For example:

- insurance broker
- ambulance driver
- school teacher
- carpet layer
- maintenance engineer
- lion tamer

What links this diverse list of roles is the notion of service. You might consider (a) what service is being given, (b) who pays for it?

Patterns of ownership

Having looked at the broadest categories to which companies can fall we shall now focus on how they are organised in terms of ownership. There are seven descriptions that cover the main types of ownership:

- sole trader
- partnership
- private limited company (Ltd)
- public limited company (PLC)
- franchise
- co-operative
- public enterprise (national, local authority)

Sole trader

An individual who runs his/her own firm who may or may not have staff. The sole trader is personally answerable for tax returns, insurance and all debts that occur.

Advantages

- Controls development and scope of firm.
- Controls decision making.
- Working time not restricted by industrial agreements on overtime rates and recommended hours of work.
- Retains all profits.

Disadvantages

- Liable for all debt if firm fails.
- Continuity of work can cause problems.
- Too much work can result in a failure to meet deadlines.
- Holidays difficult to plan.
- Raising capital can be difficult.

Partnership

This can be a natural development from sole trader as it allows some of the disadvantages found by the sole trader to be overcome. A partnership can involve up to twenty people as partners. Usually a deed of partnership is drawn up showing the rights of partners and how profits are to be distributed. Even without a partnership deed, partners have obligations covering profits and rights laid down by the Partnership Act.

Advantages

- Work can be planned to avoid a build-up of deadlines.
- Planning for continuity of work made easier.
- Capital easier to raise.
- Responsibility and supervision shared.

Disadvantages

- Disagreements within the partnership over roles and strategies.
- Unlimited liabilities for business debts.

Private limited company (Ltd)

This is often the route that sole traders and partnerships take when they wish to expand. The 'limited' in private companies refers to the investors' limited liability for company debts. The limitation of the shareholder in such a company is the extent of the money personally invested. Only that sum can be lost, unlike the position of the sole trader or partner; here bankruptcy can be declared when debts need to be recovered.

Each investor buys shares in the company, and each shareholder receives a portion of the company profits. The law lays down certain rules about the conduct of such a company – to protect the shareholder there must be a board of directors elected by shareholders and these in turn must elect a managing director.

'Private' in private limited company indicates the way capital is raised. Shares are not sold on the stockmarket, i.e. publicly, but to specific groups, organisations or individuals. Examples would be a family firm or, at the other end of the scale, the Co-operative Society whose members (shareholders) contribute to the finances of societies by the shopping loyalty that membership implies.

Private limited companies by law must carry the letters 'Ltd' after their name.

Advantages

- Liability for debt is limited to the investment placed in the company.
- Can raise capital easier than sole trader or partnership.
- Allows business development and strategy to be planned effectively.

Disadvantages

- Doesn't have the capital-raising potential of a public limited company.
- Discord at boardroom level can have far-reaching effects throughout the company.
- Domination of company practice and recruitment by family or section can result in poor management.

Public limited company (PLC)

'Public' in public limited companies means capital is generated by selling shares on the stockmarket to members of the public or financial institutions. As with the private limited company, it has elected directors and a managing director. A private limited company needing to raise capital in order to take advantage of the market will 'go public' – that is, to sell to the public shares that are based on the stock exchange's valuation of the company. This allows the company to expand, and public limited companies tend to be significantly larger than private companies.

Advantages

- Size allows more flexibility in product range.
- Bulk purchasing gives greater control of pricing.
- Downturns in profit can be offset by developments in other areas.
- Opportunities in foreign markets can be analysed and developed more effectively through greater finance.
- Wider outlets.

Disadvantages

- Possibly open to takeover through share buying by another company or organisation.
- Unhappy shareholders can influence company policy by electing directors favourable to their point of view.
- It is now law that the compulsory annual audit of the company finances must be published. This could affect share values adversely with the knock-on effect of curtailing expansion and future development.

Franchise

This is the right given by one company to another to sell its branded products or services: examples are Clarks Shoes, British School of Motoring and Tie Rack. The franchisee pays a fixed sum to the franchiser to use the products, and a further sum depending on profits.

Advantages

- The sale of a tested and successful product or service.
- Advertising and marketing done by the franchising company.
- Advice and monitoring given by the franchising company.

Disadvantages

- Ownership of business not clear.
- Powers of decision limited.
- The franchisee is liable for all losses.
- Final say resides with franchiser.

Co-operative

A group of people without a great deal of capital individually, combine their resources allowing themselves to use plant, space, labour or services not available to them as individuals. Examples: a self-build group constructing their own houses using subcontract labour; a group of small farmers hiring expensive farming machinery.

Advantages

- Capital outlay small.
- Debts likely to be small.
- Work opportunities made available that would not be available outside the group.

Disadvantages

- Decision making can be difficult with members having equal rights.
- Dissension can cause the group to break up.

Public enterprise (national)

This is a nationalised company or enterprise offering a product or service financially underwritten by the public through taxation. Within the last twelve years we have seen a number of these companies broken up into sections and sold to the public in the form of share issues. These sections or areas of the old public enterprises are now public limited companies. Two major public enterprises, however, still exist: the Post Office and the BBC.

Advantages

- Responsive to public criticism.
- Answerable to pressure brought to bear on appropriate government.
- Profits can be put back into the business to aid development.

Disadvantage

- Long-term planning is difficult owing to changes in government policy.

Public enterprise (local)

Operated by finance controlled by local government, consists of things such as: sports halls, community centres, museums, refuse collecting. There is strong pressure currently to push such enterprises into the private sector.

Advantages

- Can show a greater response to local needs than outside contractors.
- Local needs can be seen as a greater incentive than the need to make profits.
- Creates local employment.

Disadvantages

- Can be a burden on local revenue through lack of profitability.
- Absence of competition from similar organisations in the private sector can lead to complacency and inefficiency.

Organisational structures

For sole traders and small partnerships, the organisational structures tend to be rather informal. One individual, or a group of individuals, can be involved in all aspects of the operation while at the same time maintaining an overview of the whole business. For larger organisations this is no longer a practical approach. It then becomes necessary to assign particular people to relatively specialised roles and to ensure that each person is aware of his/her own responsibilities, and has the necessary information and resources to fulfil them.

On a small scale, imagine that you and a partner have an idea for producing a more efficient electric battery. You have a few tools and a workbench in a shed at home, and you work on this idea until you have produced a prototype that works as you expected. If you can persuade people to buy them, you then have a small business.

As time goes by you sell more and more batteries, until you reach a point where you can no longer meet the demand, and you have to employ someone else to handle producing the batteries. At the same time, you realise that you are spending most of your time dealing with the accounts, and employ another person to do this. Finally, in order to sell more of your product you take on a salesperson to deal directly with customers.

You have now gone from a situation in which you and your partner were developing, producing and selling the product, to one in which all these functions are performed by different people employed solely for that purpose. This arrangement forms the basis of most companies, although of course finance, production and sales and marketing will normally be the responsibility of whole departments rather than individuals.

Hierarchical organisation charts

Hierarchical means, in general terms, a condition where the rulers are at the top and the ruled are at the bottom. This is classically seen in the power structures of the armed forces where each level of rank commands all ranks beneath them, and each level of rank is answerable to all levels of rank above them.

Hierarchical management systems operate in a similar fashion, but neither the power exercised by the higher levels, nor the responsibility they demand, is so absolute as that seen in military structures. Figure 1.1 shows a simple hierarchical chart, sometimes called a line and staff chart.

What we see in this line organisation is that the managing director has authority over all employees and operations through the top executive officers. These executives are shown to have direct control over their departments. So we can see that authority flows downwards. The chart also shows, for example, that the export sales manager is answerable to the marketing director, therefore responsibility flows upwards.

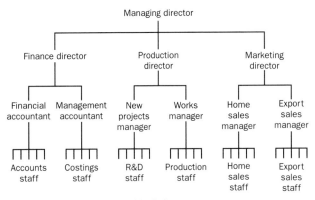

Figure 1.1 A simple hierarchical chart

Like all models, however, it does present us with certain problems, e.g. how is value to the company shown? Do the distances and spaces revealed on the chart indicate levels of salary, responsibility or authority? It has to be admitted that a company forced to adhere to the authority lines shown would swiftly collapse.

Imagine that, through some oversight, the costing staff had failed to pay a materials bill regularly, resulting in irregular delivery of materials. Do they complain to the works manager, who then complains to the managing director – all in writing of course? No, what would happen is that the works manager would contact the management accountant informally enquiring about the reasons for late delivery. This type of communication is called informal communication. On the chart such lines of communication can be shown as dotted lines (Fig. 1.2).

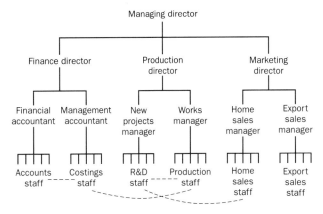

Figure 1.2 Informal lines of communication

In spite of the shortcomings of a chart like this, it does offer a description of the functions of a company, and shows quite clearly where authority and responsibility lie. It also shows the number of people directly answerable to a particular manager (Fig. 1.3).

The personnel manager here has four people under his or her supervision. This responsibility is called a span of control.

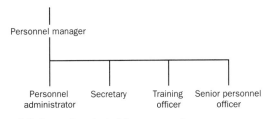

Figure 1.3 Span of control of the personnel manager

Advantages

- Offers clear paths of promotion.
- Encourages competition for posts within the organisation.
- Responsibilities are clearly defined.
- Areas of expertise clearly defined.
- Clear designation of authority.

Disadvantages

- Chain of command can be cumbersome.
- Promotion can be seen as an automatic reward which can lead to top heavy management.
- Expertise can be confirmed to departments and sections leading to inflexibility.
- Departments can be given inappropriate work in order to keep them in existence.
- Informal lines of communication are necessary for the organisation to work, this can result in ambiguity developing in the command structure.
- Communication can become unclear, travelling through the various management levels.

Flat structure

The type of organisation described above is also known as a 'tall' structure, on account of the large number of layers in the hierarchy. In contrast, some organisations, particularly smaller ones, adopt a 'flat' structure in which the number of layers is greatly reduced.

In theory a flat structure may have just one layer of management between the chief executive or board of directors and the workforce, but in practice few large businesses are simple enough to adopt this structure. However, the practice of 'delayering', that is removing whole layers of management from an organisation, has become widespread in recent years.

The reduction in the number of managers obviously gives more responsibility to individual employees, and makes decision making less bureaucratic. This approach has been adopted particularly by smaller companies in rapidly changing businesses who want to be able to adapt quickly to changing circumstances.

Advantages

- Much higher flexibility shown in staff in order to allow wider span of control.
- Increased responsibilities for wider task areas can result in greater motivation.
- Increased training opportunities given to staff.
- Stronger identification with firm's product through widening of roles.
- Reduction of management decisions and responses.
- Flexibility of workforce allows product or service to be quickly modified.
- Informal communication built into the system.

Disadvantages

- Effectiveness needs a motivated workforce.
- Training can be time consuming in achieving necessary levels of motivation.
- Restricted promotion opportunities can cause long-term discontent.
- Some aspects of flexibility can cause stress, e.g. long hours and an increased workload.
- Fewer levels of management can put excessive work on managers, particularly when there is a crisis.

Matrix system

What is a matrix system? At its base, lies the use of a multiskilled team working to produce a particular product or service. During the period of its existence, the team functions with considerable independence of the company. Usually a matrix formation would be used in a hierarchical company, but some high-tech companies can use flexible matrix structures to respond quickly to changing market needs. A good example occurred in the development of the Boeing 747, where the team developing it consisted of 4,000 Boeing employees and 500 chosen subcontractors. The immense effort was co-ordinated by a vice president of Boeing. Usually this is done on a smaller scale, but the basic strategy is the same, as described below.

The matrix works this way: a new product is needed so an appropriate team is made up. The most important figure in this team is the co-ordinator. It is the authority of this person within the company that ensures that the team is staffed by the most appropriate personnel from the various departments. The team would focus on these questions:

- *Marketing* – Is there a real need for this product?
- *Research and design* – Can we design for that need?
- *Production* – Can we make this product?
- *Finance* – Can we afford this product?
- *Sales* – Can we sell it?

The co-ordinator links the efforts of the team, draws on other expertise when necessary and finally pushes through the decisions arrived at by the team.

Advantages

- Capable of responding rapidly to complex market and environmental factors.
- Can focus expertise.
- Product and time factors generate motivation and team spirit.
- The release from departmental restraints encourages creativity.

Disadvantages

- Individuals drawn from departments cannot see clear career paths.
- Secondment from departments means two bosses, appraisals, responsibilities, etc.
- Meetings can proliferate without strong leadership.
- Departments can exert pressures on individuals, causing divided loyalties.

Engineering and commercial functions

As we have seen, most companies consist of a number of different departments, each with a specific function. For a company to be successful it is vital that all these departments do their jobs effectively. We will now look at each of the major departments found in a manufacturing or engineering company, and identify their functions. In Element 1.2 we will then look at the relationships between different departments, and see how communication *between* departments is crucial to the success of the company.

It is possible to classify functions within a company as either *engineering functions* or *commercial functions*. Obviously, our main interest is in the engineering functions, but an understanding of commercial functions is also essential, and we will look at these first.

Commercial functions

The major commercial functions in a company are as follows:

- finance
- purchasing
- marketing
- sales

We will now look at each of these functions in turn, to discover how they contribute to the overall success of the business. We will also see how engineers may be involved, because, strange as it may seem, engineers are not confined solely to the engineering departments of a company.

Finance

The finance department is responsible for monitoring and controlling all aspects of the flow of money into and out of the company, and for producing detailed accounts of the company's business once a year. There are several distinct activities that go to make up the finance function:

- budgeting
- accounting
- investment
- cash flow monitoring

Budgeting
You will have seen the Chancellor of the Exchequer presenting a national Budget to the House of Commons every year. Perhaps you got the impression that it is simply a chance for him or her to increase taxes on cigarettes or spirits or petrol. These are certainly the parts that seem to get the most headlines. In reality, however, the Budget is where the Chancellor sets out in some detail the expected government expenditure over the next year, and explains how this money will be raised. Money may be raised by direct taxation (Income Tax) or indirect taxation (VAT, duty on alcohol, etc.), or by other means.

Similarly, companies have to prepare budgets every year in which all their expenditure is matched with the expected revenue from sales. This exercise is vital to the company's plans for the future, and it is essential that estimates of sales and expenditure are made carefully. For instance, if a company is planning a major new development, such as an extension to the factory with a corresponding increase in the size of the workforce, it is essential that the sales revenue is sufficient to meet the additional costs involved.

Estimating sales figures is a notoriously difficult exercise. However, it can be done surprisingly accurately, by careful analysis of sales figures from previous years and knowledge of the trends in the market (are there going to be changes in the regulations?, what are your competitors planning?, are your major customers expanding?). The finance department will liaise closely with sales and marketing when preparing these figures.

A side-effect of setting projected sales figures in a budget, is the motivational effect it can have on staff who have a target to aim for.

Accounting
This involves keeping track of all the money entering and leaving the company, and preparing balance sheets and profit and loss accounts. Each transaction is recorded using a system known as double-entry book-keeping. For instance, payment of an invoice to a supplier for a new piece of machinery results in a decrease in the company's cash, but an increase in the value of its *fixed assets*, and an entry for each of these will be made in the accounts. Equally, when the company sells some of its products, there will be an increase in the cash, but a decrease in the value of the *stock* held in the warehouse.

Investment
If at any time the company has a large quantity of cash that it does not immediately need, it may seek to invest it on the stock market or elsewhere. This will be handled by the senior finance staff.

Cash flow monitoring
This is really the routine work of a finance department, ensuring that all bills and wages are paid, and all incoming payments are received on time. It is important to monitor this activity carefully to ensure that the company always has enough *cash* to pay the bills. Many companies that were in theory profitable, have gone bankrupt because they did not have enough cash. This may happen if customers are late paying their bills, or if too much money has been invested and is not available when needed.

Sales and marketing

Sales and marketing are really two distinct functions, but in practice the connections between them are so strong that they are frequently combined in a single department.

Sales
The job of sales representative has a rather low reputation, due largely to the activities of door-to-door salesmen selling products such as double glazing. This image is very far from the reality in most companies, where sales representatives are highly trained personnel with a detailed knowledge of the products they are selling. For example, a sales representative for a company making agricultural equipment will be able to discuss with a farmer exactly what his or her particular requirements are, and advise on which of the company's products would be most suitable, and describe its advantages. Rather than making a one-off sale for as much revenue as possible, the aim is to establish a continuing relationship with the farmer so that he or she feels satisfied with the product and service, and will be encouraged to deal with the company again in the future. For this reason, many engineering companies employ qualified engineers as sales representatives.

Marketing
The marketing function is central to the success of a particular product, and ideally the marketing department are involved with every stage of the development of new products, from the original onwards. The various components of marketing may be classified as follows:

- market research
- promotion
- feedback

Market research involves analysing trends in the market in order to determine whether or not a particular product or service is likely to be successful. One method of doing this is to ask customers and potential customers about their requirements, either via personal visits or through questionnaires. This is known as *primary market research*. The questions that might be asked include:

- Are technological developments making a difference to your requirements?
- Do you prefer our products to those of our competitors? If not, why not?
- What features do you consider to be particularly important when choosing a supplier?

A study of the responses to these types of question enables the market researcher to draw up a brief for the products most likely to succeed. This brief can then be presented to the research and development department.

In addition, companies will also carry out *secondary market research*. Rather than direct contact with customers, this involves monitoring other sources of information about the market, such as the trade press, competitors' catalogues or promotional material, government announcements and so on.

Promotion is the most visible aspect of marketing. The first aim is to publicise the product so that as many people as possible are aware of the product's existence. For consumer goods that are likely to be bought by the general public, such as cars, this may be done by a campaign of advertising on television, in national newspapers and on billboards. A lot of ingenuity goes into these campaigns, and the advertising industry is one of Britain's most successful.

However, campaigns such as this are expensive and *untargetted* (you reach everybody, not just those people who might be interested in your product), and most promotional campaigns are on a much smaller scale. For instance, a campaign to promote a new machine tool could include the following:

- *Preliminary mailshot* – This would consist of a fairly basic leaflet giving some details of the new product, including likely price and the date at which it would first be available. This would be sent to all customers who had purchased similar tools in the past, and to other companies who might be interested. It would also include a form for interested companies to return to request more information.
- *Trade press advertisement* – This would contain much the same information as the preliminary leaflet, but would be placed in all relevant trade journals and newspapers.
- *Principal mailshot* – This would be the major promotional leaflet for the campaign, and would be sent out shortly before the new product became available. It should include some detailed information about the product, including photographs of a prototype, and should emphasise the strengths of the new product relative to the existing competition. There would be a form for companies to make definite orders of the new product.
- *Follow up* – Any companies who have expressed an interest, but not made a definite order, will receive follow-up letters encouraging them to make an order.

Feedback occurs once the promotional campaign has finished and the first products have been delivered to customers. The marketing department will monitor customers' response to the new product, and the success of the various elements of the promotional campaign. This information can then be used to adjust the specification of the product, or simply its packaging, and to design further promotional activities. For example, if it seems that an important potential market has been missed out in the initial campaign, additional promotion may be required.

Purchasing

Most large companies have a centralised purchasing department to co-ordinate the buying of products, raw materials and services by all parts of the organisation. This gives the company more influence over potential suppliers, because the orders involved are bigger than if they come from individual departments.

For products or services that are needed regularly, the purchasing department will have three main concerns:

- reliability of service, both short and long term
- quality
- price

If a supplier is providing goods or services over a long period of time, they will sign a contract with the purchasing department. This will probably specify the prices to be paid, a minimum delivery time for new orders, and the exact details of the product or service required. Any deviation from the specified terms may lead to the purchasing company cancelling the contract, or claiming a refund on money paid to the supplier.

Before entering into such a contract, the purchasing department may undertake a thorough check of the supplier's accounts in order to be sure that the supplier will be in a position to maintain the relationship over a long period.

Engineering functions

The major engineering departments within a manufacturing company are:

- research and development
- production

These can each be subdivided into a number of specific functions.

Research and development

This covers all the activities involved in developing a new product, or adaptation of an existing product, including design, analysis of competing products, and testing. It is a very important part of any company, as it is essential to continually develop new products and refine existing ones in order to be successful. Companies that continue to produce the same products year after year are always vulnerable to attack from more innovative competitors.

The initial idea behind a new product may come from two sources:

- as a response to advances in technology
- as a response to changes in the marketplace

Developments in technology may make possible the production of completely new products, or improvements to new ones. For instance, advances in electronics enable companies to produce smaller and cheaper circuits for use in computers and other electronic devices.

Developments in high technology areas such as electronics can be very fast, making existing products obsolete almost instantaneously. For this reason it is vitally important for companies to keep up-to-date, and many large companies such as GEC, ICI or BP maintain large research departments of their own. This allows them to patent any important discoveries and prevent competitors from using them.

Alternatively, a new product may be developed as a response to a company's market research, or to feedback on existing products from customers.

Once an idea has been proposed, there are a number of stages to go through before the new product is made available to customers:

- design
- viability study
- prototype
- testing

The research and development department will first produce a complete design for the new product. This will include details of the manufacturing processes required, the materials to be used and the performance specification of the final product. The design process is discussed in detail in Unit 5.

From the detailed design it is possible to obtain an estimate of the cost of manufacturing the product (this is discussed in more detail in Element 1.3). This information can be combined with the estimates of sales figures from market research to assess the financial viability of the project (discussed in Element 1.2).

Once the viability study has been carried out, and the project has been approved by the management of the company, a prototype will be produced. This enables the designers to check that the manufacturing processes laid out in the design do work satisfactorily, and that the materials employed are adequate. If problems are encountered, it will be necessary to make amendments to the design and this may involve recalculation of the costs of production.

The final stage is the thorough testing of the prototype. This involves not just ensuring that the product does what it is supposed to do, but also checking its reliability over a long timescale. Often prototypes will be subjected to *destructive testing* in which they are used continuously until they break down. If the conditions necessary to induce a breakdown are far more severe than those likely to be encountered in normal operation, there is no problem. If this is not the case, it will be necessary to determine the cause of the breakdown, and amend the design to eliminate the problem.

In general, the development process can be seen to consist of a series of feedback loops in which information gathered at each stage is used to refine the original design until the best solution is reached (Fig. 1.4)

Production

The production department deals with all aspects of the manufacturing process, including:

- development and installation of equipment
- operation of production line
- maintenance
- quality control
- stock control

Development and installation of equipment

As described above, the design and development of a new product will involve a detailed description of the necessary production process, and probably production of a prototype. When the design has been finalised, the production department will be responsible for implementing this production process on a large scale, including buying the necessary machinery and employing the staff to run it.

Operation of production line

The details of the production process obviously vary from product to product, but there are three broad categories:

- assembly line
- batch processes
- one-off project

The *assembly line* is probably the image most closely associated with large-scale manufacturing. It is a continuous process in which raw materials and components pass through a series of operations until a complete product is produced at the far end. This was initially developed by Henry Ford for use with his early cars, and is now widely used in all sorts of manufacturing. Because the system involves continuous repetition of the same process it lends itself readily to automation, and modern assembly lines are staffed by relatively few operatives.

Batch processes involve, rather than continuous production as in an assembly line, production of a small number of items over a short period. They may be used for producing speciality products where only a small number are required, and the machinery or raw materials can be adjusted after a run to change the product of the next batch. This method is often used in the chemical industry.

In some industries, such as shipbuilding, there is no requirement to produce large numbers of identical products, but each individual or *one-off project* requires a huge amount of work. In such cases, the production department will develop a detailed plan for each individual project.

Maintenance

Whatever production system is chosen, all the equipment will need to be effectively maintained in order to maximise the efficiency of the process. Ideally, the emphasis should be on 'prevention rather than cure'. In other words the maintenance staff should concentrate on regular checking of each piece of equipment in order to identify and correct any faults before they become serious, rather than just repairing equipment that has broken down. There are two reasons for this:

- down time
- quality control

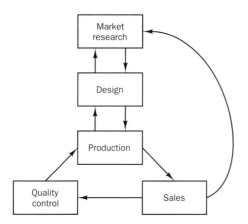

Figure 1.4 Departmental feedback loops

Down time
If any piece of equipment breaks down, the entire production process will be interrupted until it is repaired or replaced. This *down time* reduces the capacity of the factory, and can have serious consequences if it happens regularly.

A piece of machinery may function poorly for some time without actually breaking down, and this may result in a reduction in the quality of the final product, leading to dissatisfied customers.

Quality control
In an ideal world, every product produced should be identical in every respect. In practice this is never quite achievable, but the product design will specify an acceptable range within which each property may vary. In order to verify this, the production department will continuously monitor the quality of the products, and investigate any complaints from customers. As with testing of the protoype in the development stage, quality control involves checking the long-term reliability of the product, as well as its performance in the short term.

If problems are found, the causes need to be identified and corrected as quickly as possible. In general there are three possible sources of a problem:

- flaws in the product design
- unsatisfactory raw materials or components from suppliers
- malfunctions in the production process

Any flaws in the product design should have been exposed during the development, but occasionally they slip through. Correcting them will mean changes to the design, and possibly major changes to the production process, which is why testing is such an important part of the development stage.

Problems with suppliers can be solved by discussions with the suppliers, or, as a last resort, switching to a new supplier, while processing problems will be fixed by the maintenance staff. These two areas generally present less of a problem than design flaws, provided they are spotted and resolved quickly.

Stock control
It is a key aim for most businesses that they be able to satisfy any order from a customer in the shortest time possible. In order to do this it is important to be able to predict the level of demand some time in advance, so that sufficient products are in the warehouse to meet that demand. For instance, toy manufacturers will ensure that their warehouses are fully stocked for the months before Christmas; if they run out of stock in this period, they could lose a lot of business.

You might think that the solution to this would be to run the production process at full speed continuously, so that stocks of finished products were always at a maximum. However, there are two problems with this:

- extra production costs
- storage costs

If a company produces more goods than it needs, it will spend more on raw materials, components, labour and energy than necessary.

Also, and perhaps less obviously, there is a cost involved in storing products. This arises from the cost of maintaining and staffing a large warehouse, and possible loss of products due to deterioration or theft if they are stored for a long time.

For these reasons, companies are keen to reduce their stock, and try to operate a 'just-in-time' strategy, where the highest levels of production occur just before the predicted peaks in demand.

1.2 Investigate engineering and commercial functions in business

In the previous element we looked at the ways in which companies are organised, and the functions of individual departments within companies. Obviously these departments cannot operate in isolation from each other, and in this element we will look at the interfaces between them and the information flow across these interfaces. We will also look at the way in which financial factors influence the activities of these departments.

Interfaces and information flow

For a business to be successful, it is essential that everyone focuses on the ultimate goal of providing quality products and services to customers, and that all employees are aware of how their own role contributes to this ultimate aim. Communication of information between departments and individuals is therefore vital, and it is one of the functions of management to ensure that this communication takes place efficiently.

Having said that, there are certain interdepartmental flows of information that are of particular importance to a company, as follows:

- finance to/from all other departments
- research and development to/from sales and marketing
- research and development to/from production
- production to/from sales and marketing
- production to/from purchasing

We will now look at each of these key interfaces in more detail.

Finance

As described on p. 7, an important function of the finance department is the preparation of an annual budget. In order to do this, they need to gather information from all other departments:

- sales projections, and cost of promotion campaigns from sales and marketing
- costs of project development from research and development
- costs of new machinery, raw materials and components, and running costs from production

In addition, they will monitor the company's actual performance against the agreed budget. In some circumstances, it may be necessary to revise a budget during the year in order to take account of unexpectedly low sales figures, or changes in raw material prices.

Research and development to/from sales and marketing

The relationship between these two departments is crucial to the success of a new product. The principal information flows that may be involved are:

From R&D

- Idea for a new project, submitted to S&M for market research to be carried out.
- Details of the final design of a product, for S&M to include in promotional materials.
- Analysis of competing products to assist the salesforce in selling the product.

From S&M

- Market research suggesting new products or adaptations of existing ones.
- Market response to a new idea from R&D, including views on price.
- Feedback from salesforce suggesting improvements to a product.
- Information on new competing products.

The end result of these various pieces of information is a continuous flow of information between the two departments over the lifetime of a product from original concept until production is finally stopped. This is vital if the company is to produce products that are well adapted to their customers' needs, and market them successfully.

Research and development to/from production

The initial design for a new product has to include a detailed description of how it would be produced. R&D therefore need to consult with the production department at a very early stage in order to determine:

- The technical feasibility of the proposed process.
- The most efficient production method.
- What new equipment will be required.
- The implications for the number of production staff needed.
- The costs of production, including new equipment, raw materials, labour and power.

The second element of this interface is the construction of a prototype. This is a chance to assess the effectiveness of the production process, as well as the quality of the finished product. If problems are encountered, it may be necessary to revise the details of the process.

Once any problems with the prototype have been resolved, the new product will pass from research and development to the production department and go into mass production. However, there are still two ways in which the two departments need to maintain contact:

- quality control
- amendments to product

If the quality control staff in the production department identify a consistent fault in the finished products, it may be necessary for the R&D staff to amend their original design in order to resolve the problem.

Alternatively, amendments to the product may be suggested by market research or a technological development, and the two departments will need to collaborate on the best way to make these amendments.

Once again, we can see that that the interaction between these two departments spans the whole lifetime of the product.

Production to/from sales and marketing

There are two principal interactions between production and sales and marketing:

- stock control
- product quality feedback

Production relies on sales predictions from sales and marketing in order to estimate demand for products. The level of production can then be adjusted to ensure that stock levels are always high enough to meet the expected demand, without being much higher than necessary.

If market research, or feedback from the salesforce, indicates that there is a problem with the performance or reliability of the product, this information can be passed to the production department for further investigation.

Production to/from purchasing

This is a straightforward, but important, interaction. The production department have to liaise closely with the purchasing department to arrange supply of:

- raw materials
- component parts
- new machinery

In the case of raw materials and components, it is important that the contract with the supplier allows for prompt delivery so that the production process can run continuously, without building up large stocks of these. Also, the quality of raw materials and components must be specified precisely so that they meet the requirements of the production process and do not result in the production of substandard products.

Financial factors

As we have already seen, the finance department of a company has a close relationship with all the other departments. The reason for this is, quite simply, that a company needs to be financially successful in order to continue trading. Innovative design, high-quality products, brilliant marketing and excellent customer service, while all highly desirable qualities, will not make a company successful unless the finance is right.

When assessing the viability of a project, there are a number of financial factors that need to be considered, including:

- costs – capital and operational
- breakeven analysis
- make-or-buy decisions
- investment appraisal

We will now look at each of these factors in turn.

Costs

In order to assess the profitability of a project, it is necessary to know not just the revenue generated by sales of a product or service, but also the costs incurred in supplying that product or service. In crude terms, it is the difference between these two terms that gives the profitability.

The calculation of costs is a surprisingly complex operation, and it forms the subject of Element 1.3. For the purposes of this section, it is only necessary to understand that the total cost of a product can be divided between *fixed costs* and *variable costs*. Fixed costs, as the name implies, are the same however many units of a particular product are produced, whereas variable costs increase with the number of units produced.

For instance, production of a CD player may involve fixed costs of £40,000 per year, and variable costs of £45 for each player produced. Table 1.2 shows how the total cost changes as the annual production figure changes.

Table 1.2 Influence of annual production on costs

Annual production (units)	Fixed cost (£)	Variable cost (£)	Total cost (£)
5,000	40,000	5,000 × 45 = 225,000	265,000
2,000	40,000	2,000 × 45 = 90,000	130,000
1,000	40,000	1,000 × 45 = 45,000	85,000

From these figures it is clear that one effect of the fixed cost is that the total cost per unit depends on the number of units produced. For instance, the cost of producing 2,000 CD players (£130,000) is less than twice the cost of producing 1,000 (£85,000). This is reflected in the *unit cost* of the product, which is calculated by dividing the total production costs by the number of units produced, as shown in Table 1.3.

Table 1.3 Example unit costs

Annual production	Unit cost
5,000 units	265,000/5,000 = £53
2,000 units	130,000/2,000 = £75
1,000 units	85,000/1,000 = £85

This is one example of the concept of *economies of scale* – the higher the level of production the lower the unit cost, and hence the greater the profitability for a given selling price (assuming, of course, that all units produced are sold!)

Alternatively, costs may be described as *capital* or *operational*. Capital costs are large, one-off expenses incurred at the start of a project, such as the purchase of machinery. Operational costs are the costs incurred in the day-to-day running of the production process, such as raw materials, wages or maintenance.

Breakeven analysis

Breakeven analysis is a technique that may be applied to a new project as an aid to assessing its viability and the risks involved in it.

We have already seen how the production costs of a product can be a combination of fixed and variable costs. This may be represented graphically, as in Fig. 1.5, where the horizontal axis represents the total number of units produced, and the vertical axis represents the total production cost.

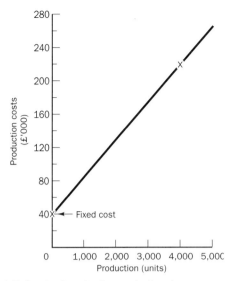

Figure 1.5 Graph of production against costs

You can see that this graph does not pass through the origin, because the fixed costs have to be paid even if no CD players are produced. The point at which the graph crosses the vertical axis (the *intercept*) represents the value of the fixed costs. From this point, the graph rises in a straight line, representing the £45 of variable costs incurred for each CD player produced. The *gradient* of the graph represents the variable cost per unit.

In order to conduct a breakeven analysis, we also need to consider the revenue from sales of the product. We can do this by adding a second line to the graph, representing the sales revenue, as shown in Fig. 1.6. In this example, the CD players are being sold for £70 each.

You can see that the revenue graph does pass through the origin (no sales means no revenue), and then rises steadily with a gradient equal to the selling price of £70. You should also see that the two graphs representing costs and revenue intersect at a point, and a little thought should tell you that this point shows the number of CD players you would need to sell for sales and revenues to be equal. This is known as the *breakeven point*, and in our example it is equal to 1,600 CD players.

The breakeven chart also allows you to calculate the profit or loss to the company of a particular level of production and sales. For quantities below 1,600, there will be an overall loss, and the extent of the loss is given by the vertical distance between the two lines on the graph to the *left* of the breakeven point. Similarly, for quantities above 1,600 (to the *right* of the breakeven point) the vertical distance between the two lines give the profit.

Obviously, for a project to be viable it is essential that the predicted sales figures are comfortably above the breakeven level, and the proposed production process is capable of meeting the predicted demand. Management can therefore use breakeven analysis in conjunction with sales predictions and production information to evaluate a project.

The technique can also be used to investigate the effect of changes in production costs, or changes in selling price. For example, if it is suggested that the selling price should be increased, it is straightforward to change the gradient of the revenue line on the breakeven chart and observe the effect on the breakeven quantity and profitability.

In our example, changing the selling price to £90 would reduce the breakeven quantity to 889. However, to evaluate the suggestion properly, you also need to estimate the effect on sales of the price increase. For instance, if there is a very similar CD player on the market costing £80, raising the price to £90 could have a disastrous effect on sales.

Make-or-buy decisions

Very few manufacturing processes involve only raw materials as starting points (the steel industry, and some chemical industry processes are examples). Far more frequently, the manufacturing process will involve the assembly of previously prepared components into a new product.

In the latter case, the company has the choice of either buying ready-made components from a supplier, or making them itself. We can perform a similar exercise to the breakeven analysis described above to assist in making this decision. This time the two graphs represent the cost of making the component, made up of fixed and variable costs as before, and the cost of buying it ready made (Fig. 1.7).

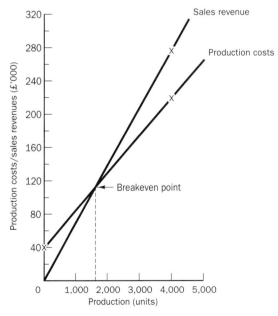

Figure 1.6 Breakeven point for the production and sale of CD players

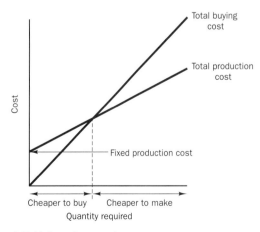

Figure 1.7 Make or buy graph

Again we have a point at which the costs are equal. At lower quantities (to the left of the graph), it is cheaper to buy, and for higher quantities (to the right) it is cheaper to make. In purely financial terms the decision would seem to be straightforward, but there are a number of other factors that need to be considered before a decision is made:

- *Availability* – Is there a suitable supplier, and can they meet the likely demand?
- *Quality control* – This is easier to achieve if components are made rather than bought in.
- *Reliability* – You need to be sure that the supplier is reliable, and unlikely to go bankrupt or experience other major problems.
- *In-house expertise* – Does the production department possess the necessary expertise to make these components?

All these factors will need to be considered carefully before the make-or-buy decision is made.

Investment appraisal

This is a longer term technique, which can be used to assess the contribution of a particular project to the company over a number of years. In particular, it is helpful in choosing between a number of proposed projects in a situation where sufficient funds are available to develop only one. There are three common methods of conducting this kind of assessment:

- payback
- return on investment (sometimes known as accounting rate of return)
- net present value

Payback

This is a fairly crude, but none the less useful, method. It takes no account of profitability, but simply concentrates on the length of time necessary for the company to recoup its initial investment in a project.

This is particularly significant for projects which require a large initial outlay, for instance on expensive new equipment. If the payback period is long, the money invested is not available for investment in other projects and the company may therefore miss opportunities for development.

Suppose that a company has £40,000 available for investment. Two projects have been proposed, each requiring an initial investment of the full £40,000. Project A is the purchase of new machinery to increase the capacity of an existing production process, while Project B is the development of a totally new product that will be produced in fairly small quantities, but will be sold at a high price.

In order to determine the payback period, we need to estimate the *cash flow* for the project for each year of the project's lifetime. The cash flow is the difference between the revenues generated by sales, and the operational costs incurred during the same period.

For Projects A and B, the cash flows over the first five years are shown in Table 1.4, from which we can see that Project A starts to generate cash immediately, because it is an improvement to an established process, and the initial £40,000 investment is paid off during the second year of the project. Project B on the other hand, generates no cash in the first year, and a relatively small amount in the second year, because a complete development process needs to be undertaken before any revenue comes in. Once it becomes established, the cash flows become much higher, and the investment is paid off during Year 4.

Table 1.4 Comparison of cash flows

Year	Project A	Project B
1	20,000	0
2	25,000	10,000
3	25,000	25,000
4	20,000	40,000
5	15,000	35,000
Total	£105,000	£110,000

Following the payback method of investment appraisal, the company would choose Project A, in order to regain its investment within two years, even though Project B is more profitable over the first five-year period.

Return on investment

An alternative method of investment appraisal takes less account of the timescale of the project, and concentrates more on the overall profitability. It works by comparing the average annual cash flow over the lifetime of the project with the value of the investment.

In the example above (Table 1.4), we can assume that the lifetime of each project is five years. They may continue to generate revenue beyond this point, but five years is the maximum lifespan we can be sure of. The average cash flow is then given by dividing the total cash flow by 5:

$$\text{Project A: Average cash flow} = 105,000/5$$
$$= £21,000 \text{ per year}$$

$$\text{Project B: Average cash flow} = 110,000/5$$
$$= £22,000 \text{ per year}$$

We can also assume that the value of the initial investment declines to zero over the five years. That is the machinery purchased at the start of Project A has no resale value at the end of the project, and the various goods and services purchased during the development of Project B also have no value after five years. This means that the average value of the investment over the five years is half its initial value, i.e. £20,000.

The *return on investment* is the ratio of average cash flow to average investment, expressed as a percentage:

$$\text{Project A: Return on investment} = (21,000/20,000) \times 100$$
$$= 105\%$$

$$\text{Project B: Return on investment} = (22,000/20,000) \times 100$$
$$= 110\%$$

Therefore, using this criterion, the company would choose Project B.

This example is rather a simple one to illustrate the method. The technique can also be used to compare projects with different life spans.

Net present value

This method, also known as *discounted cash flow*, has gained popularity over recent years. Although it is more complicated to do than the previous methods, it has the advantage that it takes account of both profitability and payback period when assessing projects.

The key factor in this technique is the fact that the real value of cash flows to the company depends on their timing, because cash received in Year 1 can be invested for four years at a good rate of interest. £1,000 received in Year 1 and invested, is therefore of more value than £1,000 received in Year 2 and invested for three years.

Alternatively, we can say that £1,000 received in Year 2 is worth less than £1,000 to the company now, by an amount related to the interest rates applying over the next five years. This is known as the *net present value* of the £1,000 (Table 1.5). For instance, suppose that the interest rate over the five years remains constant at 6 per cent, the net present value of £1,000 received in Year 2 is £943.40.

In other words, if we receive £943.40 in Year 1 and invest it at 6 per cent interest, it will be worth £1,000 at the end of Year 2. £943.40 received in Year 1 is of as much value to the company as £1,000 received in Year 2.

Similar calculations can be performed for the cash flows received in each year of the project, and these can be added to give the total net present value of the project. For an interest rate of 6 per cent, the net present value of £1 received in Years 1–5 is given in Table 1.5, and you can see from the table that every pound earned in Year 5 of the project value, is worth only 75 pence to the company now.

Table 1.5 Net present value

Year	Present value of £1
0	£1
1	£0.9443
2	£0.8900
3	£0.8396
4	£0.7921
5	£0.7473

Self-assessment tasks

1. Using Table 1.5, calculate a total net present value for each of Projects A and B described previously. Which project would a company be more likely to choose using this method of investment appraisal?
2. If a new technological development reduces the time necessary for development of Project B, the new product may become available sooner than first anticipated. Suppose that the cash flow in Year 2 is now predicted to be £20,000 rather than £10,000. What effect does this have on the result of the investment appraisal? (Assume that all other cash flows are unaffected.)

When using any of the methods of investment appraisal described above, it is important to bear in mind that all the calculations involve a lot of assumptions and estimates, and the quality of decision making based on these techniques can only be as reliable as these estimates. For this reason, accurate estimation of future costs and future sales figures is a vitally important function in any company.

1.3 Calculate the cost of engineered products and engineering services

We have seen in the previous element that the accurate calculation and estimation of costs is a key factor in a company's strategic planning. In this element we will look at the ways in which costs can be categorised and calculated, and apply these methods to both manufacture of engineered products and the provision of engineering services.

We have already seen how the total cost of a product or service can be viewed as a combination of *fixed* and *variable* costs (see p 12). It is also possible to categorise costs as *direct* or *indirect*, and we will consider this division first.

Direct costs

Direct costs are those costs which are directly attributable to the manufacture of a particular product, or the provision of a particular service. They are approximately similar to *variable costs* which vary with the level of production, but there are differences.

For an engineered product, direct costs generally consist of the following:

- direct labour
- parts and raw materials
- depreciation of fixed assets

For an engineering service, such as repair of machinery, direct costs might be:

- direct labour
- cost of spare parts and materials

We will now briefly consider each of these direct costs.

Direct labour

This is the wages paid to employees solely concerned with the product or service being considered. For instance, a machine operator working on an assembly line, or a service engineer dealing with one group of products, would have their wages included in the direct costs.

Direct labour costs are also *variable*; they vary with the number of products or services supplied.

Wages paid to staff whose work relates to a number of different products, or to the company as a whole, are treated as indirect costs, as we will see later.

Parts and raw materials

This is the most obvious component of direct costs for an engineered product. Parts and raw materials are purchased specifically for a particular process and the cost rises for each extra unit produced. This is also a variable cost.

For an engineering service, it may be necessary to supply replacement parts for machinery being serviced, and these are also direct costs.

Depreciation of fixed assets

Depreciation costs arise from the fall in value of fixed assets such as heavy machinery, servicing equipment or vehicles.

Imagine that a company pays £30,000 for a new assembly line. The assembly line is only to be used for one range of products, and the estimated lifetime of the project is six years.

As mentioned previously, when the machinery is purchased there is no change in the total value of the company, just a transfer of £30,000 from the value of the cash held to the value of the fixed assets. However, at the end of the six years the machinery is worth nothing, so the value of the company's fixed assets has decreased by £30,000 over the period. This decrease is known as *depreciation*.

In order to take account of this depreciation when calculating the profitability of the project, it needs to be counted as a production cost. If the whole amount is included as a fixed cost when calculating the first year profitability, the first year figures will look very bad. As the assembly line is used throughout the lifetime of the project, it is more realistic to spread the cost over the full lifetime as well.

There are a variety of methods for doing this, but we need only consider the simplest. This is the *straight line method*, in which the value is assumed to decline by a constant amount each year, and this amount is treated as a fixed cost for each year. In our example, the amount would be £30,000/6 = £5,000, so £5,000 would be added to the fixed costs of production each year.

Depreciation is an example of a *fixed* direct cost. It is direct because the assembly line is used only for one range of products, but fixed because it does not vary with the level of production.

Of course, depreciation also applies to buildings or machinery not directly related to the production process, and this is an *indirect* cost.

Indirect costs

Indirect costs covers all the expenses involved in running a company that do not relate directly to the provision of a particular product or service. They may also be known as *overheads*.

The list of factors contributing to the indirect costs is very large, and will depend on the exact nature of the company involved. However, some of the major factors are:

- power
- depreciation
- wages of non-production staff
- administration
- marketing
- running expenses

Power

Power is required to operate all production machinery, but also for heating and lighting of offices and factories, running computers, and operation of other machinery. In theory, the power required to run a particular production process is a direct cost, but in practice it is not generally possible to separate this element from the total power bill.

Depreciation

As we have seen, depreciation of assets used directly in the production process is a direct cost. However, depreciation also needs to be calculated for other fixed assets, such as buildings, vehicles, maintenance machinery, computer systems and office equipment.

Depreciation on these items cannot be identified with a particular product or service, and is therefore counted as an indirect cost.

Wages of non-production staff

A similar argument can be applied to wages. If you recall the description of engineering and commercial functions in Element 1.1, it is obvious that a number of departments have no direct involvement in a product, and their wages therefore contribute to the indirect costs.

Even within the production and research and development departments, there are maintenance staff and other support staff whose work cannot be identified with any one product or service, and their wages are also indirect costs.

Administration

This category includes all the expenses (other than wages) incurred by support departments such as finance, purchasing, personnel and facilities management (general upkeep of buildings and vehicles).

Marketing

This covers all the expenses involved in the marketing function, including:

- paid-for market research
- production of catalogues and leaflets
- advertisements

The marketing department will have its own budget from which to pay these expenses, and the total marketing budget forms part of the indirect costs.

Running expenses

This includes all other expenses incurred in running a company. The major components will be rent, warehousing and rates.

All indirect costs can be regarded as *fixed* costs as they do not vary with the level of production. They will, of course, vary from year to year, and for this reason they may sometimes be referred to as *semi-variable* costs.

Allocation of indirect costs

In order to take account of indirect costs when calculating the total cost of a product or service, it is necessary to distribute the total overhead cost between the various products and services provided by the company.

If the company only produces one product, this is straightforward. The overheads are added to any fixed direct costs to give a total fixed cost, and this is then added to the variable costs to give a total cost.

$$\begin{array}{l} \text{Indirect costs (overheads)} \\ + \text{ Fixed direct cost} \\ \hline = \text{Total fixed cost} \end{array}$$

$$\begin{array}{l} \text{Total fixed cost} \\ + \text{ (Variable cost} \times \text{No. of units produced)} \\ \hline = \text{Total cost} \end{array}$$

However, for most companies there are a number of products or services to consider and the direct costs have to be distributed between them. There are a number of ways of doing this, and several different methods may be used within one company.

The two simplest methods involve allocating indirect costs on the basis of direct labour costs, or on the basis of the value of fixed assets.

We can illustrate these methods with a simple example. Suppose a company produces gas cookers, and operates a repair and maintenance service for any make of gas cooker.

The production line for the cookers contains a number of pieces of modern machinery, with a combined fixed asset value of £140,000. One foreman and four employees run the process, and their annual salaries total £85,000.

There are four travelling service engineers, with a combined salary of £75,000 per year. They each have a van and a complete specialist toolkit, with a total value of £70,000.

In addition, the company employs small staffs for maintenance, sales and marketing and administration and has a small warehouse for unsold stock. The total annual costs incurred by these service departments, including wages, are £120,000, and the company must also pay rent, electricity bills and rates totalling £35,000 per year. Total indirect costs are therefore:

$$\begin{array}{r} 120{,}000 \\ + 35{,}000 \\ \hline = £155{,}000 \end{array}$$

This £155,000 now needs to be distributed between the production business and the servicing business in order to obtain the fixed costs for each business. If we base the calculation on direct labour costs, we get the following:

Total direct labour cost
= 85,000 + 75,000
= £160,000

Production direct labour as a percentage of total
= (85,000/160,000) × 100
= 53.1%

Servicing direct labour as a percentage of total
= (75,000/160,000) × 100
= 46.9%

Therefore, on this basis the company will charge 53.1 per cent of the total indirect costs to the production business, and 46.9 per cent to the servicing business:

$$\text{Indirect cost for production} = 155{,}000 \times (53.1/100)$$
$$= £82{,}305$$

$$\text{Indirect cost for servicing} = 155{,}000 \times (46.9/100)$$
$$= £72{,}695$$

Alternatively, the company could base its calculation on the value of fixed assets associated with each operation. The calculation then becomes:

$$\text{Total fixed asset value}$$
$$= 140{,}000 + 70{,}000 = £210{,}000$$

$$\text{Production fixed assets as a percentage of total}$$
$$= (140{,}000/210{,}000) \times 100$$
$$= 66.7\%$$

$$\text{Servicing fixed assets as a percentage of total}$$
$$= (70{,}000/210{,}000) \times 100$$
$$= 33.3\%$$

Therefore, on this basis the company will charge 66.7 per cent of the total indirect costs to the production business, and 33.3 per cent to the servicing business:

$$\text{Indirect cost for production} = 155{,}000 \times (66.7/100)$$
$$= £103{,}333$$

$$\text{Indirect cost for servicing} = 155{,}000 \times (33.3/100)$$
$$= £51{,}1667$$

As the above example shows, the method of overhead allocation chosen can make a significant difference to the fixed cost of a product or service. For very labour-intensive industries with large numbers of employees involved in each project, the first method makes sense. However, the increase in automation over the last twenty years or so has meant that most manufacturing industries have more money invested in machinery than in their staff, and it is now more common to allocate overheads on this basis.

Case study

As part of your evidence for this unit, you will be required to calculate direct, indirect and total costs for an engineered product and an engineering service. In this case study we will perform similar calculations for a particular product and service in order to demonstrate the principles involved.

At the end of the previous section, we introduced a company selling and servicing gas cookers in order to illustrate the allocation of indirect costs between different business operations. We can expand this example to build a complete picture of the costs for each operation.

Cooker manufacture

We need to determine a total cost for the manufacture of each new gas cooker. As we have seen, this is composed of a fixed cost and a variable cost. The fixed cost represents:

- Indirect costs allocated to the manufacturing business
- Direct labour (assuming that the labour force is constant across the year; if extra staff are needed for busy periods, their wages are a variable cost depending on the level of production).
- Depreciation of production machinery.

The initial value of the production line was £210,000, and its estimated lifetime was seven years. Using the straight line method, we can calculate the depreciation:

$$\text{Annual depreciation cost} = \text{initial value/lifetime}$$
$$= 210{,}000/7$$
$$= £30{,}000$$

If we allocate indirect costs on the basis of value of fixed assets, the total fixed cost is calculated as follows:

$$\text{Total fixed cost} = \text{Allocated indirect costs}$$
$$+ \text{Direct labour} + \text{Depreciation}$$
$$= £103{,}333 + £85{,}000 + £30{,}000$$
$$= £218{,}333$$

In this case, the variable costs associated with each additional unit of production are simply the costs of the component parts that need to be assembled. The company has taken a decision to buy all components, rather than make them themselves. The total cost of all the required components is £57.50. Therefore, the total cost of producing x gas cookers is given by:

$$\text{Total cost} = \text{Fixed cost} + (x \times \text{Variable cost})$$
$$= 218{,}333 + 57.50x$$

Table 1.6 shows this calculation for a number of levels of production.

Table 1.6 Variation of total cost with production units

Production (units)	Total cost (£)
1,000	275,833
2,000	333,333
3,000	390,833
4,000	448,333

Self-assessment tasks

1. If the cookers are to be sold at £425 each, calculate the *breakeven* figure for the manufacturing operation.
2. If the price of components is increased to £65, what effect will this have on the breakeven figure?

Servicing of cookers

We can do the same type of calculation for the engineering service. Again, we have to calculate a total fixed cost, made up of:

- allocated indirect costs
- depreciation of vehicles and servicing equipment

Indirect costs have been discussed previously, so we just need to calculate the depreciation. The vans and toolkits were purchased two years ago for a total of £84,000 and are expected to have a lifetime of six years. Using the straight line method, we can calculate the depreciation as follows:

$$\text{Annual depreciation} = \text{Initial cost/Lifetime}$$
$$= 84{,}000/6$$
$$= £14{,}000$$

Therefore:

Total fixed cost = Allocated indirect costs + Depreciation
= 51,667 + 14,000
= £65,667

Calculation of the variable cost for a service is less straightforward. Whereas all the cookers manufactured are identical and therefore carry the same cost, each servicing job is unique and carries a different cost. The elements of the variable cost are:

- labour
- travel expenses
- spare parts and materials

Labour is a variable cost in this case, because, although the total wage bill is fixed, each job occupies a different length of time and therefore absorbs a different fraction of the total labour cost.

In order to take account of this, we can calculate a labour cost per man-hour (i.e. the cost of one engineer working for one hour). Suppose that each of the four engineers works a 35-hour week (i.e. seven hours a day, five days a week), and has 20 days holiday per year plus the eight bank holidays and 52 weekends.

The total number of hours worked by each engineer is therefore:

Total hours = 7 × number of days worked
= 7 × (365 − 20 − 8 − 104)
= 7 × 233
= 1,631 hours

The total number of hours worked by all the engineers is therefore 4 × 1,631 = 6,524 hours. We have already seen that the total wage bill is £75,000, so we can calculate the hourly rate.

Cost per man-hour = 75,000/6,524
− £11.50

The other elements of the variable cost are dependent on the nature of the job and its location, and cannot be estimated in advance. However, the customer will normally be charged for any parts and materials required in addition to the service charge, so we can ignore this when deciding on an appropriate rate for service charges.

In this example, we will asume that all services are carried out in the local area, so travel expenses and travelling time between jobs are negligible.

We have already calculated:

- labour cost per man-hour = £11.50
- total fixed cost = £65,667

For this type of service, it is usual to charge customers by the hour. If we assume that all available hours are filled, we can calculate the charge per hour required to break even, and the profit made for a particular rate.

In this example, the breakeven rate is given by:

Breakeven rate = hourly labour cost
+ (Total fixed cost/Number of hours)
= 11.50 + (65,667/1,631)
= 11.50 + 40.26
= £51.76 per man-hour

If the company decides to charge £60 per man-hour, the total profit would be:

Total profit = revenue − costs
= (60 × 1,631) − [(11.50 × 1,631) + 65,667]
= 97,860 − 84,423
= £13,436

This is a relatively straightforward example, and we have made several assumptions about the company in order to simplify the calculations. However, it does illustrate the principles involved, and should assist you in compiling a similar report for your portfolio of evidence. You should also find that an assignment of this type will give you opportunities to demonstrate skills in Application of Number. The calculations are also suitable for a spreadsheet program, and using one of these would demonstrate skills in Information Technology.

1.4 Unit test

Test yourself on this unit with these multiple-choice questions.

1. In which economic sector would you place a company manufacturing mining equipment?

    (a) primary
    (b) secondary
    (c) tertiary
    (d) service

2. Why might a private company convert to a public limited company?

    (a) to raise capital
    (b) to improve its advertising
    (c) to be under government control
    (d) to reduce the number of shareholders

3. Which of the following describes a tall organisational structure?

    (a) one with a large board of directors
    (b) one in which individual employees are given responsibility
    (c) one in which teams of specialists work together on specific projects
    (d) one with numerous layers of management

4. Which of the following is a privatised company?

    (a) BBC
    (b) British Telecom
    (c) Mercury
    (d) Esso

5. Which department is responsible for budgeting?

    (a) human resources
    (b) production
    (c) sales and marketing
    (d) finance

6. Which of the following would be used in primary market research?

    (a) catalogues
    (b) trade magazines
    (c) questionnaires
    (d) design briefs

7. The purpose of a prototype is:

    (a) to ensure that a new product works as predicted
    (b) to estimate sales figures
    (c) to advertise the product
    (d) to train staff

8. What is down time?

    (a) the time for which a production line cannot be used
    (b) the time taken to manufacture each unit
    (c) the time taken for a component to fail
    (d) the time at which an assembly line is shut down for the night

9. It is decided that a product needs to be slightly modified to meet customer requirements. Which two departments will be responsible for implementing the modification?

    (a) sales and marketing, purchasing
    (b) research and development, sales and marketing
    (c) research and development, production
    (d) production, purchasing

10. Which of the following pieces of information would need to be supplied to the sales and marketing department by the research and development department?

    (a) the cost of raw materials
    (b) market research results
    (c) analysis of similar products made by other companies
    (d) what new equipment will be required

11. If the salesforce report that a product is regularly failing because of a problem with the raw materials, which other two departments will be involved in remedying the problem?

    (a) Finance, production
    (b) Production, purchasing
    (c) Research and development, production
    (d) Finance, purchasing

12. The purchase of components used in the manufacture of a television set is a:

    (a) unit cost
    (b) fixed cost
    (c) capital cost
    (d) variable cost

13. The television sets of the previous question are to be sold for £190 each. If the fixed costs are £30,000, and the variable costs are £60 per set, what is the breakeven figure?

    (a) 231
    (b) 158
    (c) 500
    (d) 342

14. Which of the following is an advantage of making components in house rather than buying them from an outside supplier?

    (a) it is always cheaper
    (b) increased control over quality and reliability
    (c) fewer staff needed
    (d) lower capital costs

15. Which factor is most important in the payback method of investment analysis?

    (a) profitability
    (b) rate of return
    (c) capital costs
    (d) cash flow

16. Four proposed projects have the projected cash flows shown. Use the return on investment method to determine the most attractive project for the company.

	Investment	Year 1	Year 2	Year 3	Year 4
(a)	10,000	5,000	6,000	5,000	4,000
(b)	8,000	5,000	5,000	4,000	3,000
(c)	5,000	5,000	4,000	3,000	2,000
(d)	12,000	5,000	5,000	5,000	5,000

17. Which of the following is a direct cost?

    (a) rent
    (b) electricity
    (c) raw materials
    (d) maintenance

18. If a company pays £20,000 for a delivery van and expects it to last for five years, use the straight-line method to determine the depreciation cost to be charged in the first year.

    (a) £5,000
    (b) £20,000
    (c) £10,000
    (d) £4,000

19. Overheads are:

    (a) direct costs
    (b) labour costs
    (c) indirect costs
    (d) operational costs

20. Which of the following forms part of the total fixed cost for a product?

    (a) allocated overheads
    (b) spare parts
    (c) capital costs
    (d) interest payments

Unit 2: Engineering Systems

Engineering products span a vast range of applications and employ many different processes in their manufacture. While being designed, engineering products will require the use of calculators, computers, etc. and when being used these products will operate according to many scientific principles.

In spite of the many differences between engineering products, they all have one feature in common; they are all *systems*, as are the machines and processes which design, produce and test them.

In this unit the student will learn more about systems and how they can be described. The student will learn how to:

- Investigate engineering systems in terms of their inputs and outputs.
- Investigate the operation of engineering systems.

The unit is further divided into sections corresponding to the performance criteria described in the GNVQ Advanced Engineering syllabus. This will help the student to produce the right kind of evidence necessary to pass this unit

After the theory covered in each element has been studied the student will be able to apply the ideas to real engineering systems, firstly by means of 'worked' examples and then using suggested further examples which can be used to generate evidence in conjunction with practical work.

2.1 Investigate engineering systems in terms of their inputs and outputs

Identify and describe the purposes of engineering systems

As we have already mentioned in the introduction, all engineering products and processes can be called engineering systems. So what are the features shared by all products and processes? You would probably agree that all products and processes do something, i.e. they have a purpose and a function.

To clarify the difference in meaning between *purpose* and *function* we can use a power drill as an example:

- The *purpose* of the drill is to make holes.
- The *function* of the drill is to convert electrical energy to rotational mechanical energy.

In other words, the *purpose* is the reason for making the product whereas the *function* describes the changes that take place to make the product operate.

To illustrate this let us consider two engineering products. Firstly, a measuring device, a voltmeter. It is relatively simple to describe this system in terms of its purpose; it is designed and made to measure voltage within a specified range (e.g. between 0.001 and 500 volts). A second product we can consider is the car. The purpose of this product could be described as many things, depending on who and what you are; for instance:

- Transporting people and luggage.
- A vehicle to go faster than other vehicles.
- A means of impressing the neighbours.

Someone like Damon Hill or Michael Schumacher would consider the car in a different light to a taxi driver or someone who uses their car for leisure only.

Now let us think about the functions of these two products. In the case of the voltmeter, the function is to convert electrical energy to light energy (if the display is digital) or rotational mechanical energy (if the display is in the form of a needle with a scale). The car becomes a very simple device if we describe it in terms of its function. Its function is to convert chemical energy from fuel into kinetic energy of the whole vehicle.

The point being made here is that the purpose of a system can often be described in more than one way. The function is usually quite clear cut, and often requires only thinking about energy changes. The best way to approach the problem of defining the purpose or function of a system is to think about the *inputs* and *outputs* of the engineering product, which leads to the next section.

Identify and describe the input(s) and output(s) of engineering systems

All systems have an input and an output. The function of the system is to process the input to produce an output. This sounds rather vague, so let us think about our two systems. Once again the voltmeter is relatively simple to analyse; the input is voltage and the output is some form of numerical display. The car, however, leaves us with a problem.

Considering first the inputs. All of the following quantities and possibly a few others, could be thought of as inputs:

- petrol or diesel
- air
- electrical energy (from the battery)
- 'human energy' (from the driver)
- cold water (for cooling)

Similarly, the following could be thought of as outputs:

- kinetic energy
- rotation of the drive wheels
- exhaust gases
- heat
- sound
- vibration

Once again the answer to the question 'What are the inputs and outputs of a system?' really depends on your viewpoint, and whether you are examining the purpose or function of the system. Also, you might consider some of the outputs to be undesirable, in which case they will not be part of the purpose of the system. For instance, the heat and noise produced by a refrigerator are definitely not desirable outputs, or part of the purpose. They are by-products of the function.

Represent input(s) and output(s) of engineering systems using block diagrams

You have most likely found the first two sections rather wordy; so would most engineers. A much clearer way of representing an engineering system is with a *block diagram*. All systems can be defined by a simple block diagram of the type shown in Fig. 2.1

Figure 2.1 A block diagram

A block diagram such as Fig. 2.2 can give all the information required about an analogue type of voltmeter, and by adding a few words of description (voltage and scale position) the system is defined without the need for a paragraph of writing!

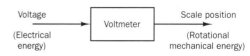

Figure 2.2 Voltmeter (analogue type)

A block diagram of a car can be seen in Fig. 2.3, which simplifies greatly our ideas about what a car does. However, as you will find out later, complex systems can often be broken down and represented by several block diagrams joined together. For now, look at the list of systems contained within a car and draw a block diagram for each:

Figure 2.3 Block diagram of a car

- ignition system (as an example, this has been done for you in Fig. 2.4)

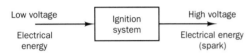

Figure 2.4 Block diagram of an ignition system

- cooling system
- fuel system
- lubrication system

If you cannot figure out these block diagrams at this stage, don't worry. In Element 2.2 you will find suggestions for some of these systems.

> **Explain the implications of input(s) and output(s) for the engineering systems**

When an engineering product is being designed the nature of the input(s) and required outputs will have implications for the system. For instance, any device which requires mains electrical power will have to be designed (in the UK) to run at 240 volts 50 Hz. A similar system used in the USA would have to run at 120 V 60 Hz. Hence the nature of the electrical input has an implication for the system.

If we use a rotary sander as an example, then there are several implications of the outputs for the system. These are:

- The size of sandpaper available.
- The amount of movement required.
- The vibration.

The size of sandpaper available will determine the size of the sanding pad and the amount of movement required will influence the design of the oscillatory mechanism. You may not have thought about vibration as you read this section, but this can have an implication for many systems, although it is usually an unwanted by-product. Products have to be designed to withstand the effects of vibration, which include:

- Nuts, bolts and screws loosening.
- Failure of components due to fatigue.
- Injury to the operator.

Real engineering systems

Now that you have studied the theory covered by Element 2.1 we can look at some real engineering systems in terms of their *inputs*, *outputs*, *purposes* and *functions*. The systems which will be analysed in detail are:

- a lathe (an electromechanical system)
- a hydraulic jack (a fluidic system)
- a battery (a chemical system)
- a mobile phone (an information/data communication system)
- a refrigerator (a thermodynamic system)

Lathe

As part of your current GNVQ course, or in a previous course, you may well have used a centre lathe to produce an engineered product (Fig. 2.5). Of course the lathe itself is also a product, and hence an engineering system. Lathes are produced in many shapes and sizes, from tiny model maker lathes to large CNC (computer numerical controlled) machines, all of which produce objects by turning the workpiece while removing material with a tool moving in a straight line.

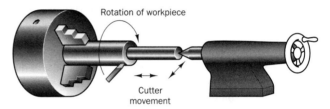

Figure 2.5 A lathe in action

Even a small, simple lathe has many components which go together to make the system. However, as we have already discussed, a block diagram will simplify things. To draw the diagram we need to be sure of:

- the input
- the purpose
- the function
- the output

If we think about the system in terms of its purpose then the block diagram might be in the form of Fig. 2.6.

Purpose: The production of products from stock material

Figure 2.6 The purpose of a lathe

In terms of the lathe's function, the block diagram will be in the form shown in Fig. 2.7.

Function: Conversion of electrical energy to rotational mechanical energy

Figure 2.7 The function of a lathe

Implications of inputs and outputs
In both diagrams the inputs and outputs have implications for the system. Thinking first about the inputs, the implications might be:

- Size of stock material will define chuck size and gear ratios.
- Supply voltage will define the type of electric motor used.

The implications of the stated outputs might be:

- Size of product will define gear ratios, bed length, type of tool, etc.
- Required chuck and tool motion will define gear ratios.

You can probably think of more implications. Some by-products of lathe operation are:

- Heat from the cutting process.
- Heat from the motor and gearbox.
- Sound.
- Vibration.

These are all undesirable; without them there would be no need for coolant, and the machine shop would have a much more pleasant environment. Designers of machine tools strive to remove or absorb vibration as it can cause reduced accuracy and a poor surface finish.

Hydraulic jack

Hydraulic systems make use of liquid pressure to transfer forces from one point to another. The advantage of hydraulic systems over purely mechanical devices, such as levers, is that force can be transmitted over long distances, round corners, etc., with very little loss of energy in the process.

Simple hydraulic jacks are used to raise motor vehicles to allow wheels to be changed, etc. Jacks consist of two hydraulic cylinders, one with a small diameter and one much larger connected by a pipe which cannot distort under the hydraulic pressure (Fig. 2.8).

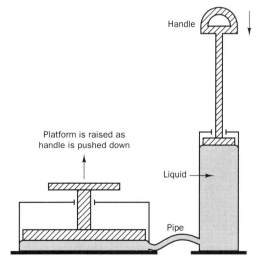

Figure 2.8 A hydraulic jack

The *purpose* of the system is to raise vehicles, the *function* is to amplify force. Two block diagrams could therefore be drawn, as shown in Fig. 2.9.

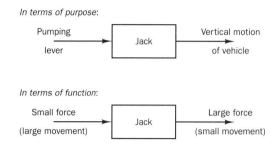

Figure 2.9 Purpose and function of a hydraulic jack

Note that although the face is amplified, the operator of the jack must move the pump handle much further than the vertical distance moved by the vehicle. The work (energy) put into the system cannot be less than the work output. As work done is given by Force × Distance (moved in direction of the face)

Force on handle × Distance moved by handle

must be greater than

Force on vehicle × Distance moved by vehicle

Why must the work input be greater than (and not equal to) the work output?

Implications of inputs and outputs
The most important implications which should be considered are:

- The weight of the vehicle to be raised (although only a proportion of that weight will be lifted by one jack).
- The height to which the vehicle is to be raised.
- The size of force which can be applied by the operator.

Of these, the first is the overriding design consideration or implication; if you wanted to buy a hydraulic jack the vehicle weight would determine the size of jack chosen.

Battery

Batteries come in many shapes, sizes and voltages, but the function of all of them is to use chemical energy to either store or supply electrical energy. (Batteries which are non-rechargeable can only supply electrical energy.)

Before drawing the block diagrams we are going to simplify further the way in which the analysis of systems is presented. This will help you when you come to present your GNVQ evidence.

System: 12 volt motorcycle battery

Function:

Electrical energy → Battery → Chemical energy (Charging)

Chemical energy → Battery → Electrical energy (Discharging)

Purpose:

Electrical energy from generator → Battery → Electrical energy to power starter, lights etc.

(Note: The purpose of the battery is not to be recharged)

Implications of the input:

- The input voltage to the battery will govern the generator voltage.

Implications of the output:

- The power required by the starter and accessories, and their frequency of use, will govern the power rating of the battery.

By-products of the system:

- Heat
- Hydrogen (when the battery is charging)

Implications of the input:

- Loudness of voice will determine the sensitivity of the phone.
- Frequency range of average voice will determine range of sound frequencies to which the phone is sensitive.
- Network signals will determine the type of signal which can be received by the phone.

Implications of the output:

- The output signal will be governed by the type of signal which is used by the phone network.
- Ear sensitivity will determine the loudness of the phone.

By-products of the system:

- None of any significance.

This method of presentation should:

- Simplify your presentation.
- Make your work more easily read and marked.
- Reduce the amount of writing you have to do.

From now on the systems used as examples will all be presented in this way, with the inclusion of diagrams/photographs.

Mobile phone

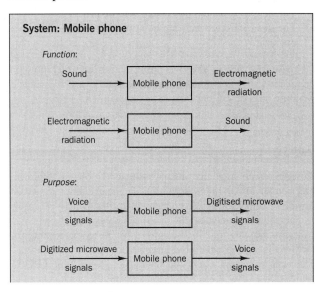

System: Mobile phone

Function:

Sound → Mobile phone → Electromagnetic radiation

Electromagnetic radiation → Mobile phone → Sound

Purpose:

Voice signals → Mobile phone → Digitised microwave signals

Digitized microwave signals → Mobile phone → Voice signals

Refrigerator

The refrigerator is an example of a fairly complex thermodynamic system which has a very simple function. The block diagram below shows the basic components of a fridge, the overall function of which is to remove (heat from the stored food) at a low temperature and release it (via the condenser) at a higher temperature. This raising of energy requires another energy input, usually in the form of electricity. You will notice that there are now two inputs shown on the block diagram for the function of the system.

Function or purpose?
Throughout this chapter so far we have drawn block diagrams to illustrate both the function and purpose. As has already been stated, the way in which you look at a system depends on who you are. Now that the systems being analysed are more complex the function of the system is becoming more relevant to us as engineers; the purpose of the system, however, should always be borne in mind as it is the only reason for the product's existence!

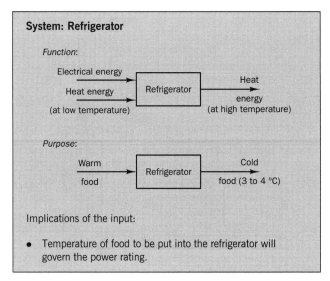

System: Refrigerator

Function:

Electrical energy, Heat energy (at low temperature) → Refrigerator → Heat energy (at high temperature)

Purpose:

Warm food → Refrigerator → Cold food (3 to 4 °C)

Implications of the input:

- Temperature of food to be put into the refrigerator will govern the power rating.

Activities for Element 2.1

The GNVQ syllabus requires five short reports looking at one system from each of the categories:

- electromechanical
- fluidic
- chemical
- information/data communication
- thermodynamic

Each report should contain descriptions of:

- the purpose of the system
- the input(s)
- the output(s)
- the implication of the inputs and outputs
- block diagrams

It is suggested that you follow the format used in the text for the last few 'worked' examples.

Although you are free to choose any systems for analysis, it would pay, at this stage, to stick to simple products. The following are suggestions; you must decide into which category they could be placed.

Electric screwdriver   Air compressor
Bicycle pump   Aquarium aerator
Cassette player

2.2 Investigate the operation of engineering systems

In Element 2.1 you studied the way in which systems can be represented using block diagrams and the ways in which the nature of the inputs and outputs influenced the design of systems.

In Element 2.2 you will be looking at systems in more detail by carrying out the following:

- Breaking systems down into subsystems.
- Describing the functions of subsystems.
- Looking at the relationships between subsystems.
- Looking at the ways in which systems can be controlled.

The last part of this element is concerned with the practical evaluation of engineering system performance. Once again real engineering examples are used to illustrate the theory, and suggestions for evidence-gathering activities are given towards the end of this element.

Describe engineering systems using block diagrams

In Element 2.1 we mentioned that many engineering systems consist of several component subsystems. Each subsystem can be treated as a separate system, and can be represented by a block diagram.

To illustrate this idea we will consider a hi-fi system consisting of a CD player, an amplifier and speakers. Even if the hi-fi system is sold as one unit (as opposed to separate components) it will contain these separate engineering systems. If the system is considered as one unit the block diagram might be as shown in Fig. 2.10.

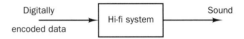

Figure 2.10 Block diagram of hi-fi system

Of course another input to the system would be electrical energy, but although necessary for the operation of the system, the electrical input is not part of the hi-fi system's purpose. You might like to think about its relevance to the overall function of the system, however. Figure 2.11 shows the separate subsystems

The next example is a different type of engineering system altogether. The structure of Fig. 2.12 represents a small manufacturing company making one-off engineering products.

The company may carry out some design and drawing work before the product is made, and the product itself may go through several processes – for instance, a cowling might require some machining – but breaking the system into subsystems, the block diagram would have the form of Fig. 2.13.

You can see that this block diagram is becoming rather large. We have included some of the 'financial' subsystems,

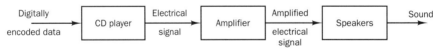

Figure 2.11 Block diagrams of separate subsystems

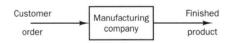

Figure 2.12 Block diagram for one-off product

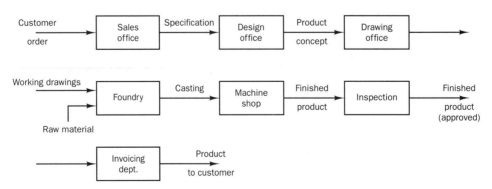

Figure 2.13 Block diagram of operations required for a one-off product

the sales and invoicing offices, and have introduced a second input at the production stage, the raw material. There may also be other inputs, and possibly outputs, depending on the type of product.

Types of subsystems

In this section we will examine the functions of subsystems by putting them into six broad categories

- energy delivery subsystems
- energy conversion subsystems
- energy utilisation subsystems
- monitoring subsystems
- control subsystems
- waste discharge subsystems

In each category, examples of subsystems will be examined, and suggestions for further study given.

Energy delivery subsystems

In many engineering systems a source of energy or power is required to allow the system to operate. In this category we will examine three energy delivery subsystems:

- power supply units
- electric motors
- combustion engines

Power supply unit (PSU)

PSUs are an integral part of many electrical and electromechanical systems, which are mains operated. The reason for using mains electricity is that it is a reliable and constant source of energy. However, many components operate at voltages considerably lower than the 230 volts supplied. In some components, such as televisions and devices containing thermionic valves (e.g. guitar amplifiers), voltages greater than 230 volts are required. In addition some devices require a direct current supply and not alternating current as supplied by the mains.

PSUs may take the form of encapsulated components ('bricks') which plug directly into a three-pin socket, or can be integral with the device they are supplying, as in personal computers.

The functions of a PSU can be summarised as:

- The supply of electric at a voltage different to the mains supply.
- Rectification (a.c. to d.c.) of the supply.
- Stabilisation of the supply (constant current or voltage).
- Safety isolation.

PSUs may not have all these functions but some combination.

Changing voltage
Voltage changing is carried out using a 'step-down' (voltage-reducing) or 'step-up' (voltage-increasing) transformer. Many PSUs use transformers which have several outputs, and offer different voltages, e.g. 12 V, 9 V and 6 V.

The voltage changing stage of a PSU can be represented by Fig. 2.14, or in the case of a multiple output transformer as might be fitted to a computer, by Fig. 2.15.

Figure 2.14 Block diagram of a transformer

Figure 2.15 Block diagram of a multiple output transformer

An important point to note here is that the voltage changing *must* be carried out before rectification takes place. Transformers will only operate on a.c.

Rectification
Many PSUs include diode bridges to convert the reduced voltage supply from a.c. to d.c. to run d.c. circuits, d.c. motors, etc. Rectifiers often include capacitors to smooth the output. This is particularly true in audio devices where any 'ripple' will cause a hum to be heard (Fig. 2.16).

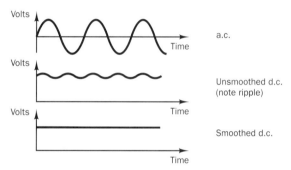

Figure 2.16 Converting a.c. to smoothed d.c.

The rectification stage of a PSU is shown in Fig. 2.17.

Figure 2.17 Block diagram of a rectifier

Stabilisation
Stabilised PSUs contain circuitry which ensures that the output voltage remains constant as the load changes. Instrumentation and audio devices are just two examples of systems which might require a stabilised supply. Cheaper PSUs are often unstabilised and care must be taken when using such devices as the output voltage will drop when the current drawn increases.

Safety isolation
Power supplies may be used to provide isolation from the mains supply, and thereby provide electrical protection for the user of electrical devices. Many power tools, particularly those used in industry, operate at reduced voltage and are plugged into PSUs. The transformers used provide further protection by isolating the mains supply from the output, so that there is no possibility of mains voltage reaching the output.

PSU block diagram

The block diagram for the complete PSU will depend on the required output. The following diagram shows a PSU for use with a radio-controlled model. The *purpose* of the device is to charge up the batteries. The block diagram (Fig. 2.18) shows its *function*.

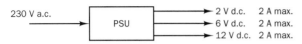

Figure 2.18 Block diagram of a PSU

Although the PSU is a complete subsystem, it can be further broken down into the components already described (Fig. 2.19).

Electric motors

The function of an electric motor is to supply kinetic energy, in the form of rotation, to a system. Motors can be categorised in many ways, but two common ways are:

- a.c. and d.c. motors
- constant speed and variable speed motors

D.C. motors
All d.c. motors are of the variable speed type, the motor speed broadly depending on the voltage supplied. However, one type of d.c. motor, the *shunt* motor, is suitable for maintaining a constant speed under conditions of varying load as its speed can be closely controlled.

Where a high starting torque is required, the *series* motor is used; however, this type of motor is not at all suitable for constant speed drives, and care must be taken that the load on the motor is not greatly reduced as it will overspeed.

Compound motors share constructional features with both series and shunt motors, and are used where a high starting torque is required but where the load may drop off considerably, at which point the motor will not overspeed. It is also suitable for use in systems where there is a fluctuating voltage supply.

A.C. motors
The speed of a.c. motors depends on the frequency of the supplied a.c. voltage. Speed variation is however possible, usually with a loss of efficiency as speed decreases. *Synchronous* motors are used in constant speed applications, whereas *induction* motors can be used where speeds below the 'synchronous speed' are required.

Torque characteristics
Apart from motor speed, the ability of a motor to supply torque is very important. In some systems, such as the traction motors on railway locomotives, a high starting torque is required to get a piece of machinery moving. In other situations, such as fan drives, the torque required increases with speed.

When selecting a motor for use as an energy delivery subsystem, reference must be made to the speed/torque characteristics which are given in manufacturers' catalogues.

Combustion engines

Internal combustion engines
Engines are used as energy delivery systems in two ways:

- As propulsive units, e.g. cars, planes, ships.
- As a means of driving other energy delivery systems, e.g. generators, compressors, pumps.

The advantage of engines over electric motors is that they are more portable in that they do not rely on a mains electricity supply. This is particularly true where large quantities of energy are required. At the lower end of the scale, battery driven electric motors are more suitable, and less expensive than engines.

The main categories of engine in use are:

- reciprocating engines (diesel, petrol, gas)
- gas turbines (kerosene, gas)

Diesel engines and turbines are particularly suitable for constant speed applications (generators, etc.) with turbines generally being limited to larger systems.

External combustion engines
Steam engines and air engines (e.g. Stirling engines) are both examples of external combustion engines, i.e. the fuel is burned outside the engine. In some cases, no fuel is actually burned, but heat is supplied from, for example, an electrical coil.

As with internal combustion engines the main categories are:

- reciprocating engines (steam engines, Stirling engines)
- turbines (steam turbines)

Other types of energy delivery system

Apart from the three types of energy delivery subsystems covered in this section, the following will also be found in many engineering systems:

- compressors (pneumatic systems)
- pumps (hydraulic systems)
- batteries (portable electrical devices)

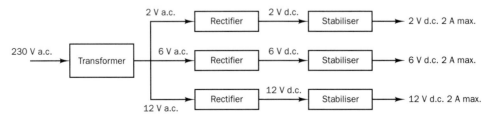

Figure 2.19 Components of a PSU

Some of these subsystems require the use of another energy delivery device, and in some cases two others. For instance, a battery-powered pump could be represented by Fig. 2.20.

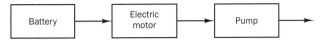

Figure 2.20 block diagram of a battery-powered pump

However, you should remember, that the function of the pump is to convert low-pressure liquid to high-pressure liquid, as shown in Fig. 2.21.

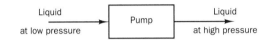

Figure 2.21 Block diagram of a pump

Energy conversion subsystems

In this section, we will look at some devices, the function of which is to convert energy from one form to another. Many devices convert energy as a means of carrying out their function, but the devices considered here do so as their main purpose.

Transducers

Transducers are usually, but not always, electrical/electronic devices which are used to sense changes in quantities, for instance, in measurement or control systems. An example of a transducer is a thermocouple (Fig. 2.22). This is a device for detecting temperature changes.

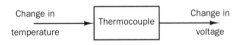

Figure 2.22 Block diagram of a thermocouple

There are far too many different types of transducers available to give descriptions of each; instead, Table 2.1 summarises the functions of some of the more common devices. Should you need a transducer for a particular purpose, reference should be made to a manufacturer's catalogue. Some of the terms used to specify and evaluate the performance of subsystems, such as transducers, are described on p. 29.

Table 2.1 Inputs and outputs of some transducers

Sub-system	Input	Output
Thermocouple	Temperature	Voltage
Thermister	Temperature	Resistance charge
Bourdon tube	Pressure	Mechanical motion
Piezo	Pressure/force	Voltage
LVDT	Displacement	Voltage
Strain guage	Small change in size/displacement	Change in resistance

Actuators

Actuators are devices which convert delivered energy to motion or force (Fig. 2.23). The input energy to actuator subsystems can be obtained from:

- pneumatic and hydraulic actuators
- electrical actuators

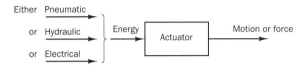

Figure 2.23 Block diagram of an actuator

Pneumatic and hydraulic actuators
The way in which pneumatic and hydraulic actuators operate is very similar, the major difference between the two systems is that the fluid pressure in hydraulic systems is very much higher than that in pneumatic systems.

According to the type of motion produced, actuators can be categorised as:

- linear
- rotary

Linear actuators The hydraulic or pneumatic cylinder or ram is the commonest type of linear actuator (Fig. 2.24). These can be seen in use on mechanical diggers (hydraulic) and automatic bus doors (pneumatic).

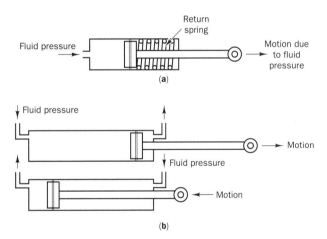

Figure 2.24 Linear actuators: (a) single-acting cylinder (fluid pressure causes motion in one direction only); (b) double-acting cylinder

Rotary Actuators Rotary actuation can be achieved in two ways (Fig. 2.25):

- Using a linear actuator with a crank mechanism.
- Using a property rotary actuator – usually more expensive.

Dedicated rotary actuators can achieve large angles of rotation in a very small space. Linear actuators with cranks, on the other hand, are bulky and involve several points of rotation and hence wear.

In both type of actuator the force or torque produced will depend on the fluid pressure. Due to the greater system

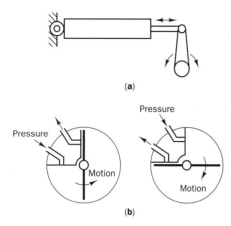

Figure 2.25 Actuators: (a) linear with crank; (b) rotary

pressure, hydraulic actuators are used where large forces and torques are required. Conversely, pneumatic systems tend to act quicker as air does not have the viscosity problems associated with hydraulic oil.

Electrical actuators
Again, both linear and rotary motion can be achieved using electrical actuators.

Linear actuators Linear actuation is achieved electrically using solenoids, as shown in Fig. 2.26.

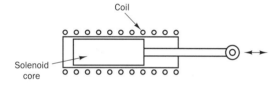

Figure 2.26 Linear actuation using a solenoid

When a current is passed through the coil the core will move either in or out of the coil, the direction of motion depending on the direction of current. For a given solenoid, the force produced will depend on the magnitude of current passed.

Rotary actuators Electrical rotary actuation can be achieved in three ways:

- solenoid with a crank (similar to the hydraulic system)
- stepper motor
- synchro

Signal-processing units

Signal-processing subsystems are used to convert electrical signals from other subsystems, such as transducers, into different electrical signals which can be used by subsequent subsystems. In other words, a signal-processing subsystem acts as an interface between other subsystems.

The diversity of signal processors available prohibits an in-depth coverage. Instead, three of the most common types are covered in some detail. These are:

- Wheatstone bridge
- amplifier
- charge amplifier

- frequency discriminator
- A to D and D to A converters

Wheatstone bridge
In each case, we will look at the applications of the subsystems and not the principles by which they operate. Wheatstone bridges convert small changes in resistance to changes in voltage, which can be displayed. Probably the most common use for this subsystem (Fig. 2.27) is as a signal processor in strain gauge systems.

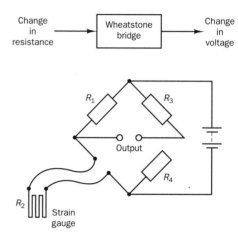

Figure 2.27 Wheatstone bridges in a strain gauge system

When the ratio R_1/R_2 equals R_3/R_4 the output voltage will be zero. However, when the component on which the gauge is mounted, is strained, the value of R_2 will change, R_1/R_2 will not equal R_3/R_4 and there will be a voltage across the output terminals, which can be measured.

Temperature compensation
With the configuration shown, an error can occur if R_1 and R_2 are at different temperatures, as increased temperatures will cause an increase in resistance. This problem can be removed if R_1 is a 'dummy' gauge, attached to a piece of material, similar to the active gauge and in the same environment. Any temperature variation will then affect both resistances and the ratio R_1/R_2 will only reflect resistance changes due to strain.

The Wheatstone bridge can be further refined, and made more sensitive by making R_1 an active gauge, but mounting it so that when the resistance of R_1 increases, the resistance of R_2 decreases, and vice versa, as depicted in Fig. 2.28. When R_1 is stretched (in tension), R_2 will be in compression. The ratio R_1/R_2 is therefore increased. This is known as a 'half-bridge'.

A further refinement utilises all four gauges in a group or 'rosette'. In this configuration the gauges must be arranged so that R_1 and R_4 increase as R_2 and R_3 decrease, and vice versa, to maximise sensitivity. This arrangement is known as a 'full-bridge'.

To further increase the sensitivity of the subsystem the output can be fed to an amplifier before the signal finally reaches the display unit, usually a digital meter, calibrated, in this case, show the strain.

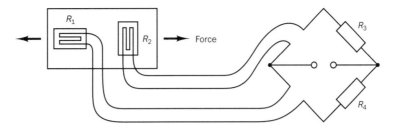

Figure 2.28 Example of a half-bridge

Amplifiers
The voltage output from many transducers is very small. In order to efficiently read the signal it must first be amplified, so that readings taken on meters (digital or analogue) are high enough to make reading errors small. If signals are to be transmitted through long lengths of wire amplification it is essential to avoid significant losses due to resistance.

The most common type of amplifier used in instrumentation is the Operational Amplifier (op-amp). This device is contained in a chip about 10 mm × 6 mm.

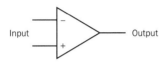

Figure 2.29 The operational amplifier

Op-amps amplify the difference between the voltages at the two inputs (Fig. 2.29). One input is known as the inverting input (denoted by a '−' sign); the signal at this input is subtracted from that at the non-inverting (+) input. The difference between the two is then amplified (Fig. 2.30).

Amplifier gain
The amount by which the input is amplified (multiplied) is known as the *gain*.

$$\text{Gain} = \frac{\text{Output voltage}}{\text{Input voltage}}$$

In the examples shown in Fig. 2.30, the amplifiers have been given a gain of 10. The gain itself is controlled by the resistors connected to the input and output (Fig. 2.31). The gain is found by dividing the feedback resistance by the input resistance, i.e. $G = R_F/R_I$.

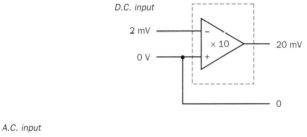

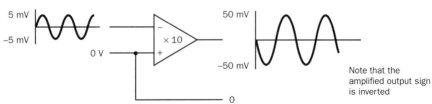

Figure 2.30 Uses of an op-amp

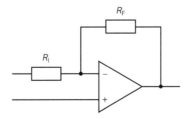

Figure 2.31 The use of resistors to control the gain

Applications

A common application for op-amps is as a thermocouple amplifier. Thermocouples are widely used as temperature transducers, but their voltage output is very low, about 1 mV for a 200 °C temperature change. The signal often has to be transmitted through long wires, from the heat source to the monitoring device, so the output from the thermocouple is amplified. Thermocouple amplifiers can be obtained as ready made units, as can other amplifiers. For instance, strain gauge amplifiers can be found in manufacturers' catalogues for use with Wheatstone bridges, the two devices often being combined in one package.

A to D and D to A converters

Most modern recording/storage devices, especially computers, store data in the form of digital information. However, the output from most transducers is analogue, i.e. continuously variable. If the information from an analogue transducer is to be transmitted to a digital device, then the signal will have to be converted by an 'A to D' (analogue to digital) converter.

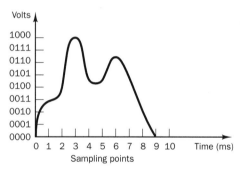

Figure 2.32 An analogue signal

Figure 2.32 shows an analogue signal. The amplitude of the signal is divided into 4 bits (binary digits). The signal is sampled every millisecond (i.e. 1,000 times per second). At each sample point the size of the signal is measured to the nearest bit. The digitised signal would be as shown in Fig. 2.33.

Clearly there is some error known as quantization error involved in the digitising process. This error can be reduced in two ways:

- Using more bits to cover the amplitude of the signal, i.e. increasing the resolution. (Table 2.2 gives the number of steps (increments) available for various numbers of bits.)

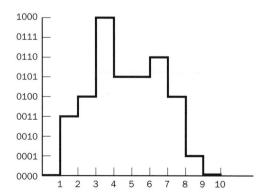

Figure 2.33 A 4-bit digitised signal

- Increasing the sampling rate; in general the sampling rate should be at least twice the highest frequency to be converted.

Table 2.2 Capabilities of different analogue to digital converts

Type of converter	Increments
4 bit	15
8 bit	255
16 bit	65,535
32 bit	4,294,967,295

If a computer is used to control an analogue device such as a valve or motor, the digital signal has to be converted to analogue. The process is similar, and once again some error will occur.

Pulse counters

Pulse counters are a particular type of D to A converter used with speed/acceleration measurement systems, using magnetic or optical sensors (transducers). A pulse counter (Fig. 2.34) receives a stream of electrical pulses from the transducer which it converts to a d.c. voltage. The magnitude of the voltage is proportional to the rate at which pulses are received.

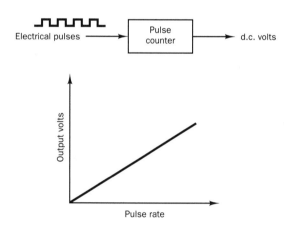

Figure 2.34 Block diagram and graph of a pulse counter

Signal modulation

Information transmitted by radio telemetry has to be encoded so that the receiver can interpret the data. The signal to be transmitted is 'transported' by a carrier wave which is modulated or changed in one of two ways – amplitude modulation (AM) or frequency modulation (FM) – by the signal, as shown in Fig. 2.35.

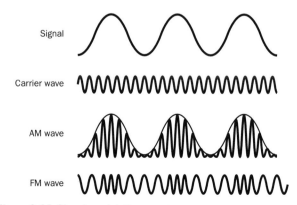

Figure 2.35 Signal modulation

In most FM systems, the signal is also digitised. The advantage of FM systems is that they are less sensitive to interference.

Monitoring systems

Monitoring systems can be split into two categories.

- Straightforward measurement systems.
- Measurement systems which are part of control systems.

The two types may be represented by block diagrams (Fig. 2.36).

The common features of the two categories of monitoring system are

- The transducer which senses the quantity being measured.
- The signal processor which converts the transducer output signal into one which can be interpreted by either a display unit or the control system.

The previous two sections covered in some detail, various transducers and signal processors. In this section we will look at the way the two components are used together. Table 2.3 gives the type of signal processing which is used with variety of common transducers. This table follows on from Table 2.1 which gave the inputs and outputs of the same transducers.

All the systems in Table 2.3 could be represented by block diagrams (Fig. 2.37), e.g. a speed measurement system using a magnetic pickup to sense the revolutions of a shaft.

Table 2.3 Typical transducers and typical signal processors

Measured quantity	Transducer	Signal processor
Temperature	Thermocouple	Amplifier
	Thermistor	Wheatstone bridge and/or amplifier
Pressure	Bourdon type	Gearing
	Piezo device	Amplifier
Position/displacement	LVDT	Amplifier and rectifier
	Strain gauge	Wheatstone bridge and amplifier
Speed	Tachogenerator	Possibly a rectifier
	Optical or magnetic sensor	Pulse counter and amplifier

Control systems

Control systems comprise several subsystems. The subsystems used will depend on whether the control system is:

- open loop
- closed loop

The type of strategy employed to control engineering systems is discussed in some detail on p. 39. In this section we will look at the types of subsystem you might find being used in open- and closed-loop systems (Fig. 2.38).

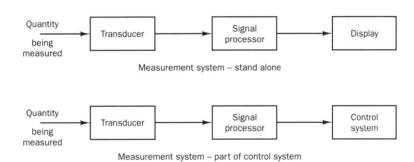

Figure 2.36 Block diagrams of monitoring systems

Figure 2.37 A speed measurement system

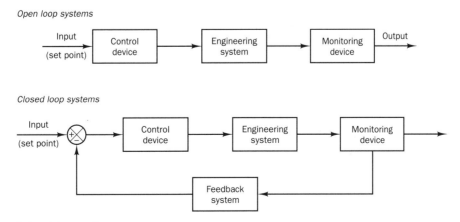

Figure 2.38 Open- and closed-loop systems

Open-loop systems

Open-loop controllers
Any type of open-loop controller must be preset by the user, i.e. the user defines the set point. Any disturbance to the output is likely to cause the system to move away from the set point, and some readjustment will be necessary.

Control of electrical and electromechanical systems
Electric heaters, motors and motor-driven devices, lights, etc., can all be controlled by varying the voltage across the device which in turn controls the current flowing (Ohm's law!). Figure 2.39 shows such a system; the voltage being varied by the variable resistor.

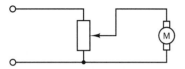

Figure 2.39 A variable resistor

This type of system can, however, only be used in low-power systems as the 'unwanted' energy has to be dissipated through the variable resistor, in the form of heat.

An alternative method of controlling, for instance, the output from electric heaters, is to switch in or out a series of identical heating elements, as shown in Fig. 2.40.

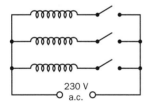

Figure 2.40 Mains electric heater (3 elements)

This type of control does not provide continuous variability as with a variable resistor, but, as shown in Fig. 2.41, changes the output in discreet steps.

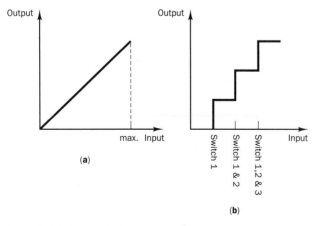

Figure 2.41 Changes in heat output: (a) control by variable resistor; (b) control by switching

A simpler form of switching is shown in Fig. 2.42. In this configuration the heater has only two possible states, on or off, and temperature control is very limited.

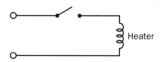

Figure 2.42 A simple form of switching

The two methods of open-loop control then can be described as either

- continuously variable or *proportional*
- switched *on/off*

These ideas can now be extended to other types of system.

Open-loop control of fluidic systems
Fluidic systems can be controlled in much the same way as electrical systems, by either providing a valve which allows continuous variation of pressure or flow rate, or by using a valve which is either open or shut.

Valves may be controlled remotely by providing actuators to provide the energy required to move the valve. Motor-driven valves provide continuous variability while solenoid valves are on/off devices, but are considerably cheaper and relatively maintenance free.

Control of engines
Open-loop control of engines is something that we probably experience every day. The speed of a car engine is controlled simply by varying, continuously, the flow of fuel/air mixture to the combustion chambers.

It is possible to use on/off control on stationary engines, of the type used to drive pumps, generators, etc. This can be achieved by switching cylinders in and out, or simply by turning the whole engine on and off. However, it is far more common to find engines controlled by governors, a form of closed-loop system.

Closed-loop systems

Closed-loop systems contain the same types of control subsystems as open-loop systems, the major difference being that in closed-loop systems the control subsystem is driven not by a switch or preset control knob, but by a signal or setting determined by another subsystem, the error detector or comparator (Fig. 2.43).

The error signal is the difference between the set point (the input which gives the *required* output) and the feedback signal which is obtained from the *actual* output.

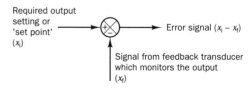

Figure 2.43 A monitoring system

Possible output conditions
- Actual output = Required output
$$x_f = x_i \quad \text{so} \quad (x_i - x_f) = 0,$$
i.e. the error signal will do nothing to the output.
- Actual output > Required output
$$x_f > x_i \quad \text{so} \quad x_i - x_f \text{ is negative}$$
i.e. the error signal will *reduce* the output.
- Actual output < Required output
$$x_f < x_i \quad \text{so} \quad x_i - x_f \text{ is positive}$$
i.e. the error signal will *increase* the output.

Electrical systems
In electrical closed-loop systems, the op-amp is frequently used as an error detector. The inverting input provides a negative sense to the feedback signal, and the amplifier acts as the difference between the two inputs, as shown in Fig. 2.44.

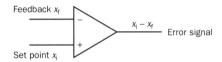

Figure 2.44 The op-amp as an error detector

The amplifier may be set with a gain of 1, i.e. there is no amplification and the device acts purely as an error detector. Alternatively, it can be used to amplify the error signal to a level which can be accepted by the control device.

Waste discharge systems

The unwanted by-products of engineering systems often require separate subsystems to deal with their removal. This section describes four waste discharge systems, but it should be remembered that there is a huge diversity of such subsystems. The unwanted by-products we will deal with are:

- heat
- noise
- chemical waste
- radiation

Heat removal

Heat is a by-product of many processes and systems ranging from chemical processes to heat engines. The removal of heat can be carried out by:

- a passive subsystem (requiring no energy input)
- an active subsystem (requiring energy input)

A good example of a passive system is the finning around air-cooled engines and compressors. The fins are designed to present a large area to the air so that it can be radiated. This can be enhanced by providing an appropriate surface finish, such as matt black paint. The disadvantage of such a system is that there is no control over the cooling process. To achieve this the system must be made active.

The simplest way to do this would be to provide a fan to force air through the fins. By either switching the fan, or continuously controlling its speed, the system can be made to respond to changes in heat output, air temperature, etc.

The system can be further refined by putting a water (or other coolant) jacket around the device giving off heat. The water flows through a heat exchanger (radiator) which may be fan cooled. This is the type of system used in most road vehicles.

The three subsystems described can be represented by the block diagrams shown in Fig. 2.45.

Noise removal

As was the case with heat removal, the removal of noise from a system may be carried out using either passive or active subsystems.

The simplest way to remove noise is to absorb it. Sound is a form of energy which can be damped out or absorbed by acoustic materials. Machinery noise can be absorbed in this way by lining the area in which the machine is operating with sound-deadening material. In cars, a similar process is used to absorb engine and road noise.

Another type of passive process is used in vehicle silencers to remove energy from the exhaust gases by forcing them to pass through, or around, a series of baffles. This has the effect of dissipating sound waves.

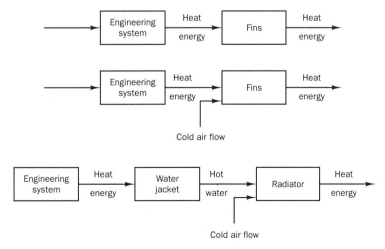

Figure 2.45 Block diagrams of heat energy systems

Active noise removal systems use loud speakers to emit sound of the same frequency as the offending noise (Fig. 2.46), but 180° out of phase.

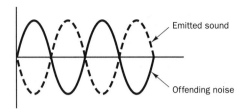

Figure 2.46 Reducing/eliminating noise

The two waves 'cancel out' and no noise is heard. In practice it is difficult to emit a sound which is exactly 180° out of phase, but very good results can still be achieved. At the time of writing this type of noise removal subsystem has been tested by some car manufacturers with some success.

The best way to remove noise is to prevent it starting. Components vibrating are a common cause of unwanted noise, and by careful design this problem can be minimised. The study of vibration control is a subject in its own right. Apart from the problem of noise vibration it can lead to structural fatigue failures, human injury and poor accuracy in manufacturing processes.

Chemical waste

Chemical waste can be very difficult to deal with. Since the 1960s we have become more and more aware of the environmental problems associated with chemical waste and, as engineers, we have a responsibility to deal with these materials in a safe manner.

There is no one way of dealing with chemical by-products, but as with any other waste, the best way to remove it is at its source, i.e. to prevent the *production* of waste. If this is not possible, then ways of neutralising the waste chemically must be found; dumping in the sea or landfill sites are *not* acceptable solutions, as sooner or later the chemicals will leak into the sea and then the food chain, or into a water course, and hence our drinking water, as can be seen in Fig. 2.47.

Radiation

Radiation, or more specifically, ionising radiation of the type emitted by radioactive materials, is dangerous to all life. Even low-level radiation in the form of α-particles can cause damage to living cells, which can then mutate, leading to cancers and other problems. Non-ionising radiation such as ultraviolet rays, and microwaves, can still be dangerous, and as scientific understanding increases, so more hazards have become apparent.

Energy utilisation systems

Many devices use an external energy source as a means of carrying out their function. The efficient use of energy is of great importance in industrial systems, where wasted energy means lost profit, and is increasingly becoming a domestic concern.

As engineers, we have a responsibility to use energy efficiently by:

- using 'low-energy' systems
- producing 'low-energy' systems

Over the last ten years or so much research has gone into producing devices which use energy, particularly electrical energy, in an efficient manner. Although the main reason for this is the reduction of operating costs, there is another, perhaps more important issue at stake.

Most of the electrical energy produced is converted from non-renewable resources such as coal, oil and gas, or from nuclear power stations, about which there is much concern regarding safe disposal of waste, and safety in general.

If man is to continue using energy, then renewable energy resources must be found. At the time of writing, a lot of research is being carried out into the viability of:

- solar devices
- wind power
- wave power

Identify and describe subsystem interrelationships

The systems used so far, whether for measurement, energy conversion, material processing or information processing, have all been of a linear nature. This means that there is only one path taken as the input is processed. The overall sensitivity of such systems can be determined, where appropriate, using the sensitivities of the component subsystems.

The sensitivity (G) of the whole system in Fig. 2.48 is $G_1 \times G_2 \times G_3$.

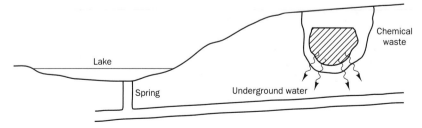

Figure 2.47 An unacceptable solution to waste disposal

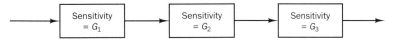

Figure 2.48 Interrelationship of subsystems

Feedback systems

Many systems make use of *feedback loops* in order to allow the system to respond to changes (disturbances) to the output. The details of such systems will be covered in more detail in the next section.

The common feature of all feedback systems, whether they are 'hard' or 'soft', is that there is a loop back from the output to the input (Fig. 2.49).

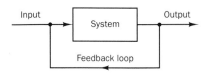

Figure 2.49 A feedback loop

We have now introduced an alternative or secondary path for the signal, so the system is no longer linear.

In a material-processing (manufacturing) system, the inspection procedure is often the point at which the 'signal' is monitored and the feedback loop starts (Fig. 2.50). It may be that components which fail inspection can be reprocessed; this is especially true of cast components.

Figure 2.50 A feedback system

Alternatively, the feedback signal may be in the form of information which might lead to a modification of the input and/or process itself. If there is tool wear the products might be consistently oversize, in which case the feedback information would lead to renewal of the offending cutter.

The constant speed of a diesel engine driving a generator would be maintained by means of a governor, which could be a mechanical or electronic device. In either case, a signal would be sent from the output to the input, and the information acted upon in such a way as to maintain the speed at the required value.

Real engineering systems

Management information systems

Most of the examples used so far, have either been electromechanical/mechanical devices, or production processes, all of which are engineering systems. However, none of these could exist or take place without a 'human system', i.e. the system which produces or carries out the systems.

In an engineering business many different human systems operate at the same time. Some of these are:

- design systems
- production systems
- financial systems

All such systems are controlled by a management system. In the section on closed-loop control, we talked about signals; the type of signal which passes through a management system is information.

To illustrate the idea of management systems, we will consider an open-loop and a closed-loop system.

Open-loop system

The block diagram (Fig. 2.51) shows a hypothetical human system.

You have already seen a similar system in Element 2.1. In this diagram the flow of information takes place in one direction only. In reality, there would also be a flow of information backwards, and two of the purposes of management systems are (a) to act as the feedback path for comparing the output of a process with that required, and (b) to modify the system to control the output.

For the human system shown in Fig. 2.51, we could add the three feedback paths shown in Fig. 2.52.

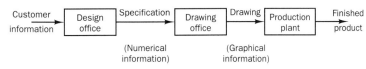

Figure 2.51 Block diagram of a management information system

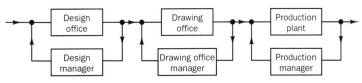

Figure 2.52 Feedback loops in a management information system

Remember, this diagram represents the flow of information, and is really a greatly simplified version of the block diagram, which would describe the management information system in a real engineering business. The main point to make is that in human systems, as in any other, the only way to maintain control of the output is to provide feedback paths, so that the actual output can be compared with the required output, and adjustments made accordingly.

Conveyor systems

Conveyor systems used in continuous production processes are examples of electromechanical systems. We will look at two different processes, starting with frozen peas. When you open a bag of frozen peas, you take it for granted that you will see thousands of individual peas and not a large green mass! However, peas are awkward to freeze, as they are blanched by immersion in very hot water or steam before freezing, which means that even after drying they still contain a lot of moisture. If the peas are to maintain their freshness, they must be frozen quickly so there is not much time for drying.

Having been blanched, the peas pass onto a chain freezer belt through which very cold, dry air is blasted, so that the peas 'float' above the belt. If the belt travels too slowly then the layer of peas is too deep, and the air cannot 'fluidise' the product. If the belt speed is too fast there will be large gaps between the peas, and they tend to stick to the belt rather than float. There is also the problem that, at high belt speeds, the belt is more likely to jump off one of the drive cogs or even break. Figure 2.53 shows the block diagram for the system.

Figure 2.53 Block diagram of a conveyor system

As it stands, this is an open-loop system, but it would be normal to provide some form of feedback by inspecting the frozen peas as they come off the belt, and adjust the speed accordingly. The feedback in this case is not automatic, i.e. not part of the conveyor system and, as such, the system cannot be referred to as closed-loop, although it could be represented by the loop diagram in Fig. 2.54.

The error detector in this case is the person carrying out the inspection. Human feedback systems are quite common in production processes of this type, as automatic monitoring of the output is not always technically feasible.

A major problem which can occur in the control of belt systems, is the time lag between input and output. In the case of the freezer belt, this can lead to a considerable amount of waste before a problem is rectified.

Identify and describe system control strategies and techniques.

Control

The purpose of controlling any system is to maintain the output at the required level, for a given input. The study of control is a subject in its own right, frequently involving complex mathematics to analyse the behaviour of control systems. However, we can look at some simple control systems, and the strategies used to maintain control without the need to delve into mathematics.

Open-loop control

A system with open-loop control has no feedback loop. An example of this is shown in Fig. 2.55. An electric motor is used to drive a pump at a specified speed. The pump is connected to the motor via a gearbox, as is shown in the block diagram.

The motor has been selected so that it will run at a constant speed. The speed is reduced by a fixed ratio in the gearbox, so that the pump also runs at a constant speed. Under normal conditions, this set-up operates satisfactorily; however, if the feed to the pump 'dries up' the resistance felt by the motor will suddenly disappear and it will speed up. As there is no monitoring system to provide feedback, this situation cannot be prevented. In other words, the system cannot *automatically* respond to disturbances to the output.

One way in which this system could be controlled without automating would be to add some form of pump speed indicator which would sound a warning if the pump speed fell below, or rose above, predetermined limits. Somebody could then investigate and adjust the system accordingly.

On/off and continuous control

Control over a system can be maintained in one of two ways:

- 'Switching' the input on or off, i.e. the input is either zero or at a specified level.
- Varying the input continuously between specified limits.

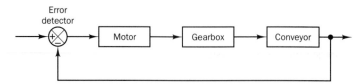

Figure 2.54 Feedback loop on a conveyor system

Figure 2.55 Block diagram of an open-loop system

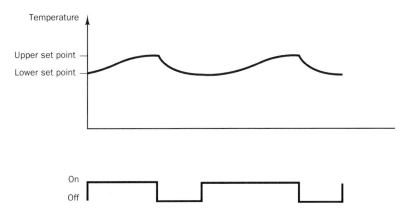

Figure 2.56 Automatic control of a heating system

Domestic central heating systems are examples of an on/off system. Room temperature is monitored and controlled by a thermostat which switches the flow of hot water *on* when the room temperature falls, and *off* when it rises above the specified value. To prevent continual switching, there is a tolerance (built into the thermostat) on the temperature. The method of controlling the temperature is demonstrated in Fig. 2.56, which shows the behaviour of the thermostat switch. It is worth noting at this point that the input changes digitally whereas the output responds in an analogue form.

Closed-loop control

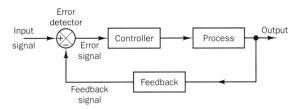

Figure 2.57 Block diagram of a closed-loop feedback system

The fundamental difference between open- and closed-loop control is that, in the case of the latter, the output quantity is monitored and the system responds when there is a difference between the monitored (feedback) value and that required by the input.

Figure 2.57 shows a general block diagram for any closed-loop system. To put it in an engineering context, we will return to the motor/pump unit used in the open-loop section, and approach the problem of control by modifying the original system. Just to remind you, the original block diagram is repeated as Fig. 2.58.

The first thing to point out is that, from the point of view of controlling the system, the whole of this diagram can be regarded as the process that is to be controlled, and in order to achieve control, the pump speed must be monitored, using some form of speed transducer, such as a magnetic pickup, which would provide electrical pulses, the frequency of which would be proportional to speed.

The signal from the transducer will have to be processed in such a way that it can be compared to the input signal, which specifies the required speed. This signal processing is carried out by the feedback system.

The processed feedback signal is then compared with the input signal, i.e. the required speed is compared with the actual speed and the *error* between the two is communicated to the controller.

The controller acts on this *error signal* to return the pump speed to that specified by the input signal. If the error signal is zero, then the output will not change.

A point of confusion, which might arise here, is the idea of the motor/pump unit continuing to run at constant speed with no input. In this context, the input is *not* the electrical energy required to drive the motor; it is an instruction or signal which controls this energy.

At the end of this element, we will look in more detail at several controlled systems, which should help to clear up any confusion.

Control strategies

The purpose of controlling a system is to maintain the overall system output at the level required. If the output is disturbed by events outside the system (external influences), the control system attempts to restore the output to the required level. The degree of success with which the system achieves this depends on the strategy chosen by the designer. This section will examine some of the techniques employed, without going into the mathematics of control, which can become very complex.

Open-loop systems

The speed of small d.c. electric motors can be controlled by the voltage applied to the terminals. In an open-loop system, the voltage would be set at a particular value, and the motor allowed to run. If any disturbances to the output occur, e.g. the load on the motor changes, there are no means of changing the applied voltage to correct the speed.

Figure 2.58

The block diagram for an open-loop system is shown in Fig. 2.59.

Figure 2.59 Block diagram of an open-loop system

Now compare this diagram to Fig. 2.60, which describes our voltage-controlled motor.

Figure 2.60 Block diagram of voltage-controlled motor

Note that in Fig. 2.59 the input signal for the controlled system is called the *reference* signal.

The important disadvantage of an open-loop system, is that the signal passing through the system only travels in the forward direction. This means that if the motor speed reduces because of an increase in the load, there is no way of sending a signal back to the input to say 'increase the voltage to bring the motor back up to speed'.

Closed-loop systems

The motor control system could be improved by employing someone to monitor the motor speed, by watching a tachometer, and altering the applied voltage when changes to the speed occur. In this way a backward path would be provided, i.e. information about the output could be sent back to the input to allow changes to be made (Fig. 2.61).

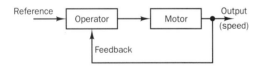

Figure 2.61 Monitoring the speed in a motor control system

To visualise the operation of this system, put yourself in the position of the operator. The reference signal tells you the required motor speed, and the feedback signal tells you the actual speed. Your brain processes this data, which results in a change in the motor voltage if necessary. The way in which you change the voltage will depend on the difference between the required and actual motor speeds.

Automatic closed-loop systems

If the operator is replaced by a subsystem which can automatically respond to changes in the output caused by disturbances, then full control can be achieved. The general block diagram for an automatic closed-loop control system is shown in Fig. 2.62.

There are some new features in this diagram:

- the error detector
- the error signal

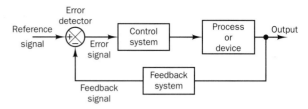

Figure 2.62 An automatic closed-loop system

The error detector, usually just part of the control system itself, compares the reference signal with the feedback signal, i.e. it is indirectly comparing the actual output with that required. The error detector subtracts the feedback signal from the reference signal, and sends the resulting error signal to the controller, which acts on this information to correct the output. Because the feedback signal is subtracted from the reference signal, it is known as negative feedback.

The feedback signal is usually supplied by a transducer, a device which responds to the output by emitting a signal, often electrical, which can then be sent back to the error detector. Transducers are discussed in more detail at the end of this chapter, in which real systems are analysed.

The fundamental difference between open- and closed-loop systems can be summarised:

- In open-loop systems the output has no effect on the input (reference signal).
- In closed-loop systems changes in the output have an effect on the input.

Control strategies

From now on in this section, we will be discussing only automatic, negative feedback control systems. The control strategy is the way in which the controller responds to the error signal.

On/off control

This is the simplest type of response, which can be observed in thermostatically controlled heating devices. When the heater is switched on the dial is rotated to the position at which the desired room temperature is indicated. When this temperature is reached, the contacts on the thermostat open, and the heater coil is switched off.

As heat is no longer being supplied, the room temperature starts to drop until the point is reached where contraction of the thermostat coil causes the contacts to close again, and the heat to be switched on again.

The whole process can be summarised graphically as in Fig. 2.63.

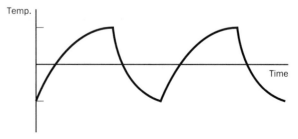

Figure 2.63 Graphic form of an on/off control system

The temperature changes between upper and lower limits or set points. The gap between these limits will depend on the construction and adjustment of the thermostat mechanism, and is known as the 'dead band', i.e. the band of conditions or values that the output can have without causing the control system to respond.

If the dead band is too narrow, then the control system will be switching the output on and off continually, a situation which can lead to wear and tear of switchgear, and other components, and which, in dynamic systems, can lead to instability.

If the dead band is too wide, then the system will suffer from hysteresis, i.e. the control system response will lag behind the changing output. In static systems, this may be of little consequence, but in dynamic systems instability again becomes a problem.

To summarise, on/off Control is often an inexpensive option, but its use is limited to systems where any changes to the output occur slowly.

Continuous control

Proportional control
Proportional control is the simplest form of continuous control. In a proportional control system, the amount of correction applied to the output is proportional to the size of the error signal (Fig. 2.64).

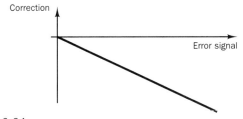

Figure 2.64

Put another way, the amount of correction depends on the amount by which the output has deviated from its specified value. Figure 2.65 shows the relationship between a disturbance to the output, and the resulting correction which is applied by the control system. Note that the correction occurs in the opposite direction to the disturbance. If it did not, there would be no correction.

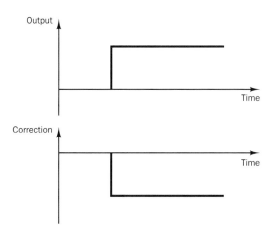

Figure 2.65 Disturbance/correction relationship in control system

Many simple control systems, especially those in which disturbances occur slowly and which have a fairly wide tolerance on the specified output, operate well with proportional control. However, more sophisticated control systems employing two other control strategies provide 'tighter' control on the output. These control strategies are *derivative control* and *integral control*, which are used in conjunction with proportional control. The detailed study of these strategies involves some complex mathematics, but an idea of the purpose of each one can be obtained by looking at the way each one operates in words.

The family tree of control systems (Fig. 2.66) shows the ways in which derivative and integral control are used with proportional control.

Derivative control

If the output disturbance is large, it is advantageous for the control system to react quickly to prevent the output going out of control. In a derivative control system, the rate at which a correction is made is proportional to the size of the disturbance.

Closed-loop gain

In a closed-loop system, there are two signal paths:

- The forward path through the engineering system itself.
- The reverse path through the feedback system.

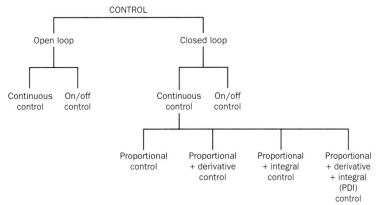

Figure 2.66 A family tree of control systems

The overall sensitivity or gain of the system will depend on the gains of both the forward and reverse paths (Fig. 2.67).

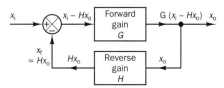

Figure 2.67 Forward and reverse paths in a closed-loop system

Remember that

$$\text{Gain} = \frac{\text{Output}}{\text{Input}} = \frac{x_o}{x_i}$$

So the feedback signal is Hx_o.
The error signal is

$$x_i - x_f = x_i - Hx_o$$

The output signal is

$$x_o = G(x_i - Hx_o) = Gx_i - GHx_o$$

So

$$x_o + GHx_o = Gx_i$$
$$x_o(1 + GH) = Gx_i$$
$$x_o = \frac{G}{1 + GH} x_i$$

Hence

$$\text{Gain} = \frac{x_o}{x_i} = \frac{Gx_i}{1 + GH} \cdot \frac{1}{x_i}$$

$$\text{Closed-loop gain} = \frac{G}{1 + GH}$$

Evaluate the performance of a given engineering system

If you were asked to assess the performance of a system, your first response should be to ask which aspect of the system behaviour was to be evaluated. Extremely complex systems can be described by many variables; a car's performance, for instance, could be assessed in terms of:

- economy
- speed
- comfort
- acceleration
- pollution
- space
- resistance to corrosion

It is common practice to assign values to these performance variables. In this way, comparisons can be easily made. In this section, we will examine some of the more 'technical' variables used to evaluate the performance of systems. Some particular systems have a unique set of performance variables; the car is an example. However, many other systems can be described by the variables covered in this section.

In this section we will define and expand on some of the terms used to specify and describe systems and subsystems. These terms are most frequently used when referring to measurement systems and control systems, although their use is not limited to these areas.

In some cases, calculations can be made of system performance. Where example calculations are shown, you may take it that similar questions might occur in the GNVQ examination.

Sensitivity

The function of a subsystem (or complete system) is often given as the *sensitivity*:

$$\text{Sensitivity } (G) = \frac{\text{Output}}{\text{Input}}$$

If you are wondering why the letter G is used as the symbol for sensitivity, the answer is as follows. The sensitivity of electronic subsystems is often referred to as *gain*, e.g. amplifiers, hence the symbol G.

Measurement systems often have sensitivity quoted as part of their specification. For instance, a thermocouple temperature-measuring device has the block diagram shown in Fig. 2.68.

Figure 2.68

The sensitivity of the device can be calculated if values for the input and output are known. If the thermocouple device has been calibrated so that at $0\,°C$ the output is zero volts, then the sensitivity will be:

$$\frac{\text{Output}}{\text{Input}} = \frac{70\,\text{mV}}{35\,°C} = 2\,\text{mV}/°C$$

When electric motors are selected two important quantities to be specified are the torque developed by the motor and the current drawn at that torque. This information can be obtained if the sensitivity of the motor is known from test results. Figure 2.69 shows an electric motor which develops $x\,\text{N\,m}$ of torque at a current of y amps.

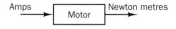

Figure 2.69

$$\text{Sensitivity} = \frac{x\,\text{N\,m}}{y\,\text{amps}} = z\,\text{N\,m/amp}$$

Function description

It would be difficult to give a definitive list of subsystem functions from which a description could be selected. However, a generalised block diagram can often be drawn, which covers a particular range of systems. Using again, the example of measurement devices, all such systems can be shown.

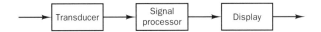

Figure 2.70

Figure 2.71 Subsystems in a thermocouple device

The function of each of the three subsystems in Fig. 2.70 is clearly defined.

- The transducer converts the quantity to be measured to a 'signal' which can be understood by the rest of the system.
- The signal processor converts the signal to one which can be understood by the display unit.
- The display unit shows the user the value of the quantity being measured.

Our thermocouple device could be broken down in this way, as shown in Fig. 2.71. Each of these will have its own sensitivity.

Time lag

When a signal passes through a subsystem a finite time will elapse between the input and corresponding output. If the system is static or in a steady state (i.e. the input is not changing) then this log will be of little consequence. However, if there is a disturbance to the output or the system is dynamic (input changing) any time delay will put the input and output out of phase. This can give rise to *hysteresis* effects.

Put simply, hysteresis occurs when the output lags behind the input. Hysteresis also occurs if the system behaves differently, when the input is increasing to the way it behaves with the input decreasing. This is shown graphically in Fig. 2.72.

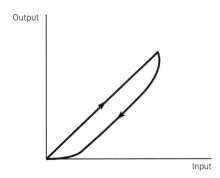

Figure 2.72 An example graph of hysteresis

System response

Sooner or later, while a system is operating, the output will be disturbed. In the case of a fridge, for example, the output could be disturbed by placing warm food inside it. The response of the system will be to return the output to its specified value. This is particularly true of systems fitted with subsystems which control the overall operation.

Types of response

The way in which a system responds to a disturbance depends on the amount of *damping* in the system (Fig. 2.73). In mechanical systems damping is often achieved by fitting a 'damper' but in many other systems damping occurs due to overall system design. Too much damping will make the system 'sluggish' (Fig. 2.73), whereas insufficient damping will make it difficult to maintain a constant output (Fig. 2.74).

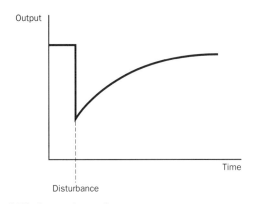

Figure 2.73 An overdamped response

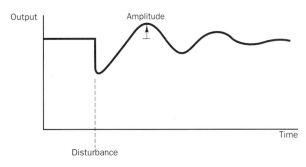

Figure 2.74 An underdamped response

When a system is underdamped, the output will overshoot its original value when responding to a disturbance, and then oscillate at a constant frequency, but with reducing amplitude. The less damping, the longer it will take for the oscillation to die away. You might like to consider what would happen if there was a further disturbance while the output was still oscillating, and furthermore, what would happen if there were repeated disturbances occurring at the same frequency as the response oscillation. This problem will be discussed in the section on 'stability' below.

Figure 2.75 may appear to be the same as Fig. 2.73. However, when a system is critically damped the time taken for the output to return to its original value is the fastest possible without any overshoot.

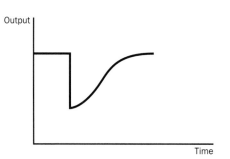

Figure 2.75 A critically damped response

Response time
This is the time taken for the output to return to a value 'within specified limits' after a disturbance. It would be impossible to hold the output at an absolutely constant value, and so, in most systems, especially control systems, the output value is defined by lower and upper limits known as *set points*.

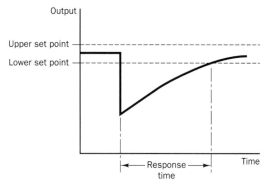

Figure 2.76 Response time in a damped system

Figure 2.76 shows how the response time of a damped system may be determined. The response time of an underdamped system is trickier to work out, because of the oscillation already discussed.

The response time is the time taken for the oscillation to be contained within the boundaries of the set points, as shown in Fig. 2.77.

Stability
A stable system is one which is able to maintain an output between set points and respond quickly to disturbances to return the output to a satisfactory level.
Some of the factors which affect stability are:

- system damping
- system natural frequency
- time lag or hysteresis
- the gap between the set points (dead band)

An underdamped system will clearly have trouble maintaining a steady output, and if a repeated disturbance is applied at a frequency close to that of the oscillating output, a phenomenon known as *resonance* can occur, in which the amplitude of the output oscillation will increase and the system will go out of control.

A lag in the system can lead to a delay in the response to a disturbance which again, when the disturbance is repeated, can result in instability due to the 'wrong' response occurring for the type of disturbance (see Fig. 2.78).

If the set points are too close together, the system will have to respond to the slightest output change and will 'hunt', i.e. the output will oscillate at a constant amplitude. A domestic central heating system may suffer from this problem if the thermostat is too sensitive. The result is that the heating is switched on and off every few minutes, and a lot of energy is wasted.

Accuracy

This term tends to be used mainly when describing measurement systems. Accuracy is defined as:

$$\frac{\text{Indicated value} - \text{True value}}{\text{True value}}$$

As part of the production process, measurement systems are calibrated – i.e. checked against known standards to assess and check their accuracy. Calibration may also have to be carried out, for instance, at yearly intervals if accuracy is to be maintained. This is particularly true of equipment used in precision engineering measurement.

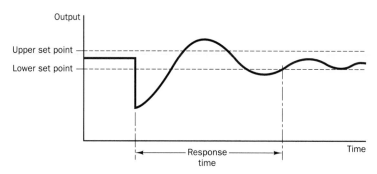

Figure 2.77 Response time in an undamped system

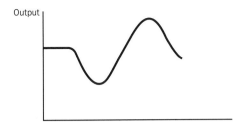

Figure 2.78 A delay in response to a disturbance

Accuracy can be expressed in two ways:

- A percentage of the reading made.
- A percentage of full-scale deflection (% f.s.d.).

The latter method is often applied to electrical meters. The following example will illustrate the two methods of expressing accuracy.

A digital multimeter has a stated accuracy of:

$$\pm 0.8\% \text{ of reading plus } \pm 0.2\% \text{ f.s.d.}$$

Let us consider a reading, taken on the 0–20 volt range, of 14 volts. The possible error would be:

$$\pm 0.8\% \text{ of } 14\,\text{V} = \pm 0.112\,\text{V}$$

$$\text{plus} \quad \pm 0.2\% \text{ of } 20\,\text{V} = \frac{\pm 0.040\,\text{V}}{\pm 0.152\,\text{V}}$$

The ability of a measurement system to maintain accuracy over many readings is known as its *repeatability*.

2.3 Unit test

Test yourself on this unit with these multiple-choice questions.

1. The *purpose* of an electric heater is to:
   (a) reduce the amount of gas used
   (b) convert electrical energy to heat energy
   (c) provide heat to raise the room temperature
   (d) operate efficiently

2. Considering the *function* of the system, the input to a car battery, when the output is electrical energy, is:
   (a) chemical energy
   (b) electrical energy
   (c) sulphuric acid
   (d) connecting wires

3. Figure 2.T1 shows a block diagram for a pressure transducer.

   Figure 2.T1

   The sensitivity of the system is:
   (a) 0.25 bar/mV
   (b) 11.5 mV/bar
   (c) 17.5 mV/bar
   (d) 4 mV/bar

4. One of the inputs to a car engine is petrol. An implication of this input for the system is:
   (a) that a locking fuel cap is required
   (b) an ignition system is required
   (c) a fire extinguisher is necessary
   (d) a fuel gauge is required

5. A hydraulic jack is an example of:
   (a) a mechanical system
   (b) an electromechanical system
   (c) a fluidic system
   (d) a thermodynamic system

6. Instability in a system might be due to:
   (a) loose nuts and bolts
   (b) a blown fuse
   (c) hysteresis
   (d) analogue inputs

7. A transducer is an example of:
   (a) an energy delivery system
   (b) an energy conversion system
   (c) an energy storage system
   (d) a control system

8. The purpose of a dummy gauge in a Wheatstone bridge is:
   (a) to counteract the effect of variations in temperature
   (b) to calibrate the instrument
   (c) to increase the sensitivity
   (d) to reduce hysteresis

9. Figure 2.T2 shows the response of a system following a disturbance to the output. The response time of the system is:
   (a) 45 ms
   (b) 115 ms
   (c) 130 ms
   (d) 160 ms

10. The dead band of a system is:
    (a) the area in which the system does not function
    (b) the magnitude of the feedback signal
    (c) the time taken for a system to respond to a disturbance
    (d) the area between the upper and lower set points

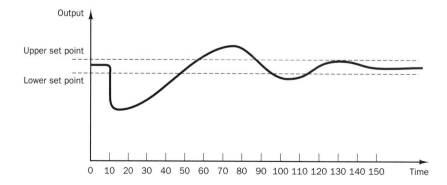

Figure 2.T2

11. In a control system, the purpose of the *error detector* is to:

    (a) find mistakes
    (b) provide positive feedback
    (c) compare the input to the feedback signal
    (d) compare the actual output to the required output

12. A meter is described as having an accuracy of ±1.5% f.s.d. It is used to measure a current of 5.5 amps on a 0–10 amp range. Calculate the range of values within which the measured current lies.

    (a) 5.4175–5.5825 amps
    (b) 5.35–5.65 amps
    (c) 5.5–5.65 amps
    (d) 5.35–5.5 amps

UNIT 3: Engineering Processes

Engineering processes cover vast topics and are evident in everyday life in all that we see or use. Very little thought is given as to how products are manufactured or maintained. This unit will offer a greater understanding of what is required to manufacture electromechanical products to a specification and the consideration given to maintaining the performance of the product throughout its working life. This is to be achieved while maintaining an understanding of the health and safety requirements.

In this unit the student will learn how to:

- Select processes to make electromechanical engineered products.
- Make an electromechanical product to specification.
- Perform engineering services to specification.

After reading this unit the student will be able to:

- Recognise a range of processes suitable for making and servicing electromechanical products.
- Evaluate manufacturing and servicing processes for an electromechanical product.
- Describe specific techniques and sequences in order to perform a range of processes.
- Select suitable processes and tools on the basis of given parameters.
- Select the most appropriate techniques required to perform a selection of tasks.
- Devise a sequence of the processes required to manufacture a given product while recognising the impact of safety procedures and equipment to perform each selected process.
- Recognise the importance of maintaining tools, equipment and working area during and after processing.

3.1 Select processes to make electromechanical engineered products

Topics covered in this element are:

- Identification of manufacturing processes.
- Specific techniques required to perform manufacturing processes.
- Evaluation of the manufacture of a given product.
- Safety equipment and procedures.

The manufacture of a given component requires a substantial amount of prior knowledge, understanding and skill before a product may be made.

Initially the recognition of the processes suitable for manufacture have to be realised. These may take many different forms and may be dependent upon specific criteria such as quality, cost, quantity, materials and tolerances. However, it is in the initial stages that we must first look towards answering typical questions such as:

- *Material removal* – Just how much can be removed in one cut?
- *Shaping of materials* – Can the product be moulded into shape?
- *Joining and assembly* – Are the products to be permanently joined?
- *Heat treatment* – How can the hardness be increased?
- *Chemical treatment* – Will the product require corrosion protection?
- *Surface Finish* – How is consistent quality obtained?

Many of these processes are taken for granted or are mainly overlooked because we often only consider the final product or process based on past experience.

> **Self-assessment task**
>
> Before reading on, write brief notes on how you think the following common components are made:
>
> 1. A spanner
> 2. An engine block
> 3. A bike frame
> 4. An electric plug
> 5. A water tap

Identifying processes for making electromechanical engineered products

Material removal

Three of the most common forms of removing material from a component are by:

- turning
- milling
- drilling

It can often be difficult to decide which particular method to adopt, when a similar outcome could be achieved by more than one process. If this happens to be the case then a further set of questions have to be considered, some of these being:

- How many have to be produced?
- What are the intended tolerances?
- Is any one process quicker than another?
- Is there a difference in cost?
- Will one process give a better quality of product?
- Is this the best process to use with this type of material?

In order to make an informed choice on the manufacture of a particular product, there is a need to be familiar with not only the general material removal processes but also with some of the more specific techniques available, for example:

- A typical turning process may consist of cylinders, tapers, holes, threads.
- A typical milling process may consist of flat surfaces, slots, curved surfaces.
- A typical drilling process may consist of through holes, blind holes, counterbores, countersinks.

> **Self-assessment task**
>
> By which process do you think the following components are generally made?
>
> | | Nuts | Bolts | PCBs | Brake discs | Screws |
> |---|---|---|---|---|---|
> | Turning | | | | | |
> | Milling | | | | | |
> | Drilling | | | | | |

Shaping of materials

Not all components require machining for them to function and some components are too complicated to be machined. Certain processes can change the physical shape of the material/component without actually removing material, and the three main areas that adopt this approach are:

- casting
- forging
- forming

Each process has the ability to change the shape of the component without loss of material.

Joining and assembly

Any product containing more than one part must be assembled, that means joining has to take place, for the main aim of assembling is to place separate components together in such a way as to form a working whole component. Joining may be permanent or temporary and could involve fastening together many differing materials, therefore careful consideration has to be given to the selection of the methods to be employed.

Heat treatment

The properties found in many materials often require modification during or after manufacture in order to enhance the function of the component or process. The main method of modification uses controlled heat usually in a furnace, applied to the material and often just as important a controlled cooling rate.

For steels the following four main techniques are used:

- annealing
- normalising
- hardening
- tempering

Each provides a specific change to the properties and hence the practical working life of a component. Basically it is the application of heat and the subsequent cooling period that gives one of these forms.

> **Self-assessment task**
>
> List four common household items that need to be hard to withstand wear and four that need to be tough to withstand shock.

Chemical treatment

When differing materials come into contact with each other they often form a chemical reaction. By carefully utilising certain reactions it is possible to alter a component to provide an economic enhancement that may otherwise prove too costly or impossible to achieve by any other method. Two of the most common forms of chemical treatment are etching and plating: one basically removes material from a component and one effectively coats it.

Surface finish

The surface finish of a component is a result of the process used. It may be from the cutting tools on a lathe or milling machine or it could be obtained from the casting or forging process. Surface finishes can be added to a component by coating the surface with another, as in painting, to offer surface protection.

The measurement of the finish is obtained by making a comparison against a known standard. The comparison is made of the small almost undetectable peaks and troughs that are associated with any surface, in general the smaller the difference between these peaks and troughs the finer the finish of the surface.

Identifying specific techniques to make engineering products

Turning

Turned components are manufactured predominantly on a centre lathe (Fig. 3.1), and the main function of turning is to produce cylindrical shapes and flat surfaces to a high degree

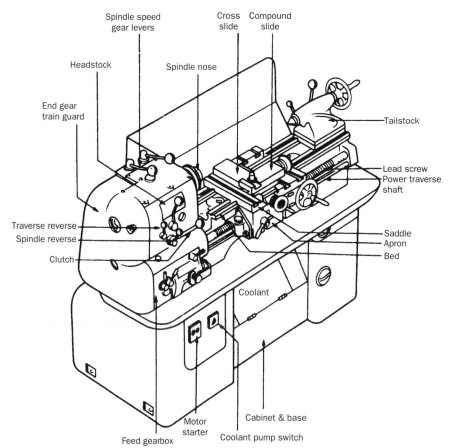

Figure 3.1 A centre lathe

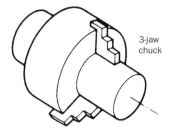 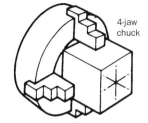 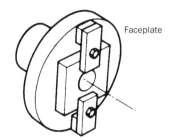

Figure 3.2 Work-holding devices

of accuracy. This is achieved by locating and clamping the workpiece in a work-holding device (i.e. 3-jaw chuck, 4-jaw chuck, faceplate; Fig. 3.2), this is then rotated about a horizontal axis. A cutting tool is moved relative to this axis as the spindle rotates and makes cutting contact with the workpiece removing material so that the desired shapes may be generated.

Work-holding devices allow a variety of components to be held, the most common forms are:

- 3-Jaw self-centring chuck – for regular cylindrical shapes.
- 4-Jaw independent chuck – mainly for irregular or square shapes.
- Between centres – generally for long continuous cylindrical shapes.
- Faceplate – for large or irregular shapes.

Taper turning

Tapers play an important part in engineering components; they are used for concentric location, guides and driving devices. There are four main methods of producing a taper on a centre lathe, these being:

- Using the compound slide.
- Offsetting the tailstock.
- Using a taper-turning attachment.
- Using a form tool.

Compound slide method
This taper-turning process involves rotating the compound slide of a lathe to half the included angle of the required taper (Fig. 3.3) – i.e. set at 15° to produce a full 30° taper. This offers the advantage of using any angle and is very quick to set up; however, there is no provision for power feed.

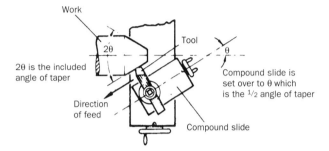

Figure 3.3 Taper-turning processes

Offset tailstock method
While set up for turning between centres, a taper is achieved by offsetting the top half of the tailstock away from the fixed lower half (Fig. 3.4) – which is a movement perpendicular to the axis of the centre lathe. This then changes the central axis of the component slightly from the central axis of the lathe. An advantage of this method allows long small-angled tapers to be machined using power feed, but only small angles may be produced.

Taper-turning attachment method
The cross-slide of the lathe is disengaged from its driving mechanism, and thus has free movement along its slide way. Behind the cross-slide there is a taper-turning attachment (Fig. 3.5), which is a metallic bar set at a required angle. The free cross-slide is fastened to this bar and as the saddle moves along the axis of the lathe the cross-slide copies the angle of the attachment. An advantage of this method is that power feed may be used and repeatability for batch type production is enhanced. A disadvantage is that the attachment is often quite large, difficult and time consuming to set up correctly.

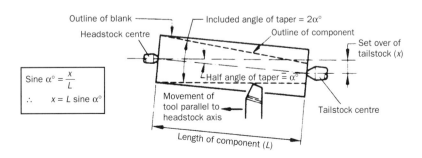

Figure 3.4 Set over of centres

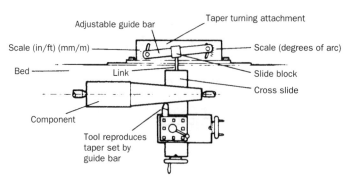

Figure 3.5 The taper-turning attachment

Form tool method

This simplest of taper-turning methods comprises a cutting tool with the desired angle ground at the cutting edge; this is then presented to the workpiece and the resulting cut takes on the form of the cutting tool (this is known as forming, Fig. 3.6). This is a cheap and quick approach to producing tapers; however, only short tapers may be produced effectively by this method.

Screw-cutting threads

When making a thread on a lathe it is important that consideration be given to the shape (form) of the thread and to the pitch. The shape (form) is often governed by the tool cutting the thread being set square to the workpiece. The pitch is achieved by having a connection from the lead screw to the drive spindle via a gearbox (Fig. 3.7). It is the selection

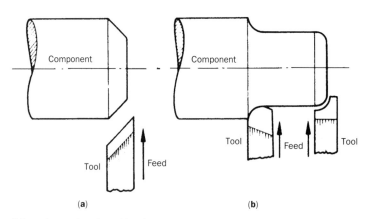

Figure 3.6 Taper-turning process: (a) cutting a chamfer with a form tool; (b) radius cutting with form tools

Self-assessment task

Identify the taper-turning methods used to produce the shapes in Fig. 3.T1.

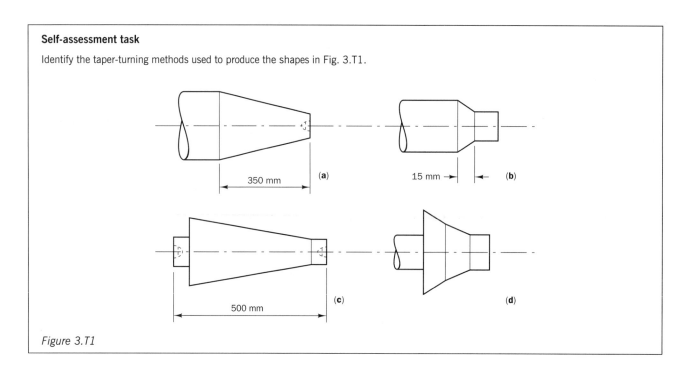

Figure 3.T1

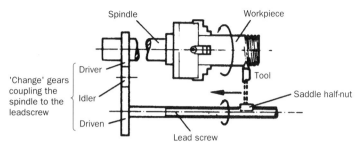

Figure 3.7 The 'change' gear train

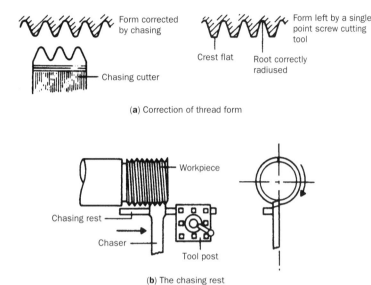

Figure 3.8 Thread chasing: (a) correction of thread form; (b) the chasing rest

of the gears in the gearbox that enables differing pitches of the thread to be cut. The set-up for single point screw cutting involves rotating the compound slide over to half the included angle of the thread to be cut, while maintaining that the cutting tool remains perpendicular to the axis of the spindle. By feeding the tool into the work (depth of cut) via the compound slide, it allows the tool to take progressive cuts on only one flank of the thread, making chip removal smoother.

This process may not give a full thread form due to the tool not having the full profile to cut a thread crest, however, this may be overcome by using a chaser. A chaser (Fig. 3.8) is a full pitch-shaped form tool which follows the form already set by the single-point cutting method but it enables the form to be corrected by removing only small amounts of material until it has the same full pitch shape as the chaser.

Cutting tools

Three considerations must be given to cutting tools for them to perform effectively.

- Type of cutting tool
- Method by which the cut is made.
- Tool height.

Type of cutting tool
There are many sizes and shapes of cutting tools and they also may be made from different cutting tool materials. The profile often shows generally just which application the tool is appropriate for.

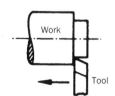
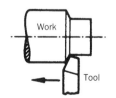

(a) The cutting edge is perpendicular to the direction of feed. Useful for producing a square shoulder at the end of a roughing cut

(b) The cutting edge is inclined to the direction of feed. Most efficient form for rapid metal removal

Figure 3.9 Orthogonal and oblique cutting: (a) orthogonal cutting; (b) oblique cutting

Method by which the cut is made

Two general forms of cutting take place: orthogonal and oblique (Fig. 3.9). Orthogonal cutting has a tool cutting perpendicular to the direction of feed, while oblique cutting has a tool cutting at a slight angle to the direction of feed. The forces are spread over more of the cutting face than for orthogonal, therefore if removing the same amount per cut it uses less power and gives a longer tool life.

Tool height

For cutting to be effective the choice made from the two considerations above has to be supplemented by positioning the tools cutting edge on-centre height. Failure to do this alters the cutting angle characteristics of the tool (Fig. 3.10). This will affect the overall cutting conditions and may also transform the profiles of tapers and screw threads.

Holes

Holes may be produced on lathes by using standard drills. These are fine for achieving quick material removal, but as they often give poor finishes and lack accuracy, other methods have to be adopted to obtain better quality. One such method is reaming. This requires that the hole to be produced is drilled slightly under size and then a reamer acts as a finishing tool removing very little material which offers a high degree of roundness with a good surface finish.

There are drawbacks to using reamers:

- Standard sizes only can be achieved.
- The reamer follows the axis of the original drilled hole and may reproduce any wavering that was present while drilling.
- They may only be used along the spindle rotation axis.

Boring

This requires great care, for usually boring tools tend to chatter and deflect due to the length of overhang from the tool post and the general small cross-sectional size of the tool itself in order to allow it to produce small bored holes. Most of the material should first be removed by drilling and then bored out. It is a process which may obtain a high degree of positional accuracy, but it is very difficult to obtain good surface finishes.

Milling

There are three basic types of milling machine, each uses multitooth cutters which allows rapid material removal:

- *Vertical* or *horizontal* (Fig. 3.11) – So named because the spindle axis is aligned either vertically or horizontally.
- *Universal milling machine* – This is similar to the horizontal milling machine only it has a table that may swivel through a prescribed angle.

Work holding

There are a variety of methods to secure work ready for machining, often the choice depends upon the size and shape of the workpiece and the type of operation to be carried out. The most common method for large work is to clamp the work directly to the table. For smaller regular shaped work a machine vice is used (Fig. 3.12). This is located axially by tenons in the base to give a speedy alignment and is then clamped to the table of the machine. It has a fixed jaw and a moving jaw (operated by a screw thread) perpendicular to each other. This vice has a limitation in that it will only allow the machining of surfaces at right angles to each other, but this limitation may be overcome by using a swivelling machine vice.

Rotary table

When there is a requirement to machine arcs, a rotary table (Fig. 3.13) may be used on a vertical milling machine, the rotary table is like the machine vice clamped to the machine table and the component is fixed on to the table of the rotary table itself. Table rotation is achieved by the use of a worm and wheel mechanism operated by hand, the handle of which has graduations marked on it to give up to 5 minutes of arc. Note that a rotary table often has a 3-jaw chuck clamped to it to allow a quick method of work holding regular shaped work when batch production is required.

The dividing head

When an equal number of slots or flats are required on a cylindrical component or ones that need to be set at precise angles, then indexing becomes very important. Indexing is achieved by the use of a dividing head (Fig. 3.14). This works in principle like the rotary table with a worm and wheel mechanism, but gives greater accuracy; it may be hand operated or power driven via a gear train linking the lead screw of the table to the worm of the dividing head (used for helical milling).

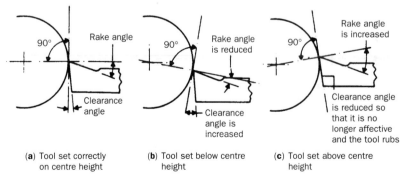

(a) Tool set correctly on centre height
(b) Tool set below centre height
(c) Tool set above centre height

Figure 3.10 Effect of tool height on turning tool angles

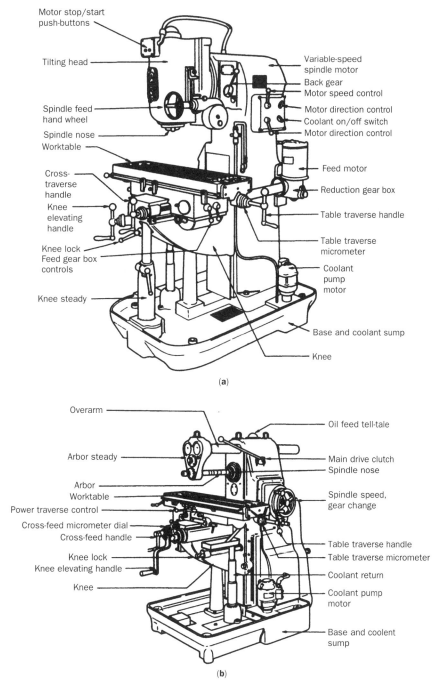

Figure 3.11 Milling machines: (a) typical vertical machine; (b) typical horizontal machine

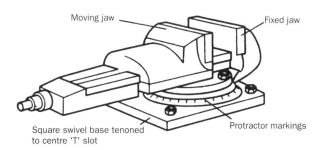

Figure 3.12 Machine vice with swivel base

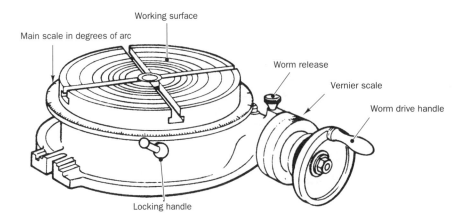

Figure 3.13

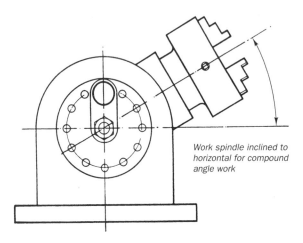

Figure 3.14

Material removal by milling

Generally a milling cutter may remove material in one of two ways (Fig. 3.15):

- Up-cut milling (conventional type milling) by allowing the cutter rotation and the feed of work to act in opposite directions.
- Down-cut milling (climb milling), thus having the cutter rotation and the feed acting in the same direction.

Up-cut milling can be used on older machines which may have worn slideways, as there is no tendency to grab the work underneath the cutter (as there is in climb milling). However, it produces the poorer finish of the two methods due to the bouncing effect of the cutter tooth leaving the workpiece on a full size cut. This tendency may be reduced by using a milling cutter which has a cutting edge in the form of a helix, thus ensuring cutting contact throughout the whole rotation.

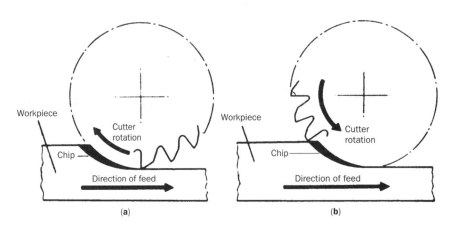

Figure 3.15 Milling techniques: (a) up-cut milling; (b) down-cut milling

Shaping of materials

Sand casting

This is the process by which a pattern (a slightly larger copy; Fig. 3.22) is made of the required component. This is usually in wood because it is cheap and easy to work with to make the desired shape.

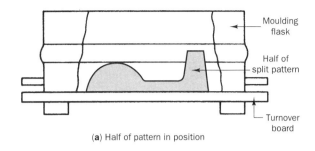

(a) Half of pattern in position

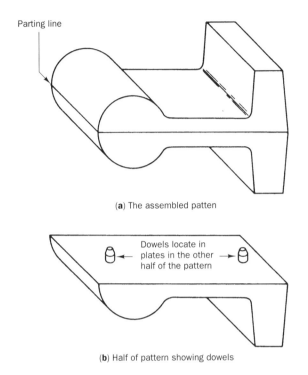

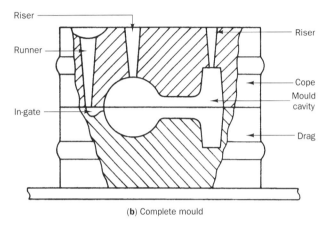

(b) Complete mould

Figure 3.23

Figure 3.22

This is placed on a flat turnover board and in a mould box (which will become the bottom half of a two-part mould and is known as the drag); it is then dusted with parting powder (to evenually aid removal from sand) and packed with fine green sand. Once full and packed the pattern is carefully removed, leaving an impression behind. The whole process is then repeated to make the top half of the mould (the cope), only in this half two extra pieces are used, a *runner* to allow pouring of the molten metal to reach the cavity, and a *riser* to allow gases to escape and show when the mould is full of molten metal. (They also act as reservoirs from which the casting can draw molten material as it cools and contracts.) Once the two halves have been made and the pattern removed (Fig. 3.23) the impression is filled with molten metal until metal appears in the riser. It is then allowed to cool. On solidification the metal contracts, which is the reason for making a slightly larger pattern.

If there is a requirement for the casting to be hollow or to have hollow sections, then the manufacture of a core using sand and linseed oil to mimic the hollow shape has to be made. The pattern also has to take into account that a core is to be used, and this is often achieved by adding steps or lugs to the pattern so that when it has made an impression in the sand the core itself has a location in which to be placed.

This process is very good for allowing any material that may be melted to be cast and for intricate shapes to be made; however, the sand mould itself is not reusable, only the sand, and has to be made each time a casting is required, which is a time-consuming process. The process produces castings which, although complex, are not dimensionally very accurate and if any machining is required then an allowance for machining must be made when designing the pattern.

There are also defects that may arise from sand casting, such as:

- Distortion due to the differing contraction rates of thick and thin sections causing massive internal stresses.
- Cold shuts may occur when the solidification of thin sections within the casting stop further flow of molten metal.
- Blowholes are very common in castings. They occur because gasses are trapped in badly designed castings or because the sand has not been vented (tiny holes pierced into the sand to allow gases to escape).
- Scabbing occurs when sand is dislodged from the mould (usually during pouring) and forms a disfiguring shape within the cavity.

Die casting

Two of the main drawbacks of sand casting – the remaking of the mould for every component and the accuracy of the finished casting – are addressed successfully with die casting. This is achieved by the use of permanent reusable steel moulds (dies; Fig. 3.24) and the resulting casting in general is left in its finished state with no further machining requirements. This speeds up the casting process and thus lowers the costs; however, there is a price to pay as it predominantly casts only non-ferrous materials, principally aluminium and zinc-based alloys. Although steel and iron have been cast, the high melting temperatures required for these materials attack the expensive dies as it is poured and this results in their rapid loss of finish and accuracy.

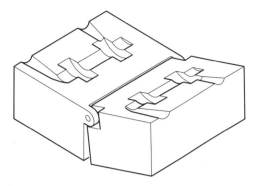

Figure 3.24 Steel moulds used in die casting

Gravity die casting

Using a metal die, this process resembles the principles of sand casting in that the molten metal fills the mould through the force of gravity (Fig. 3.25). If cores are required, they may be made to collapse to allow them to be withdrawn from the casting. If this proves to be problematical then, as with sand casting, a sand core may be employed. This casting process is not very suitable for zinc-based alloys as it tends to promote a coarse grain.

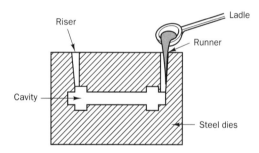

Figure 3.25 Gravity die casting

Pressure die casting

As the name implies, this casting process is influenced by the introduction of pressure. The pressure is applied to the metal as it is fed into the dies (by use of a plunger; Fig. 3.26). This ensures that the metal makes contact with the walls of the casting and, by maintaining the pressure during cooling, leaves a sharp well-defined casting, which may require little or

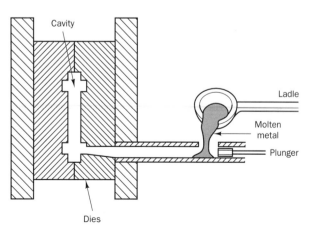

Figure 3.26 Pressure die casting

no machining. If the dies are water cooled, the components solidify more quickly and may be removed while solid yet still hot, enabling the process to start again offering a much faster turnaround of components.

A limitation of this process is the cost to manufacture the dies, which makes pressure die casting only economical if large numbers of components are to be made.

Investment casting (lost wax process)

If there is a need to make a complex component from a material which is very difficult to machine and which requires a high degree of accuracy or finish, then the processes mentioned above would prove unsuitable. Investment casting allows all these attributes to be addressed; it achieves this by making a mould out of a fine refractory material that can withstand tremendous heat ranges and can offer dimensional accuracy to match, if not better, a steel mould. This involves the following procedures (Fig. 3.27):

1. Make a wax pattern of the components (if large numbers of components are required then a mould would be made to make the wax pattern).
2. Coat the wax pattern with a refractory slurry (this is usually sprayed on, or the pattern is dipped into a slurry bath); as the slurry sets it forms a hard brittle shell (die).
3. Melt out the wax by placing the mould in a heated furnace. This burning out of the melting wax leaves a cavity and also helps to set the refractory mould.
4. Pour the molten material into the cavity and allow it to solidify, then break off the refractory lining of the mould.

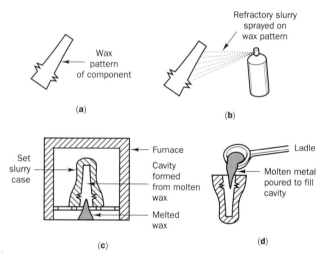

Figure 3.27 Investment casting system

This offers castings to be manufactured from any material that may be melted and give a high degree of complexity and accuracy. It is, however, expensive as the process is painfully slow.

Self-assessment task

State which casting process is most suitable for the production of the following components:

(a) An engine block.
(b) A toy car.
(c) A jet aircraft propeller blade.
(d) A water pump housing.
(e) An electrical motor housing.

Forging

This process is usually chosen when the strength of the material needs to be used to give maximum effect. This is achieved by allowing the grain structure present in the material to be aligned in a manner that takes advantage of natural grain flow of the material (Fig. 3.28).

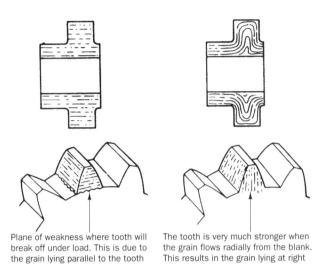

(a) Machined from bar — Plane of weakness where tooth will break off under load. This is due to the grain lying parallel to the tooth

(b) Machined from forging — The tooth is very much stronger when the grain flows radially from the blank. This results in the grain lying at right angles to the tooth

Figure 3.28

The process itself involves applying pressure to a material – usually metal – often while it is red hot, because it then requires less force. This pressure may be manually or mechanically applied. The manual pressure usually derives from hitting the component with a hammer while it rests on an anvil.

It is easier to manipulate the component when hot, so to obtain this heat the component is placed in a heated hearth (usually gas or oil-fired; Fig. 3.29). The extra intense heat is generated by blowing air into the heated hearth through a water-cooled pipe called a tuyere. As the component will be too hot to handle, tongs – which have mouths of varying shapes to accommodate differing workpieces – have to be employed to hold the workpiece. This manual manipulation is limited and means that only relatively small forgings can be made by this method. There are, however, still several basic forging techniques (Fig. 3.30) that may be used, for example:

- *Punching or piercing* – In this method square holes may be punched out by punching a pilot hole half way through the component, then turning the component over to complete the punch from the other side. The hole produced is then gradually opened up to the desired size. This gives an advantage of the grain flow swelling out around the eye of the hole, avoiding weakening the component.
- *Swaging* – This process refers to making a round or part round bar out of a square bar. This rounding off is done by placing the heated bar between a pair of swages resting on the anvil and striking the top swage with a hammer. This slowly rounds off the square bar (it is sometimes prudent to flatten the square bar into an octagonal shape prior to swaging as this requires less force).
- *Upsetting* – If there is a need to increase the thickness of a bar for a short length (i.e. the head of a bolt) then the bar is heated only where the upsetting thickness is required. It is then placed end on to the anvil and struck along its axis, which tends to force out the heated part of the bar as it is more plastic than the cold section, and further forging processes may then be carried out.
- *Drawing* – This process is the reverse of upsetting in that the thickness is reduced but the length is increased. The tools used are *fullers* which draw out the material and *flatters* which smooth off the uneven surfaces left by the fullers.
- *Bending* – This is probably the most common forging operation. Heat is applied to enable easier bending around the sides and back of the anvil, but as the material bends it has a tendency to thin out and may, therefore, first require upsetting at the position of the bend to compensate for this thinning.
- *Cutting* – A special part of the anvil is left soft for this process. As the hard cutting edge of a chisel breaks through the workpiece to be cut it will come into contact with the softer material on the anvil and thus the hard but brittle cutting edge of the chisel will not be damaged. This may also be achieved by using hardies, which are chisel sets that sit in the hardie hole on an anvil allowing the blacksmith to have more control.

Drop forging

If large forgings are required, then a machine is used to exert the forces needed. One such method is the drop forge (Fig. 3.31) which houses two halves of a die, the bottom half is fixed securely while the top half is raised mechanically, attached to a tup weighing up to 2,200 kg (giving a blow energy of 4,500 kgf m). They are then allowed to drop (hence the term 'drop forging') under the force of gravity on to the bottom die (this may occur up to 70 times per minute depending upon the length of stroke), thus squeezing the hot billet. As it is impossible to gauge just how much material is required from a billet to make the complex component, any excess material is squeezed out as thin metal known as 'flash' between the two dies and is then trimmed or clipped.

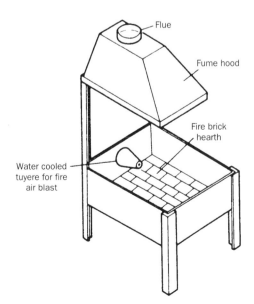

Figure 3.29 A hearth

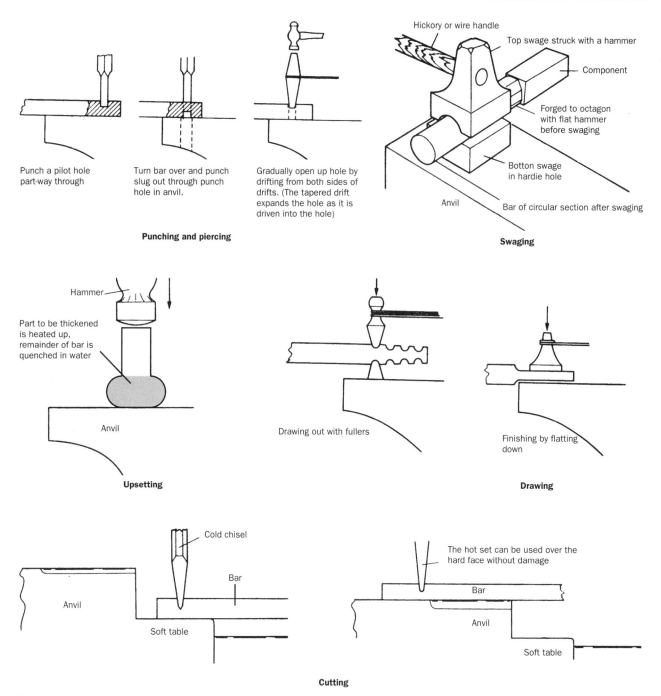

Figure 3.30

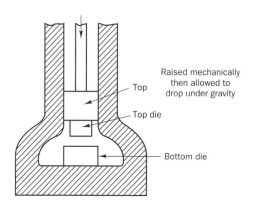

Figure 3.31 Drop forging

Pressure forging

The mechanical pressure for this process (up to 7,500 kgf m) is derived from a forming press (Fig. 3.32) in which two split dies are set. Half of the required shape is moulded in each die. The two dies are brought together under immense pressure with the red hot billet sandwiched between them. This pressure is so great the metal billet is squeezed and forced to take on the shape of the die impression.

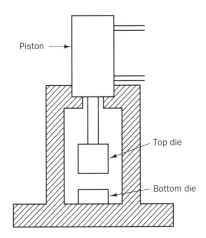

Figure 3.32 Pressure forging

Forming

Bending

When material requires a bend to be formed this bend has to have a significant radius otherwise the material would crack and split. To achieve this bend it is often rolled. During any bending the material itself stretches on the outside of the bend while it contracts on the inside of the bend, so that when an accurate bend is required an allowance must be made to avoid placing the material under undue stress and strain to allow it to form properly. This allowance is commonly known as the *bend allowance* and may be calculated as follows:

$$\text{Bend allowance} = \frac{\text{Angle of bend}}{360°} \times 2\pi \times \text{Average radius}$$

i.e. for a 90° bend of outside radius 100 mm in a 2 mm thick material, the following calculation would give the bend allowance:

$$\text{Bend allowance} = \frac{90°}{360°} \times (2 \times 2) \times 100$$

$$= \tfrac{1}{4} \times 4 \times 100$$

$$= 100 \text{ mm}$$

To actually achieve this bend the material would have to pass through a rolling machine which has three hardened steel rollers. Two front rollers are geared together and pinch the material which is then fed through these rollers (Fig. 3.33) to the third adjustable back roll. This is offset so that as the material passes this roll it is deflected gradually and starts to produce a curved shape.

If the desired bend cannot be obtained in the first pass then the back roll is offset further and the material is once again passed through the rolls. This process continues until the required radius is achieved.

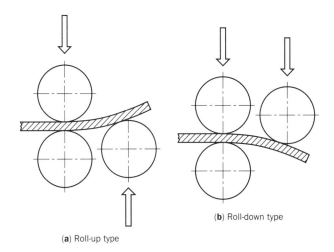

Figure 3.33 Basic arrangement of the rolls in 'pinch-type' rolls: (a) roll-up type; (b) roll-down type

Injection moulding

As polymers (plastics) become liquid when heated and solidify when cooled, they may be formed into specific shapes by injecting (Fig. 3.34) the liquid polymer into a mould and on solidification final mouldings are obtained

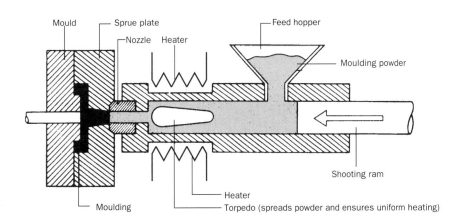

Figure 3.34

The raw plastic used is generally in granular form and is hopper fed into a heated cylinder, the heat (approximately 200 °C) plasticises the material and this is then subjected to a high pressure to force it into the mould cavity. The moulds are often water cooled to enable a more rapid cooling cycle. This process, apart from the removal of the component, may be fully automatic and may have a turnaround time of about 40 seconds per cycle, most of which is the cooling time.

This gives the advantages of the plastic component being made to the finished shape and size with a high degree of repeatability. It has a very high production rate and if added strength is required (as in the need for accepting screw threads) then metallic inserts may be placed in the mould prior to the injection of plastic.

The disadvantages of this process are:

- It has a high initial capital cost for the injection machines.
- The dies are very costly due to the accuracies and finishes required.
- It is very difficult to mould components with large variations in wall thickness.

Vacuum forming

This is a process in which a sheet of plastic is clamped over a mould, heated and then drawn into contact with the mould of the required shape by creating a vacuum between the sheet and the mould (Fig. 3.35). To enable the vacuum to be formed the mould contains many tiny drilled holes to allow air to escape. This process is suited to the manufacture of shallow thin wall parts and is widely used in the packaging industry. Larger, thicker components may be made by this process but it needs the assistance of a former of the component (often made from wood) to help to shape it.

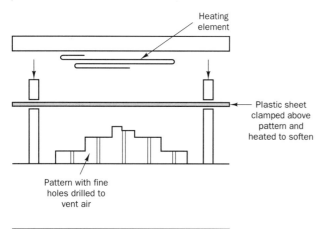

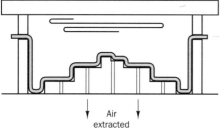

Figure 3.35 The vacuum-forming process

Bath tubs, car bumpers, boats and refrigerator liners are made by this method. There are, however, some limitations of this process:

- It is very difficult to control the wall thickness, especially near bends.
- Compared to injection-moulded plastic components, this process offers very little dimensional accuracy.
- It is limited in the intricacy of component it can produce.

Extrusion

If long straight components with complicated cross-sections are required in a ductile material that needs little or no machining, then extrusion offers a speedy and relatively low-cost process. If you can imagine how toothpaste is dispensed from its tube, then you have just imagined the process of extrusion. A heated bar of material, placed in a cylinder, is forced by a ram out through a die and forms the shape of the die (Fig. 3.36). This process offers the advantages that it is quick and produces good surface finishes with high accuracy. The only disadvantages are the size of extrusion and the limiting stresses set up in the tooling.

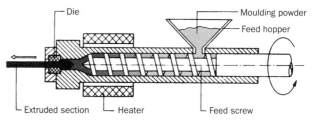

Figure 3.36

Drawing

Stretching can describe the drawing process, which conjures up an image of a material being pulled longer and longer while it gets thinner and thinner. That is basically the process of drawing (Fig. 3.37), but in reality it is much more sophisticated and tends to rely on machinery to do the drawing. This process is ideal for the manufacture of wire, as follows.

1. A heated billet of a ductile material (i.e. copper), narrowed down at one end, is too big to go totally through a die; the narrow end is fed into the die allowing some of it to protrude at the other end.
2. This protruding material is gripped by a clamping mechanism (dog) which is in turn attached to a drive arrangement and is pulled through the die.
3. This reduces the thickness of the heated billet, but it is stretched.
4. The process is repeated with ever-decreasing diameters of die until the required size is reached.
5. As the billet turns into wire it gets longer and longer, thus the need for the wire to be coiled as it has passed through the die.
6. Due to the stresses set up within the process, the material soon becomes work hardened and very brittle. It no longer becomes suitable for drawing as the ductility which enable drawing has gone.
7. This ductility may, however, be restored if the material is annealed, allowing further drawing to take place, which is essential when producing thin wire and may take several annealing stages.

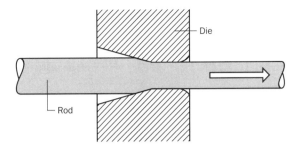

Figure 3.37

> **Self-assessment task**
>
> List four common objects manufactured by each of the processes mentioned below:
>
> (a) vacuum forming
> (b) injection moulding
> (c) extrusion
> (d) drawing

Joining and assembly

Crimping

Crimping could best be described as permanent clamping of electrical wires, it is often used within the electrical assembly industry for the fastenings of lugs or connectors onto the ends of the electrical wires so that a mechanical-fastening device (i.e. a screw terminal) may then form a non-permanent connection (Fig. 3.38). The actual connector is called a crimp fitting and is placed around the wire and compressed onto it (crimped) by the use of a crimping tool which resembles wire strippers, except that the jaws are shaped to form the crimp when squeezed. Flat edge connectors are used extensively in the electronic and automobile industry and are fastened by this method. For mass production, pneumatic or hydraulic crimping tools are used.

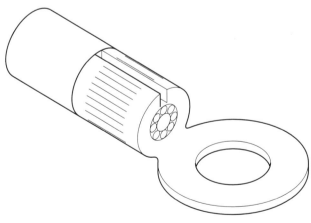

Figure 3.38 A crimp fitting

Soldering

This could also be a description for *brazing*, as both processes are closely related in that they join metal components by the use of a filler metal that has a melting point lower than the two parts to be joined. This joint is achieved by applying heat via a heated soldering iron which melts the solder (lead and tin), which then wets the surfaces to be joined and spreads by capillary action with the aid of a flux over the soldering area. The flux also acts as a cleaning agent for the soldered joint. The heat is then removed and the solder solidifies, bonding the joint area.

Joint design is of paramount importance as a large area of contact is required to give maximum strength to the joint.

Electrical soldering

This usually takes the form of making a soldered connection of a wire to a terminal. An important feature of this process is that the solder must solidify rapidly when the heat source is removed, which eliminates waiting time and misalignment of components on solidification. This form of solder is known as *Tinman's Solder* (it solidifies very quickly when the heat source is removed) and to assist in the electrical soldering operation a flux is used. This flux can be built into a stick of solder (cored solder; Fig. 3.39) or it may be a paste to be applied before soldering.

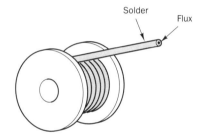

Figure 3.39 Cored solder

Only non-corrosive fluxes may be used, and they have two main purposes:

- To protect the metal from oxidising when heat is applied.
- To remove any grease or film from the surface.

Heat is applied through a heating element inside a soldering iron by passing an electric current through it. At the end of the iron is a bit, which is then heated through thermal conduction (it is this bit that makes contact with the soldered surfaces). The bits themselves may have many shapes to enable access to the soldering area and they may be iron plated to prevent rapid wear. It is important that bits are kept clean and smooth to aid the flow of solder as they soon become pitted and covered in oxide. (This cleaning can be simply done by using a file to remove unwanted material or by rubbing the bit on a wire brush or sponge pad.) Iron-plated bits must not be wire brushed; a sponge pad must always be used.

Wetting is a term given to how good a solder flows between surfaces and is dependent upon the heat of the surfaces, the presence of flux and the cleanliness of the surfaces.

Tinning refers to placing or coating solder on components or wires prior to actually making the joint. This then means that no solder has to be held during the soldering process, leaving a free hand to hold or place the component as it solidifies.

Good joints will have little electrical resistance but bad joints with too much or too little solder may initially have low resistance but cannot maintain this. If the joint is dirty, solder has difficulty flowing and there is a strong possibility that a dry joint may be made (if, for example, insufficient heat is applied or the components are moved before solidification).

Also avoid leaving the heat source for protracted periods on the joint as it may affect the component being soldered.

> **Self-assessment task**
>
> Sketch the sheet development of a 100 mm long × 100 mm wide × 50 mm deep open-top tin box prior to bending and soldering (pay particular attention to the joining tags), then describe in your own words, after bending, the steps taken to solder all the joints.

Welding

When a permanent metal joint is required designers often turn to a welded joint and – depending upon the types of materials, quality of finish, cost to manufacture, the quantity needed and the tolerances required – a choice would be made mainly from one of the following fusion processes:

- oxyacetylene
- arc (MIG or TIG)
- resistance/spot

Oxyacetylene welding

This, like any form of welding, requires the use of complex skills in a safe and competent manner, as a lot of specialised equipment is needed (Fig. 3.40).

Once the equipment is set up and the gas pressure has been adjusted to the correct level, the acetylene control valve is slightly opened and the fuel gas is ignited. The control valve is then further opened up until the flame stops producing a sooty carburising flame, then the oxygen valve is opened, slowly, until a clearly defined cone appears at the tip of the nozzle (this is an indication of burning equal amounts of oxygen and acetylene). The velocity leaving the nozzle is between 60 and 200 m/s and reaches a temperature of approximately 3,000 °C. By controlling the amounts of gases, three types of flame may be produced, each with a particular function. The neutral flame is, as mentioned above, a flame produced from an equal amount of oxygen and acetylene mixture and is in general the predominant welding flame. If a surface needs to be hardened then a slightly richer acetylene flame is used, called a *carburising* flame. When bronze or brass has to be welded then a slightly richer oxygen flame is required. In the welding

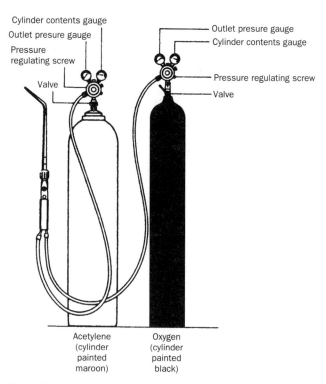

Figure 3.40 Oxyacetylene welding equipment

process itself it is important that the flame is directed at the area of the weld to be made and that it is pointing in the direction of the weld. This heat melts in a localised area the parent metals of the component, which then tend to run together and form a weld pool about the area of the joint. At this point a metal filler rod is inserted into the weld pool which, in turn, melts and joins (fuses) with the parent metal.

To form a weld run, this process must have lateral movement to give a leftward or rightward welding direction, which may be achieved by slowly rotating or swinging the nozzle from side to side about the weld area. This helps to preheat the weld area and, at the same time, melts the local parent metal. During this time the filler rod is continuously added to the weld pool. As the nozzle and rod progress left or right the temperature of the previously deposited weld drops, and the fused material cools and solidifies forming a welded joint.

Leftward welding (Fig. 3.41), as the term implies, means welding while moving to the left. Starting at the right-hand side of the joint to be made the flame is held so it points in the direction of travel (left) and is rocked from side to side

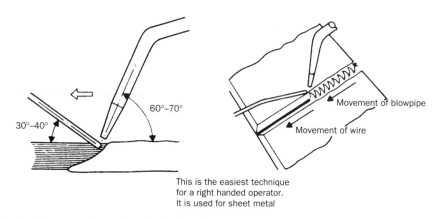

Figure 3.41 Gas-welding technique: the leftward method of welding

about the weld area, heating the parent metal. As the metal melts the filler rod is added by placing it in an 'in-and-out' movement in the weld pool. This offers an uninterrupted view of the joint and as the filler rod leaves the weld pool intermittently, it gathers oxides which it then transfers to the weld pool. This allows a welding thickness of up to 5 mm.

Rightward welding (Fig. 3.42) commences on the left-hand side of the joint and progresses to the right. The nozzle points left, back along the direction of the weld and in a series of gentle loops begins to preheat and melt the weld area. Once again the filler rod is added to the weld pool and as both nozzle and rod progress along the weld seam, the previously made hot pool cools and solidifies. This is a more common preference to leftward welding because less filler rod is used, the speed is greater and less gas is used with a lower risk of oxidation.

Arc welding

There are three general processes associated with arc welding:

- manual metal arc
- tungsten inert gas (TIG)
- metal inert gas welding (MIG)

Manual metal arc

As with each process, the heat in this process (Fig. 3.43) is generated electrically as a high intensity spark when two parts

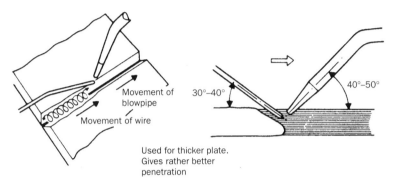

Figure 3.42 Gas-welding technique: the rightward method of welding

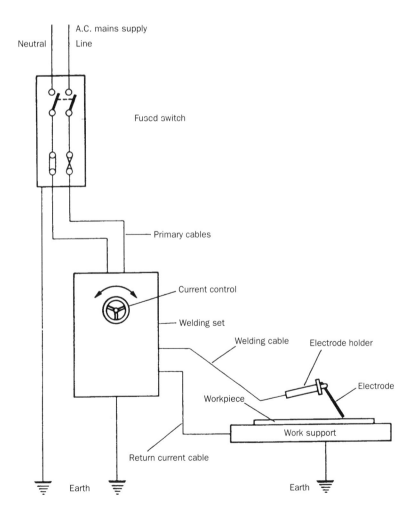

Figure 3.43 Manual metal-arc welding circuit diagram

of an electrical circuit are brought together and slowly parted. This may reach temperatures of up to 3,600 °C and is contained in a very localised area. The component forms one part of this electrical connection while the filler rod (electrode) forms the other.

The electrode is both filler rod and flux, on striking an arc (the term given to starting the weld process as the electrical connection is made and then broken almost in one continuous movement) the heat generated melts the workpiece to form a weld pool, while at the same time the electrode melts helping to fill the weld pool and the flux melts and forms a gas around the weld to protect the molten pool from the oxygen and nitrogen in the air. This requires tremendous dexterity and skill as obtaining a constant gap to create and then maintain the arc over the length of weld is not easy.

Tungsten inert gas
This method (TIG; Fig. 3.44) uses a more sophisticated approach to welding as it employs an electrode that does not melt (it is often water cooled). A shielding gas (argon) is fed through the weld torch and the filler metal is applied by hand as in oxyacetylene welding. This is an ideal process for the welding of thin sheet metals (up to 4 mm).

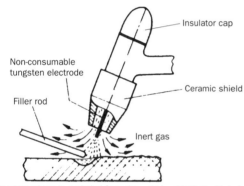

The atmosphere is excluded from the weld by shielding with in inert gas. No chemical reactions take place.

Figure 3.44 Protection of the weld – TIG process

Metal inert gas
This process (MIG; Fig. 3.45) is very similar to TIG welding but uses a consumable electrode wire (approximately up to 1.6 mm diameter), which is fed through the torch at a controlled rate, while the shielding gas is also piped through the torch. There is a variety of choice of shielding gases each with their own characteristics that enable a range of materials like aluminium, copper, nickel, stainless steel and steel to be welded by this process.

Resistance welding (spot welding)

This process relies on the resistance to an electric current between two materials. The two materials to be joined are pressed together between two electrodes, which forms an electric circuit, and as an electric current is passed through them there is a high resistance to the current at the joint, which causes intense heat and, in turn, melts the metals locally (Fig. 3.46).

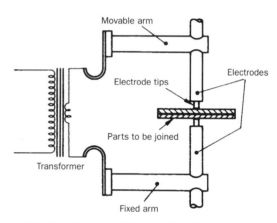

Figure 3.46 Electric spot-welding machine (schematic diagram)

As the weld area (nugget) is heated under pressure from the electrodes it becomes fluid, the pressure is maintained so there is now less solid material stopping the electrodes from coming together. As the pressured electrodes squash the softer material and start to close, the current is then automatically switched off and the fused weld solidifies. This basic principle lends itself to a variety of techniques (Fig. 3.47) for example:

- *Seam welding* – The electrodes are circular and rotate as the work passes through to be welded giving a long continuous weld seam, hence the term. These welds often overlap and form a water-tight joint.
- *Projection welding* – This relies on components having local projection which form the point of contact and hence the weld area. This process is used extensively in the manufacture of cars by welding nuts and fastenings to thin body panels.
- *Butt welding* (or flash welding) – A resistance welding process that joins the ends of bars by clamping them close together, allowing current to flow and arcing takes place between the ends. When enough heat is generated the circuit is broken and extra pressure is applied axially to the bars, resulting in a forged weld.

These processes offer the following advantages:

- little skill required
- no filler material required
- very quick
- easily automated
- little distortion as heat is very localised

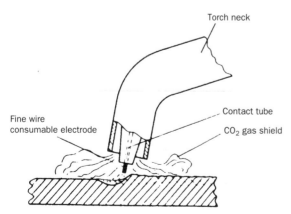

Figure 3.45 Protection of the weld – MIG process

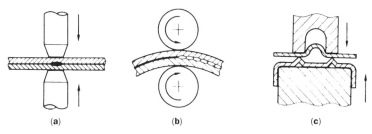

Figure 3.47 Welding techniques: (a) spot welding; (b) seam welding; (c) projection welding

> **Self-assessment task**
>
> Discuss which process is best suited to weld the following:
>
> (a) A bike frame.
> (b) An air-conditioning duct.
> (c) An aircraft landing gear strut.

Adhesives

It sometimes proves difficult to permanently join together two or more components which are made from differing materials by any of the processes mentioned above. Thus adhesives offer a solution to the engineer which is very often neglected, yet the process has a very good track record and has developed into a major versatile form of permanent joining. Often the key to a good adhesive joint is preparation of the surfaces to be joined. Many adhesives set by evaporation, and may be broken down into two distinct groups: water-based adhesives and solvent-based adhesives.

Water-based adhesives are generally made from methyl cellulose, which includes common brands of adhesives such as Polycell (for wallpaper), Copydex and Prittstick.

Solvent-based adhesives offer good resistance to water and are generally made from polystyrene solutions, polyurethane solution or phenolic resin solution.

With both types of adhesive at least one of the surfaces to be joined has to be porous so that it may absorb the water or solvent from the adhesive.

Hot melt adhesives are thermoplastic polymers – usually shellac, bitumen or wax – that have good adhesion and fluidity when hot, but turn to a solid when cool. They are often applied from a hot melt gun and are only suitable for light bonding, as their strength is limited.

One-part polymerisation adhesives fall into two categories, those that set without air (*anaerobics*) and those that set by a reaction to the water vapour in the air (*cyano-acrylates*).

Anaerobics made from acrylic ester or polyester have very little sticking properties but are used in threadlocking compounds by filling any spaces with a solid resin which then increases the friction.

Cyano-acrylates are commonly known as instant adhesives due to their rapid setting times, i.e. Loctite. They must be handled with care as skin tissue bonds just as easily as metals.

Two-part polymerisation adhesives consist of a resin and a hardener which react when mixed together, such as pheno–formaldehyde, urea–formaldehyde or epoxide–amine, tend to be more of a permanent setting process. In some cases one surface may be coated with the resin while the other is coated with the hardener, and upon joining the two the chemical mixing occurs. Another form is that of combining the two mixtures, applying them and allowing them to set, such as Isopon car body fillers.

As with any adhesive process the effectiveness of the bond is dependent on the type of adhesive, the effect of the adhesive on the materials being joined and how it is affected by temperature or the atmosphere. Many adhesives give off toxic or flammable fumes, therefore it is necessary to have good ventilation and to use protective clothing. You should always wash thoroughly after using adhesives.

Wiring

Often when two or more single-core wires need to be joined together to form a strong electrical connection, the individual wires are wrapped before soldering. There are many wrapping techniques, such as the straight twist joint (Fig. 3.48), in which approximately 25 mm of the two wires are wound tightly around each other keeping each turn closely to the next and then soldering over the twists.

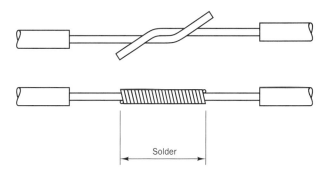

Figure 3.48 Straight twist joint

Another form of this joint is the tee twist (Fig. 3.49). In this method, remove approximately 50 mm of insulation without cutting the conductor, remove 50 mm from the end of a connecting tee wire, bend tightly around the through wire, then solder.

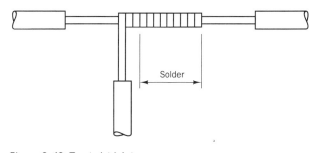

Figure 3.49 Tee twist joint

If a wired joint between two wires is to be placed under tension, then the *Britannia joint* (Fig. 3.50) would be used. This uses a third piece of wire to wrap around the two wires to be joined. The two joining wires, usually tinned, have a 90° bend about 5 mm from the end to form a stop and are placed alongside each other with their bend at opposite ends. The third wire is then wrapped around both wires often just past the bends, and is soldered.

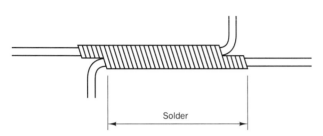

Figure 3.50 Britannia joint

Multi-cored wires require different techniques. For a straight connection a *married joint* is used (Fig. 3.51). This involves stripping approximately 120 mm of insulation from each cable, again leaving the conductor intact; splaying out approximately 75 mm of the twist strands; bringing the two cables together and wrapping the intermingled strands half a turn at a time tightly together. Each side is wrapped in opposite directions and may require tightening or crimping with pliers.

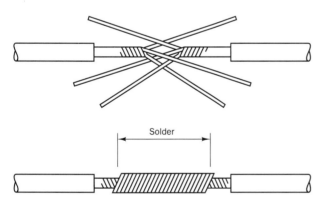

Figure 3.51 Married joint

To achieve a *wrapped tee joint* (Fig. 3.52) in stranded wire, approximately 75 mm of insulator has to be stripped from the straight connector and a similar amount from the tee or tie wire. Splay open the tie wires and divide into three sections: one section is wrapped around the tie strands to stop further splaying while the other two are wrapped equally towards the left and right of the through wire. Once again, tighten with pliers.

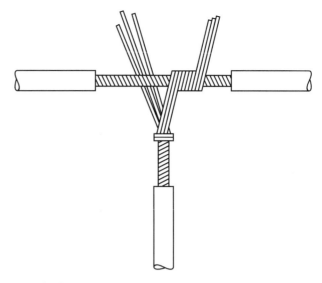

Figure 3.52 Wrapped tee joint

Heat treatment

Annealing

Metals may *work harden* or be *quench-hardened*; that is to say, the crystal structure changes and the metal becomes brittle. An example of work hardening occurs when a piece of metal is continuously bent backwards and forwards; it will harden and eventually break. Another example of work hardening is in the drawing of copper wire, where the annealing process is used to return the wire to a softer, more workable state.

When a steel has to be softened so that it is easier to work with, it is *annealed*. The annealing process requires heating a steel to the temperatures shown in Fig. 3.53 – which depends upon the steel's carbon content – and allowing it to cool slowly, often in a furnace which has just been turned off.

This restructures the grains of the component so that they become bigger and softer. Another form of annealing is *subcritical* annealing, also known as *stress relief* annealing. This involves heating work-hardened steels to between 630 and 700 °C, holding at this temperature allowing grain growth and recrystallisation, then cooling slowly. This offers a lower cost and quicker means of obtaining some grain growth.

For steels which lack ductility – usually those with more than 0.4 per cent carbon – it is doubtful if they will ever be work hardened; however, they may have been quench hardened and require softening so that further machining may be carried out. If this is the case then spheroidising annealing could be employed in preference to full annealing performed at a lower temperature to enable grain refinement. This gives all the advantages of subcritical annealing using less heat, but if cold working is required then a full-annealing process is necessary.

Normalising

When a steel requires stress relieving to remove distortion and to allow subsequent machining to take place, but does not require a full anneal to make it very ductile, *normalising* is used (Fig. 3.54). The material is heated to the same range as full-annealing, but instead of cooling within the switched-off

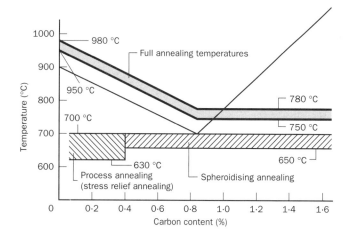

Figure 3.53

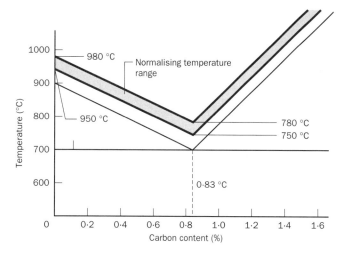

Figure 3.54

furnace, the component is brought out into the air to cool naturally. This gives s faster rate of cooling, which then also gives a finer grain structure and is still suitable for machining, but lacks ductility to be further cold worked.

Hardening

This process involves heating a steel up to 30 °C above its upper critical point (Fig. 3.55) and quenching it rapidly in oil or water. This restructures the grains by not allowing the grains to grow substantially through the normal slow-cooling process, and gives a structure of many small grains with hard grain boundaries, hence increasing the hardness.

The hardness is dependent upon the amount of carbon present in the steel and the rate at which cooling takes place. The differing rates of cooling (quenching) are achieved by using a quenching medium of water (cheap and plentiful) or quenching oil (costly, but through differing viscosities absorbing heat at different rates allows a range of cooling times, hence a range of hardness). In general the quicker the steel is cooled the harder it becomes; however, care has to be taken that the steel is not cooled too rapidly or it will crack or distort. The component to be quenched should be immersed axially in the quenching medium and agitated to allow even cooling.

Tempering

Often a component is required to be hard and strong at the same time, so it is hardened but the material the component is made from still has to be formed into the desired shape (i.e. a scriber or screwdriver blade). This hardening makes the component too brittle to change shape easily. In order for this shaping to occur the material must be ductile, yet the finished component has to be hard, strong and tough. It is therefore made in the desired shape in the ductile mode and the properties are then changed by heat treating the component. This involves hardening, which first heats the component then quenches to give it the hardness requirement. This treatment alone, however, would make it too brittle for practical use, so it is reheated to a much lower temperature which allows some grain growth, and is then allowed to cool slowly in air or sand, which removes some of the hardness and imparts a little toughness. This process is known as *tempering* and is used to form a compromise between hardness and toughness of steels. As it is very difficult and costly to ascertain the tempering temperature using thermocouples, use is made of a very simple colour coding of the oxide film which shows itself on newly polished steel when heated (Table 3.1).

Figure 3.55 Hardening of plain carbon steels

Table 3.1 Tempering temperatures colour code

Component	Temper colour	Temperature (°C)
Edge tools	Pale straw	220
Turning tools	Medium straw	230
Twist drills	Dark straw	240
Taps	Brown	250
Press tools	Brownish-purple	260
Cold chisels	Purple	280
Springs	Blue	300
Toughening (crankshafts)	—	450–600

Self-assessment task

Describe, in your own words, the process of heat treating a pair of 0.9 per cent carbon steel shear blades so that they obtain a hard but tough cutting edge.

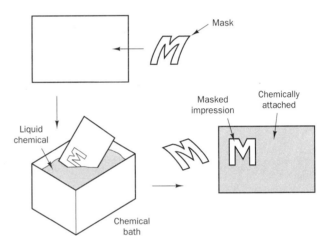

Figure 3.56 The etching process

Chemical treatment

Etching

This involves removing material by making a mask (a copy) of the desired shape, placing this mask onto the component, then subjecting the unmasked visible material to chemical attack. The component is thus changed by the erosion of the visible material until the desired feature is obtained (this may be dimensional or a type of surface finish requirement). The component is then washed to remove the chemical residue leaving the masked impression (Fig. 3.56).

The following lists typical etching fluids for various materials:

- Steel: 1 part nitric acid and 4 parts water.
- Bronze: 100 parts nitric acid and 5 parts muriatic acid.
- Brass: 16 parts nitric acid and 160 parts water.
- Aluminium: 4 ounces of alcohol, 6 ounces of acetic acid, 4 ounces of antimony chloride and 40 ounces of water.

Etching may be used to produce markings or lettering on steels (i.e. as on a steel rule); this offers very fine work to be achieved (0.003 mm depth) at very low cost on a variety of materials such as steels, ceramics and glass. It is also used extensively in the manufacture of electronic printed circuit boards (PCBs).

A typical sequence for making a PCB could resemble the following:

1. Plan out the circuit and prepare the circuit board.
2. Apply a dry transfer of the circuit onto the copper (the mask).
3. Place the board into an etching tank; keep checking the progress but take care because of the hazardous chemicals.
4. When etched, wash away any residue of chemical contamination in running water.
5. Remove mask transfers and clean up the surface to expose the copper tracks (usually done with a fine emery cloth).
6. Drill out component location holes and solder components in place.

Plating

In this process, which is the reversal of etching, a component is enhanced by coating the surface with a different material (i.e. chromium for chrome plating). This is predominantly required to ward off atmospheric corrosion, on what would normally be a very corrosive material, or to give a superior surface appearance which may be tarnish resistant. Plating or, more correctly, electroplating is based on two fundamental laws laid down by Faraday, in which he stated:

> 'The process itself requires an electrical circuit (Fig. 3.57) to be formed from an anode, which usually is the part that deposits material onto the component(s), as they are suspended in an electrolyte (a liquid conducting medium). The atoms on the anode are positively charged. The cathode is the part that will receive deposited metal from the anode and is thus the component to be plated; it is negatively charged.'

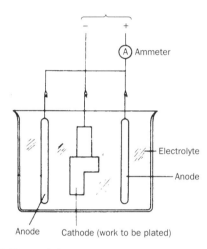

Figure 3.57 Electroplating

A typical chrome-plating procedure may be summarised as:

1. Degrease the components with a solvent.
2. Mount the components on a racking system for economic production (it is rare to have a 'one-off').
3. Clean components in an anodic alkali bath for up to 5 minutes.
4. Wash off any residue in boiling water.
5. Immerse briefly (2–3 seconds) in a 20 per cent hydrochloric solution.
6. Wash in water.
7. Place components in a chromium-plating bath for desired length of time.
8. Wash off any residue.

Chromium has several benefits in that:

- It has a very low coefficient of friction.
- It resists attack by organic compounds.
- It has a high melting point.
- It produces a light adherent oxide which clings to the surface and is dense enough not to allow oxygen to penetrate, thus maintaining a long-lasting clean surface finish.

Galvanising

This process involves the coating of mild steel with zinc and may be achieved by two methods: electrolytic galvanising similar to the plating process above and the more common hot dip galvanising.

The hot dip method galvanises by immersing a suitably prepared component in molten zinc. It ensures a low-cost process for providing a corrosion-resistant coating on steels. It is extensively used by the construction industry as corrugated or flat sheet and as weather-proof nails. The amount of coating varies from 50 to 150 μm and is dependent upon the thickness or surface of the steel.

Anodising

This is also an electrolytic process and conforms to the general methods mentioned earlier of cleaning, degreasing, washing, immersing in an electrolyte, passing a current through the component and plating. It is used generally on aluminium and may provide extra protection from atmospheric corrosion. It may also be used to harden the surfaces and to offer different coloured components. A typical household product of anodising is a knitting needle.

The type of electrolyte employed affects the finish provided, such as:

- Sulphuric acid as an electrolyte gives a colourless finish, so a dye is used to give any particular colour requirement.
- Chromic acid is used extensively in the aircraft industry as it produces an excellent key for subsequent painting to be carried out.
- Oxalic acid gives corrosion resistance which is harder and more wear resistant and is used in structural work, i.e. national grid power lines.

Surface finish

Polishing

To many the term polishing means cleaning up or making a surface shine, and this is often the outcome; however, to achieve this the surfaces to be cleaned are polished by actually marking with tiny scratches made by an abrasive material. This abrasive is generally a very fine hard powder, and is often present, immersed in a liquid. The scratches may also be made from a softer material in a disc or wheel form, such as, cotton cloth, felt, wool, etc. These are generally known not as polishing wheels but as buffing wheels and they carry the abrasive particles.

Polishing is often seen as a cleaning operation and is usually an intermediate process between the primary processes of casting, forging, machining and a finishing process, buffing.

A common method of polishing using a batch production approach is barrel polishing (tumbling; Fig. 3.58), although

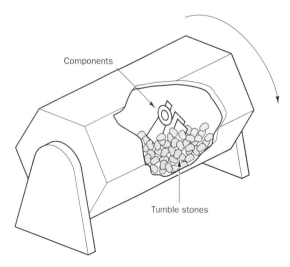

Figure 3.58 The barrel polishing system

used extensively as a deburring operation it is also an effective method of surface polishing. It involves rotating a box (barrel) containing water, fine abrasives, chunks of stone or ceramics and the components to be polished. The whole box is sealed and allowed to rotate, causing the components, abrasives and stones to rub against each other while the water in the barrel helps separate the abrasive from the component. Another polishing method is electropolishing, which operates by reversing the electroplating process. The workpiece is made the anode and material is removed from its surface, rather than coating it. This often gives a surface which is smooth and glossy in appearance.

Coating

This process has two main functions:

- surface protection
- surface enhancement

The most common form of surface coating is painting and it may fulfil both of the above functions at once, i.e. it may enhance the surface using various colours and at the same time protect it from atmospheric oxidation.

A general word of caution: to obtain the best results from a painted surface there are several points to consider:

- Good preparation makes for good adhesion.
- The amount of protection offered by painting is proportional to the thickness of coat.
- More cost is spent on applying the paint, therefore do not use poor quality paint (one with a high spreading rate and thin film).

Paint is formed from very fine powder pigments and is held in a binder (organic liquid), thinned by solvents, to obtain a consistency suitable for spreading. These solvents evaporate on application and form no part of the finished film, but this film is affected by absorbing oxygen, from the air, which hardens it through a chemical reaction. In the main, these air-drying paints are based on alkyd resins or epoxy resins and are better known as cellulose or acrylic paints. They are used for painting equipment which requires a hard solvent-resisting finish.

As air-drying paints take considerable time to harden, a quicker process which is often used is *stoved* painting. These paints contain binders which harden through a chemical change in minutes when heated, and as this often occurs in ovens or stoves they are termed stoved paints. Newer techniques have enabled this process to be activated faster by irradiation, and may also be achieved at room temperature by the use of a catalyst to trigger off the curing mechanism.

Typical paints from this process are:

- Alkyd-amino resins (urea-formaldehyde resins), used on washing machine and refrigerator finishes.
- Acrylics used on car bodies.
- Epoxies (polyurethanes and phenolics) used as a tough chemical-resistant film.

There is also a group of paints, known as lacquers, that dry solely by the evaporation of the solvent with no chemical reaction to the atmosphere.

Typical paints include:

- Nitrocellulose for car bodies.
- Shellacs for french polishing.
- Chlorinated rubber used on concrete and other lime-containing surfaces.

Another group of paints are water-based, and they form two general types:

- Paints which depend upon the drying out or evaporation of the water.
- Paints based on water-soluble resins.

The film is formed when the polymers that are suspended in the liquid solution change properties when the water evaporates. As all polymers have a *glass transition point* at around $20\,°C$ – which enables them to change from a brittle solid to a soft plastic – it is then unadvisable to paint in conditions which are very cold.

Paints that require a catalyst to instigate curing are commonly known as two-pack paints and form four distinct types:

- *Epoxy resins* – The two parts are mixed (pigment resin and hardener) and have a limited time of use before they set too hard to spread. They offer excellent protection as a weather-resistant coating.
- *Polyurethanes* – These are also mixed but may have to be thinned with solvents (in which case avoid inhaling fumes). The range of polyester resins are so great that one could coat a floor, requiring hard-wearing properties, or a rubberised beach ball. This shows a very good durability and high resistance to chemicals and solvents while giving a high gloss finish.
- *Urea–formaldehyde* – This provides extremely tough finishes but must be used quickly as it only has a pot life of approximately 8 hours.
- *Polyester paints* – These tend to give clear finishes and are extensively used to coat domestic radios or television cabinets.

Grinding

Obtaining a high degree of surface finish while demanding a close tolerance may be achieved by the use of the grinding process. It basically is a sophisticated polishing process, in which the abrasive materials that cut are held together by an adhesive (bond) often as a solid stone or wheel. This in turn is rotated rapidly to allow many of the abrasive particles (cutting edges) to come into contact with the component. Because there are many of these tiny abrasives, and material is removed very slowly and in tiny amounts, this then can provide a very smooth and accurate surface.

There are three basic reasons for employing the grinding process:

- The removal of surplus material.
- To obtain a certain high degree of surface finish.
- Machining very hard materials.

The wheel cuts with an abrasive held in a bond, but as with any cutting edge it will soon become blunt, and when this occurs the forces acting at the cutting edge then become so large that the bond is no longer able to hold the abrasive grain. It is therefore broken off or torn out, and in doing so reveals a new sharp cutting grain, giving the grinding wheel what is commonly known as a self-sharpening effect.

There are two main abrasives used in this process:

- Aluminium oxide (for use on materials with a high tensile strength).
- Silicon carbide (for use on materials with a low tensile strength).

Either of these abrasives may be held in a variety of bonds to suit differing purposes. For example, a vitrified bond allows a high rate of metal removal, while a resinoid bond allows the wheels to run at higher speed. By varying the size of grain or the amount and type of bond, there is a tremendous range of grinding wheel combinations to choose from.

Therefore, any choice should be made with these factors in mind:

- The material to be ground.
- How much material is to be removed.
- The finish required.
- The condition of the grinding machine.
- The speed of the wheel.
- The shape of the wheel.

At this point it is well worth mentioning surface finish or surface texture. All surfaces, to a greater or lesser extent, are the same: they consist of hills and valleys – the better the surface finish the smaller the distance between the tops of the peaks and the bases of the valleys. As this distance may be very small, it is measured in micrometres (one-millionth of a metre). There are special instruments to measure surface texture which employ the use of an electronic stylus to travel over the peaks and valleys. This movement is recorded and greatly magnified, enabling a calculation to be made based on a centre-line average (CLA) method, to give an overall finish rating.

In order to achieve a good surface finish and high accuracy without a polishing operation, one of three methods of grinding may be used:

Surface grinding machine (Fig. 3.59) This is similar to a horizontal milling machine in that the table movements and placement of the cutter (in this case grinding wheel) are the same. It generally is used to generate flat plain surfaces, and as it removes significantly less material than milling, and as the cutting forces are lower, it may employ a magnetic chuck as a means of work holding.

Cylindrical grinding machine (Fig. 3.60) As its name implies, this generates cylindrical surfaces and is similar to a centre lathe in appearance. The main difference is that the tool and work both rotate; the tool is a revolving grinding wheel which makes contact with the work and cuts.

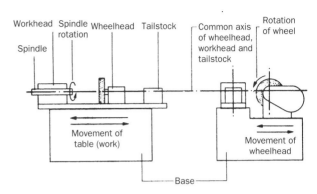

Figure 3.60

Centreless grinding machine (Fig. 3.61) This type caters generally for uninterrupted cylindrical shapes such as pins and does not require any mounting of the workpiece. This means that the work may be fed into the grinding area by hopper feed. The principle of operation is that the components pass between two axially parallel wheels: a conventional grinding wheel and a rubberised regulating wheel, inclined at a slight angle to facilitate axial rotation (like a screw thread movement).

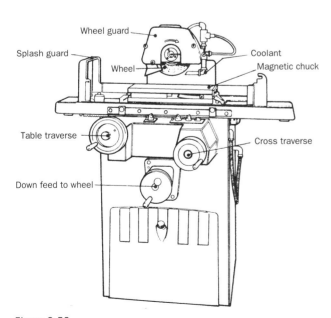

Figure 3.59

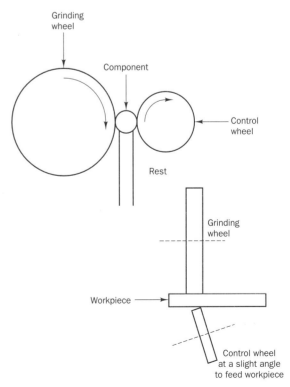

Figure 3.61 Centreless grinding machine

There are two other surface-finishing techniques used: *lapping* and *honing*. The reasons for using these techniques instead of grinding is that grinding may distort or burn the surface of the work by generating too much heat. As this would give a poor surface finish, lapping and honing tend to be used as they operate at much slower speeds and produce very little frictional heat.

Lapping uses a soft material (i.e. wood, brass, etc.) impregnated with abrasive material, and is then rotated by hand or machine over the workpiece, causing tiny scratches.

Honing is a very similar process to lapping but is confined to holes only. Not only will it remove material and give a good surface finish but it can also be used to correct out-of-roundness and tapers.

Self-assessment task

Make a comparison of the following machining techniques with reference to surface finish, and place them in a ranking order with the best finish at the top.

Also try to give some indication as to the typical roughness values obtained.

(a) turning
(b) grinding
(c) honing
(d) milling
(e) lapping

Evaluating the suitability of manufacturing processes

It may prove to be very difficult to decide how to manufacture or assemble components, especially when there is more than one method to choose from. How, therefore, may a successful choice be made? What are the factors that affect our choice?

Regardless of the type and function of a process there are *five* key areas to consider for evaluating suitability:

- The materials available or used.
- The level of quality required.
- Cost limitations.
- The quantity to be made.
- Manufacturing tolerances.

The evaluation of the use of materials within a process centres around the distinct properties of those materials, such as:

- strength
- hardness
- ductility
- malleability
- conductivity (electrical and thermal)

These and many more have to be taken into account so that the best use of a material's properties are used. An example of this is the combination of the low density and high strength offered from an aluminium alloy used on aircraft structures, which allows for a light but strong structure.

The ability to meet a customer's specification is an aid to determining the quality – the quality level, for example, may depend upon costings, finish, product life span and the type of material used. Quality is often a negotiated term which is agreed upon at the start between customer and supplier, with various safeguards in the form of guarantees ensuring satisfaction.

When there are several production processes to choose from the relative cost of each needs to be given careful consideration so that a combination of an economical and suitable method may be adopted.

This may involve employing different scales of production such as one-off, batch or mass production, and so the quantity also influences the choice of process. The quantity is often driven by trying to make economies of scale, but this may lead to an increase in capital equipment, stock, overheads and labour costs.

Tolerance is often linked to quality as a higher tolerance is seen as obtaining better quality. However, tolerance should primarily be seen as a method of obtaining the correct fitness for purpose within an economic range.

It is clearly uneconomic to use a process that takes twice as long as another to produce the same product to the same tolerances at a similar cost.

How may we go about making an informed choice of evaluating the most suited process?

A tried and tested method is to compile a tally chart, listing all the attributes required and allocating marks to various processes, e.g. the fixing of a radio grill face to the main body of the radio box may require the considerations given in Fig. 3.62.

		Screws	Adhesive	Moulded clips	Rivets	Weld
1	Ease of attachment	2	4	5	3	1
2	Ease of detachment	4	2	5	3	1
3	Security of fit	3	4	1	2	5
4	Time to complete process	3	4	5	2	1
5	Cost of process	4	5	1	3	2
6	Specialist requirements	4	2	5	3	1
7	Weight of attaching device	3	4	5	2	1
8	Aesthetics	3	4	5	2	1
	Totals	26	29	32	20	13

Figure 3.62 A tally chart

Marks are awarded in the various categories: in this case, 5 for the most appropriate method and 1 for the least appropriate. As can be seen, it looks as though the use of moulded clips would offer the best overall possible solution for fixing the grill to the box.

However, the final choice would heavily depend upon the number to be made for any one category, i.e. cost can outweigh all the others. In this case if only one radio was to be made, the cost of producing plastic moulding clips (i.e. the cost of the dies for injection moulding) would be uneconomical. If, however, thousands were to be made, then the cost would be spread over all of these and the cost savings would then be made in the time taken to assemble, thus making it a viable choice.

Sand casting versus die casting

Let us assume that a water pump casing has to be cast and a particular process has to be chosen. The two leading alternatives are sand casting and die casting. Each year, 10,000 casings are required to be manufactured in batch sizes of 2,500. The casings have an expected life span of 5 years. Table 3.2 gives some idea of the variables that affect the decision-making process.

Table 3.2 Cost comparison of sand casting and die casting

Process	Sand cast			Die cast		
	Cost of item	Frequency	Unit cost	Cost of item	Frequency	Unit cost
Tooling	£3,000 (pattern)	1/50,000	£0.06	£20,000 (die)	1/50,000	£0.40
Material	£0.12/kg	6 kg	£0.72	£0.40/kg	2 kg	£0.80
Casting setup	0.30 h @ £5/h	1/2,500	£0.00	4.0 h @ £5/h	1/2,500	£0.01
Casting labour	0.08 h @ £5/h	1	£0.40	0.04 h @ £5/h	1	£0.20
Machining set-up	£30 (5 ops)	1/2,500	£0.01	£15 (3 ops)	1/2,500	£0.01
Machining labour	0.05 h @ £5/h	1	£0.25	0.03 h @ £5/h	1	£0.15
Total unit cost			£1.44			£1.57

3.2 Make an electromechanical engineering product to specification

Topics covered in this element are:

- Devising sequences to aid production.
- Identification of specific techniques.
- Safe working practice.
- Maintenance of tools and equipment.

Devising sequences to aid production

Once a clear statement of the product requirements has been made known as the design brief or customer specification, the following are considered before production proceeds:

- time given to make it
- performance
- cost
- amount of manufacture
- strength and other specifications

The work to be carried out depends critically upon the results developed at this stage, and getting this right can save a lot of time and money trying to solve the wrong problem.

Before any manufacturing commences a clear understanding of what is required to do the job in hand is essential. Often this understanding may be broken down into three distinct areas that rely on each other for a successful product to be produced.

- planning the job
- schedules
- diagrams and drawings

Planning the job

This requires a fair degree of knowledge and experience of the types of processes available, in order to make an informed choice of:

- manufacturing methods
- skills required
- equipment needed
- time to completion
- working function

The initial form of planning may use a variety of techniques to arrive at a possible solution, such as brainstorming, group work, case studies and research studies. Once a decision is made on the method of manufacture then detailed planning may follow.

This detail can be broken down into elements that can give accurate times for individual completion of process, and a requirement of tooling. This is to enable an orderly sequence to be formed and a manufacturing schedule to be drawn up.

Schedules

The planning sequence or schedule is very useful in maintaining a sense of time, as it may provide at a glance the current or projected state of a project.

To give an example, a simple radio is to be manufactured which will probably consist of the schedule presented in Fig. 3.63 (which is also known as a Gantt chart).

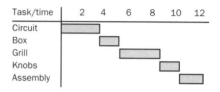

Figure 3.63 A simple schedule

Self-assessment task

Four cars require a day's schedule to be devised so that they may all be efficiently serviced. The actual service offered consists of:

(a) lubrication
(b) mechanical service
(c) electrical service
(d) a road test

Work is completed in four separate areas matching the fixed servicing order of (a) then (b) then (c) and then (d).

An estimation of the times for each job in each area is given below. Your task is to compile a schedule of the day's work. (Remember the fixed order of job (a) then job (b), etc.)

Area	Car 1	Car 2	Car 3	Car 4
(a)	60	0	40	30
(b)	120	175	30	60
(c)	40	35	90	40
(d)	20	20	20	20
Total	240	300	180	150

Diagrams and drawings

In manufacturing, drawings and diagrams play an important role as they can show, using pictorial views, items that are very difficult to visualise by the use of words alone (written or verbal). They take many differing forms but all have the same purpose – that is, to assist communication. Here are some examples.

Concept sketches (Fig. 3.64) These are very difficult to manufacture from, as they contain insufficient detail; however, they may give a general idea as to how the components fit together.

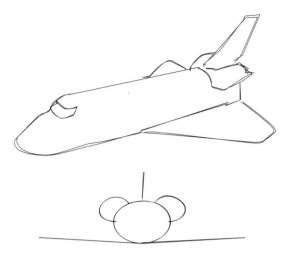

Figure 3.64 A concept sketch

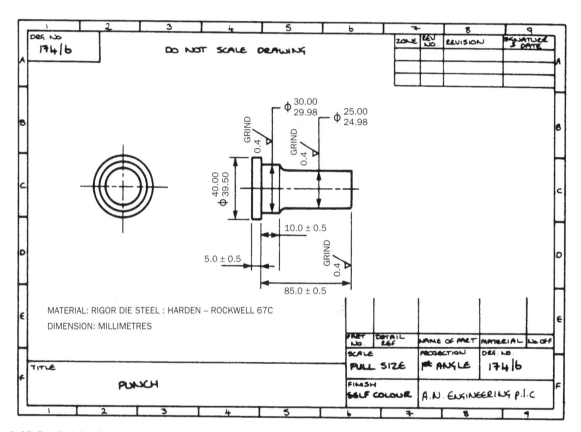

Figure 3.65 Detailed drawing

Detailed drawings (Fig. 3.65) These are very precise drawings made to scale, comprising individual components with dimensions in orthographic mode and may include manufacturing and material detail.

Subassemblies (Fig. 3.66) These are made from putting together a series of detailed drawings to show their arrangement (i.e. the brake caliper system of a car).

Assemblies (Fig. 3.67) These show the overall structure of the completed items in their working places. They often show very few dimensions as this would require too much detail, but reference is made to the individual detailed drawings and each component is identified, which is often the source of generating a bill of materials (BOM).

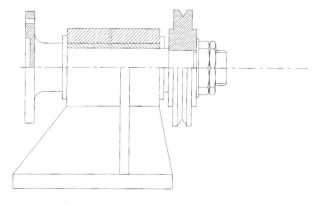

Figure 3.66 Subassembly drawing

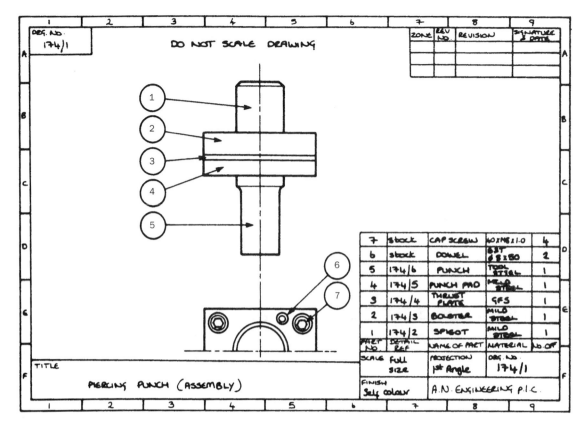

Figure 3.67 Assembly drawing

Schematics (Fig. 3.68) These often look like a young child's drawing, in which components are represented by little boxes or specific shapes. These are joined together by lines and may appear very simple, yet they provide a wealth of information so that an understanding of the process is easier to interpret (i.e. a PCB diagram). These are invaluable as wiring diagrams. Schematics are drawn this way to save tedious time recreating individual components or wires (each to be drawn 4 mm thick?), when a simple representation is just as good to interpret if not easier to follow.

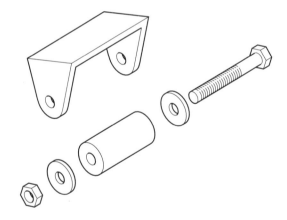

Figure 3.69 An exploded view

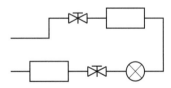

Figure 3.68 Schematics

Exploded view (Fig. 3.69) As the term implies, the details of the individual components appear to have exploded, intact, away from an assembly and frozen in space, so that they appear close together. This may be very useful in identifying all the parts pictorially and showing where they fit in relation to each other (like a DIY kitchen cabinet unit). It is, however, a time-consuming drawing to produce, but looks very effective for presentations.

Identification of specific techniques

Once an operating sequence has been formed and the general manufacturing processes have been identified (i.e. a lathe is to be used to turn a taper component) then a further decision determining which specific technique is to be employed has to be made (in this case offset tailstock, compound slide, taper attachment or form tool).

> **Self-assessment task**
>
> The workshop walls of an aircraft hangar are separated by a floor space 500 metres. There is now a need to traverse this floor space using an overhead crane. Your brief is to develop a range of crane alternatives, taking into account cost, maintenance, ease of use, time taken to cross and at least three other attributes, by creating an evaluation chart. Put forward a proposal for your selected choice outlining the reasons for choosing it.

> **Self-assessment task**
>
> Each of the types of extinguisher mentioned above has a special colour code to help distinguish it. Make a list of these colour-coded extinguishers, indicating its colour, contents and correct usage for use in the following working areas:
>
> (a) A general engineering workshop.
> (b) A computer workshop.
> (c) A car maintenance garage.
> (d) A television repair workshop.
> (e) The waiting room of a hospital.

Safe working practice

The Health and Safety at Work, etc., Act 1974 made provision for:

- The health, safety and welfare of people working.
- The protection of others at work with regards to health and safety.
- Safety regulations concerning hazardous substances.
- Atmospheric emissions.
- A medical advising service.
- Restructured building regulations.

First aid

All injuries, must receive attention and be reported. This may prevent the injury from becoming more serious and allows a record to be kept of the event so that similar incidents may be prevented. In larger establishments there should be adequate provision of first aid boxes, with suitably trained first aiders or nurse on call.

Fire exits

All fire exits should be clearly marked as such and be free from any obstruction. There should be defined assembly points well away from buildings, and positioned so that they will not block any route the emergency services may wish to use.

Portable fire extinguishers

The main categories of portable extinguisher are:

- Colour coded RED which use WATER on solid materials.
- Colour coded CREAM which use a FOAM on solid materials and small liquid spillage fires.
- Colour coded GREEN which use a GAS or VAPORISING liquid on solid materials, particularly electrical.
- Colour coded BLUE which use a DRY POWDER on flammable liquid fires
- Colour coded BLACK which use CARBON DIOXIDE on electrical fires.

Each has its own specialised use; for example, it is pointless employing a water-type extinguisher on an electrical fire as it could further damage the equipment it is trying to save. If water were to be used on an oil-based fire the consequences could be disastrous; it would only ensure that the fire would spread, as oil is lighter than water, and it would have a tendency to be carried away, burning, on the back of the water increasing the risk of further danger and damage.

Personal safety clothing

What is worn as personal safety clothing may vary depending upon the specific process in use. In general, well-maintained overalls, safety shoes and safety glasses are a minimum requirement. If a specific hazardous process is employed, it may require additional safety equipment – for example, during the casting process of pouring molten metal, where heat-resistant gloves are essential, the extra safety equipment required may consist of a leather apron worn over the overalls and spats over the shoes. If the process is in a noisy environment, ear protectors should also be worn.

Emergency systems

An emergency system that is often overlooked is the reporting procedure of damaged or faulty equipment. There should be in place a formal document procedure to record such occurrences so that a pattern may be detected and rectified before serious damage occurs. All hazardous machinery should have emergency cut-off switches, which are usually designated as a large bright red button clearly visible, sometimes surrounded by a square yellow patch to highlight it further. Once it is pressed it automatically cuts the power to that piece of equipment, while leaving others within the workshop free to carry on functioning. To restart the machine the button has to be reset; this gives an additional safety feature which ensures that the machine cannot restart accidentally. This cut-out safety system is also employed in a safety ring within workshops and laboratories. If any one of this ring of connected safety switches is pressed, then power to the whole area is cut. This enables a person some distance away from a hazardous situation to react quickly, by depressing the nearest of these switches in order to cut power. If, within modern CNC machine tools, a guarding door is left open, the machine will not operate as power is automatically cut until the door is closed. This is known as an interlocking guard.

Guarding

The guarding of dangerous machinery has 17 classification groups and further detailed information is given in BS 5304 'Safeguarding of machinery'. However, these groupings can be condensed into two distinct sections:

- The protection of rotating parts (i.e. lathes, grinding wheels, etc.).
- The protection of reciprocating parts (i.e. press tools).

Guards are positioned to offer maximum protection to the operator, but with minimum interference to their safe working practice. Basically they are designed so that they eliminate the need for operators to approach dangerous areas.

Maintenance of tools and equipment

Taking care of the equipment used will often prolong the working life of that equipment or allow it to be utilised to its full potential. Failure to take care of equipment may lead to early breakdowns or even premature scrapping of often valuable and essential equipment. The following offers some advice on the maintenance of equipment.

Storage plays an important part in the scheme of manufacturing, for if equipment is not stored correctly then early damage to the equipment could occur giving a less than useful tool life, or too much time is spent searching in a poorly organised storage system. Both of these cost extra money and are inefficient, therefore it is very important to take into account the safety, storage and maintenance of individual tooling.

A clean working area which is free from obstacles (i.e. bags, extra tooling, etc.) should always be maintained. Even if a working area is only used spasmodically, it should not be left in a state of disrepair or untidyness. A typical example of keeping equipment clean is the maintenance of a paint brush after use. If it is not cleaned the paint hardens and sets very quickly; it may, however, be possible to clean the brush later for further use, but this will require extra expense in the form of cleaning fluid and extra time to wash the brush. If, however, the brush had been cleaned immediately after use, before the paint had set, then a wash in a cheap dilution of soapy water is all that would have been required to make the brush reusable.

This kind of process may be applied to many pieces of engineering equipment.

Manufacturers' equipment often has a limited useful life span, and if the equipment is to be used to its full potential it will require regular maintenance. This may take the form of a rigorous overhaul, including a detailed strip down and replacement of parts, or a simple quick visual inspection (very much like a car's MOT). What is of importance here is that a maintenance schedule is used, allocating timescales as to when and what should be done to look after the equipment (i.e. if lathes were not cleaned regularly or have their slideways oiled frequently they would soon seize up through corrosion; this applies not only to complicated machinery but also to the simplest of tools).

3.3 Perform engineering services to specification

Topics covered in this element are:

- Identification of procedures to perform a service.
- Identification of safety procedures.
- Selection and maintenance of tools and equipment.

The function of an engineer does not cease when the components have been designed and manufactured, for these components will have to be maintained to give an optimum working life, which requires servicing.

It may well be possible that the service required is a simple task and instructions are sent out with the packaging of the equipment as it is sold, or it is clearly labelled on the component itself, as in the case of equipment requiring regular lubrication (oil here, is often a common sight). So this will have to be considered at the design stage of the component, otherwise a product modification feedback system is required.

An important aspect to performing an engineering service is the range of tools for the job. These may vary from a very large kit of tools requiring trucks to transport them, to just a few items carried in a holdall.

Generally service engineers carry standard tools (i.e. screwdrivers, spanners, etc.) to form the basis of their kit and any special tooling for specific products or processes (which may take years to accumulate).

Identification of procedures to perform a service

Procedures

In order to perform an engineering service to a specification, certain procedures have to be identified. These procedures are formed from four categories, depending upon when they occur within the servicing cycle:

- diagnosis
- inspection
- execution
- evaluation

Diagnosis

This is the initial stage of a service and probably one which requires the greatest patience. Patience because quite often a service is called upon when something has gone wrong with the equipment. There may be many reasons for this malfunction – wear of moving parts, breakage of weak components, maltreatment or abuse of the equipment through deliberate acts, or ignorance in the use of the equipment. All of these scenarios, and many more, have to be taken into consideration from the outset, eliminating each possibility until an accurate diagnosis of what caused the malfunction in the first place is arrived at.

In order to achieve a possible diagnosis certain procedures are adopted. An example of some of these procedures is given below.

An inspection of the equipment is made to ascertain that it actually requires a service or is due for a service. If the service is of a common component (i.e. a photocopying machine), there may already be in place a list of common faults that occur (based on past experiences), or procedures to run a simple diagnostic check laid down by the manufacturers.

This may take the form of a flowchart or servicing booklet, instructing on the possible problems or service requirements and procedures.

Flowcharts help diagnose troublesome problems and offer a step-by-step method of rectifying these problems. They work by the use of joining a simple series of mainly four different box-like shapes containing notations. They are joined together in a particular fashion so that it forms a code. This code is simple to draw up when developing a maintenance procedure and is invaluable when it comes to reversing a stripdown.

The four forms of boxes are:

- The process box, which is rectangular in shape and contains instructions to be carried out.
- The junction box, which is in the form of a diamond and relates to a question or a decision. If the answer to the question is 'yes' then follow a line from the side of the diamond; if the answer is 'no' then procede down through the bottom of the diamond.
- Input/output parallelogram-shaped boxes indicate that some form of interaction is required, i.e. numbers displayed or required as an input.
- Termination boxes, which are shaped like an elongated ellipse, start and end a flowchart. (A typical example of a flowchart can be seen in Fig. 3.70.)

Self-assessment task

Draw up a flowchart for the replacement of a 13 amp fuse in a domestic electrical plug.

Inspection

The use of a flowchart may even form part of a quality assurance policy, one in which an after-sales service scheme is offered (in the case of the photocopier, say after 200,000 copies have been made).

Here there are three levels of quality assurance in operation:

- The personal quality requirements of the operator servicing the equipment.
- The quality requirements of the customer.
- The quality procedures laid down by the manufacturers.

All three have to be satisfied in order for a successful service to be performed. The cost of the service also has to be diagnosed very early to allow the service to actually commence. That is to say, a decision as to whether it would be too costly to service or repair has to be agreed with the customer before any expense in either time or money is used. The components may well be serviceable but require specialised tooling or manufacturing, and this may make the overall cost too much to be a viable option.

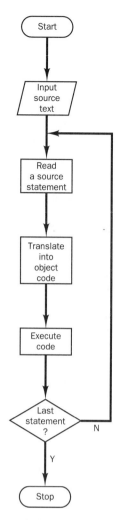

Figure 3.70 Flow chart of a typical interpreter and compiler for computer operation

Execution

This is often the most time-consuming part of any service, especially if the service involves an unfamiliar product. Throughout the inspection period which comes before the actual servicing, careful notes may have to be taken to enable any parts removed for easier access during the service to be replaced in their original positions. A useful aid is a Polaroid camera to take snap shots of the set-up prior to dismantling, or a note pad to record settings or a list of procedures, so that steps may be retraced if required. There should be, at hand, a comprehensive range of tools to ensure that time is not wasted in going back and forth to a central tool storage area.

Establish a standard (possibly a manufacture's test specimen) from which a series of tests may be compared.

Complete a series of dummy runs establishing reliability. Do not allow the customer to be the first to use the serviced equipment without trial runs. These may involve new and revised customer instruction, so be very precise about the specification. Run a set of safety checks, testing the equipment to its prescribed limits.

Throughout all of the above procedures take notes and offer these back to the manufacture, so that a detailed log of service and repair may be formed. This will help future designers to avoid costly design faults, or help them create designs to incorporate simpler or maintenance-free servicing.

Evaluation

The evaluation of a service is often the service engineer's nightmare; reports and form filling seem like boring chores. However, without these any future service may require more work than is necessary, for the same servicing mistakes will be repeated if they are not on record to warn engineers that they exist. The following shows some of the type of documentation required for a service evaluation.

Results of performance are recorded and checked against a standard, so that over a period of time this may indicate that a service may only be required after every 12 months instead of the initially prescribed 6-monthly interval.

Certification that a product has successfully met a required standard may be obligatory in some instances (i.e. the MOT of a car or the air worthiness of an aircraft). In such cases the service engineer may have to hold authorisation that they are competent to carry out such maintenance. This may even have to be reviewed or updated at regular intervals.

Part of any evaluation must be formed from checking the customer specification to ensure that it conforms to the requirements initially laid down. An example of this could be that when a second-hand car is sold, it is subject to a special service after 10,000 miles, in which the engine is to be retuned. An ordinary service would not cater for this, so documentation has to be provided as a check that it has actually been completed and that it was successful.

Inspecting, for example, a dishwasher at regular intervals may enable the identification of parts that are not performing properly or the detection of an item soon to perish. The items listed in Table 3.3 could form part of an inspection/service manual to aid the maintenance of a dishwasher.

Table 3.3 Typical inspection/services for a dishwasher

Inspection feature	Time scale
Main filter	Weekly (depending on usage)
Spraying arms	Weekly (depending on usage)
Rinse dispenser	Weekly (depending on usage)
Salt container	Weekly (depending on usage)
Door seals	Every 6 months
Valve filters	Every 6 months
Pump hose	Every 6 months
General hoses	Every 6 months
Level of machine	Yearly
Plug and connectors	After each repair
Taps	After each repair

Each of the inspected parts could have special instructions to aid maintenance; for example, the level of the machine could have a note advising you to check that the machine does not rock and is levelled by using a spirit level.

Identification of safety procedures

Many safety procedures have already been addressed; however, while performing a service it is essential to use all safe working practices. Often a service engineer will have a wealth of experience within a particular area, but will still have to maintain equipment that has new features. This requires additional training and familiarisation of all the aspects of the equipment, including safety procedures. An example of this would be the first time a recently developed piece of electrical control system has to be serviced. Apart from the routine isolation procedure, the equipment may well have to have certain elements isolated in a particular order. All of this requires that the engineer conforms to regulations, procedures and systems. The following takes a closer look at these areas.

Regulations

Service engineers not only require an understanding and detailed knowledge of how to perform a service, they also must be able to achieve this using safe techniques, so that they do not endanger themselves or others. In order for them to do this, their skills and knowledge have to be updated regularly. This may be in the form of reading material or notices, or could well be achieved though training courses, some of which offer certification of competence. The type of updating may be very simple, as in the case the installation requirements of a new design of lathe chuck guard, or it may be extremely complex and require several weeks of intensive training at the manufacturer's offices, i.e. the decommissioning of a power plant.

Procedures

Certain safety procedures have to be adopted by everyone, but the service engineer requires more than these. That is because, under normal conditions, the system is set up to work in a safe manner. In certain circumstances, to gain access to the equipment being serviced, the service engineer may have to remove some of the safety features designed to protect normal working conditions. With this in mind the service engineer has to employ the following procedures:

- If working alone, the service engineer has to let someone know exactly where he/she is and when he/she is due to finish.
- He/she needs to check the equipment before commencing with a service and needs to run a safety check on completion.
- This service should be logged to record any hazards or faults found, and should also include which, if any, components were replaced or repaired and any resetting of equipment.

All of these add to the safe use and useful lifespan of the equipment. It is also essential that only the correct equipment is used within the service – for example, if a torque setting is specified on the fastening of equipment, then the use of a normal spanner is not only insufficient and potentially unsafe, but it may even prove to be lethal.

Other safety procedures to be adopted include an awareness of what is required in the event of an accident or injury. Notices should be prominently positioned indicating whom to contact and how to report it. There should also be a log book to record all damage and faults in equipment and first aid boxes, fire alarm switches and fire extinguishers should be prominently displayed.

Emergency systems

The understanding of systems, especially emergency systems, could save lives, so it is not to be treated lightly. As mentioned above, in the event of an injury a system comes into force. This follows a set of procedures leading, hopefully, to a happy conclusion.

While performing a service, engineers have to be fully conversant with all the emergency systems in operation. It is important that they know the location and proper use of fire alarms and emergency stop switches. Not only must they have a detailed understanding of general safety procedures, but they also need to be fully aware of specific safety procedures surrounding control systems such items as limit switches, power plant, computerised controls, heat sensors, etc.

A general guide to safety procedures when servicing utilises the following points:

- Read any documentation concerning the servicing or maintenance of the equipment thoroughly before starting work.
- Make sure any electrical appliances are isolated before any inspection or repair commences.
- Take time to consider the problem and allow enough time to complete the work safely.
- Employ a methodical approach to stripping down, taking notes if necessary.
- Check everything twice.
- If in doubt ask for assistance.
- Consider your own safety and that of others.
- Use your common sense.
- Tidy up after you finished working as it makes for a safer place.
- Do not sacrifice safety by making temporary repairs.

Maintenance

Throughout the performance of a service, general maintenance should be evident. It should show itself in the form of using routine cleaning procedures to maintain a clean and safe working environment; swarf removal and the cleaning of electronic switch gear should be commonplace.

Maintenance will not be confined to just replacing or repairing equipment, it may also involve adjustment – for example, belt-tensioning devices, calibration of oscilloscopes or the lubrication of bearings. Each of these and many more features have to be considered while still maintaining an eye for fault recognition.

> **Self-assessment task**
>
> Draw up a list of safety procedures for any one of the following:
>
> (a) The servicing of a car wheel-balancing machine.
> (b) The installation of a new vertical milling machine.
> (c) The routine service and maintenance of a computer.
> (d) The regular checking of fire extinguishers.

Selection and maintenance of tools and equipment

There are many factors affecting the choice of tools required to complete a service. Here are but a few of them:

- What is the physical size of the task?
- Are there moving or static components?
- What accuracy or tolerances are required?
- How accessible is the area to work in?
- How many components are there to be maintained?

Each of these may have some effect upon the choice of tooling required or on how the task is to be attempted. As mentioned earlier, a service engineer would have a range of tooling which may have taken years to accumulate; however, within this tooling range a standard tool kit (and maybe a smaller emergency tool kit) would be formed. Descriptions of components within a typical tool kit are given below.

> **Self-assessment task**
>
> Sketch and describe the use of the following specialised spanners:
>
> (a) A chain wrench.
> (b) A pair of stilsons.

Spanners

Spanners range from the common open-ended type to a sophisticated torque wrench. The spanner has a length that is proportional to the size of its jaws. The spanner jaw must fit the nut exactly otherwise damage to either spanner or nut will occur.

Other forms of spanners are:

- *Ring spanner* – When the position of the spanner over the nut is difficult to maintain, this type offers more control than open-ended spanners.
- *Combination spanner* – This has one end as an open-ended spanner and the other end as a ring spanner. As this usually only allows both ends to be the same size, more spanners are required to complete a set.

These types of spanners are generally used when the service is a common occurrence; that is to say, the general maintenance or service requires the same tools on a regular basis, i.e. in the servicing of a washing machine there will be only a small number of different bolt sizes, therefore only a few specific spanners are required.

- A *socket wrench*, with a series of sockets, offers a similar type of spanner to the ring spanner. The handle remains constant and the sockets are changed to suit the nut. A ratchet mechanism is often employed in the head of the spanner, making it quite costly.
- *Adjustable* spanners offer greater flexibility for a single tool but attempting to apply the correct force demands greater skill than using a correct sized spanner.
- *Torque wrenches* are similar to socket wrenches in that they have a single stock handle and a series of sockets are attached to the handle for use as a spanner. The actual force applied to the socket may be calibrated to give an exact setting so that a nut may be tightened to prescribed values.

These spanners would be most useful for servicing components which use a wide variation on bolt sizes as they would cater for unusual sizes quickly.

- Hexagonal *Allen keys* or socket spanners are L-shaped hexagonal bars made in differing sizes to form a set, and the shorter leg of the L is inserted into the socket of a bolt.
- *C spanners*, or lock spanners as they are also known, are generally C-shaped with a long handle protruding perpendicular to one end of the C, while the other end has a small lug or spur which mates with a corresponding slot or hole in the nut.

These spanners would best suit a very specific servicing product, such as the removal of large locking nuts. They are often used because of limited access due to design and manufacture.

Screwdrivers

These may come in many sizes but there are only two main types of screwdriver:

- A flat blade screwdriver used on slotted screws.
- A cross-head screwdriver used on cross-recessed screws.

With the flat blade type it is very important that the thickness and width of the blade be correct for the screw head otherwise damage may occur to either the blade or the screw head. The cross-head screwdriver has various sizes of crosses and, as with the blade type, it must mate properly with the screw head.

There are screwdrivers known as *impact* screwdrivers, which operate not by twisting the handle but by hitting the handle with a hammer. They have a spring inside the handle holding the blade, and when it is compressed by a hammer blow it rotates the blade.

Hammers

There are a variety of shapes and sizes available so that a shock load can be applied by an extended weight on a shaft. The following are but a few of the various types: claw, ball pein, cross-pein, pin, sledge, lump, soft (hide mallet), welder's scaling hammer.

The correct choice of hammer is important within servicing; too large a hammer may impart too heavy a blow and damage the component, too light a hammer may not impart a sufficient load to free the component and the choice of contact face has to match the working method – i.e. soft face used when damage to the product surface is to be avoided. Scaling hammers have to be pointed to dig into the scale it is trying to remove.

Pliers

These, which often prove to be a useful extension to the engineers hands, are used for gripping, pulling, cutting and squeezing. They apply additional pressure through the use of levers.

Many different applications involve the use of pliers, and they have different types of jaws to suit particular purposes, i.e. bent ends of pliers may be used on circlips, while long slender ends may help grip work in a tight confined space. Gear pullers (pulley extractors; Fig. 3.71), which may be classed as a form of plier, consist of two or three pivoting arms with lugs or spurs on the end. These are connected to a central hub which has a screw thread running through the centre. The lugs are positioned around the pulley gear or bearing, while the screw is turned, tightening against the shaft of the pulley gear.

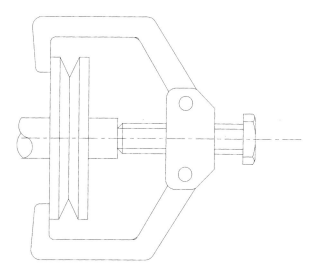

Figure 3.71 Pulley extractor

Levers

These operate like one leg of a pair of pliers in that they transmit a magnified force through the extended length of a bar. They are generally used for lifting or prising.

Typical type of levers are: the crow bar, a pinch bar and a podger bar (often used as a location lever to align two holes).

Measuring instruments

The maintenance or service engineer will have to be fully conversant with a variety of measuring instruments, and would probably have the following selection within a tool kit:

- A spirit level to check that items are horizontal or vertical. The spirit level is often used to help form datum surfaces.
- Rules, micrometers and verniers to establish size and position of components (often used when precision is required).
- Protractors or clinometers to check angles.
- A set of feeler gauges to establish fine clearances.
- A dial test indicator to measure concentricity.
- A multimeter to measure electrical voltage, current or resistance.

Drills

A comprehensive set of drills would be carried within a standard tool kit, along with some specialist drilling tools such as countersink drills, reamers and trepanning tools.

Files

These are hardened steel bars with teeth cut on the body of the bar and are mainly used by hand. They are identified by four key elements:

- The length measured from the point to the shoulder.
- The type of cut, i.e. single cut, double cut or rasp.
- The grade of cut, which is dependent upon the spacing of the teeth. Common types of cut are rough cut, bastard cut, second cut, smooth cut and dead smooth cut.
- The shape of the file, which offers the biggest range of choice of the four key elements of a file, for the amount of material to be removed or the intricacy of the component often dictates the shape of the file to use.

The following give an example of some of the files in common use:

- A flat file is tapered in width and thickness and has double cut faces with single cut edges.
- A ward file is a small file in size and is tapered in width but not thickness; it has double cut faces and single cut edges and is used for filing narrow slots.
- A half-round file is shaped as the name suggests, almost half round; it is used to file concave surfaces and has a double cut flat face and a smooth cut curved surface.
- A hand file is parallel and has faces that are double cut; however, one of the edges usually has a single cut while the other is left uncut to form a safe edge.
- Square, round and three-square files are specialist files and are shaped as their names suggest. They are used to open up holes or provide sharp corners.

A safety note: All files must have a handle to eliminate the danger of the tang of the file slipping into the hand.

Hacksaws

These often vary in size only and are all based on the same operating principle. A cutting blade is fitted into a frame with the teeth pointing away from the handle. The blade has teeth which are set so that they produce a cut which is slightly wider than the thickness of the blade, this helps stop friction occurring on the sides of the blade and the slot being cut.

Electrical tools

A basic tool kit for a service engineer should contain at least a small amount of electrical equipment. This could include, a multimeter for checking values of circuits and components, an electrical soldering iron for repairs to circuits, wire strippers and crimpers and a set of electrical screwdrivers.

Sundry tools

Although these may be optional, many service engineers find them invaluable and often have them at hand. They include items such as:

- Markers for identifying parts and settings, especially useful when reassembling equipment.
- Shears and chisels for cutting material often for access.
- Charts and tables to be used as reference material.
- Treading tools such as taps and dies, also pipe benders for conduit work.

> **Self-assessment task**
>
> Devise an emergency tool kit, comprising no more than 40 items, to be carried by a photocopying service engineer.

3.4 Unit test

Test yourself on this unit with these multiple-choice questions.

1. Which of the following heat treatment processes enables a steel component to increase its ability to withstand wear:
   (a) annealing
   (b) normalising
   (c) hardening
   (d) tempering

2. On a centre lathe a chamfer of 45° taper is to be produced, which of the following taper turning methods is most suitable:
   (a) using a compound slide
   (b) offsetting the tailstock
   (c) using a taper turning attachment
   (d) using a form tool

3. Which of the following work holding devices employed on milling machines is used to securely hold small regular shaped work:
   (a) a machine vice
   (b) a rotary table
   (c) a dividing head
   (d) vee block and clamps

4. When drilling a trapanning tool or tank cutter is used to:
   (a) countersink a hole
   (b) counterbore a hole
   (c) cut a large diameter hole in sheet metal
   (d) spot face a hole

5. In the process of sand casting a wooded pattern of the component is made slightly larger than the desired finished component, this is because:
   (a) wood is cheap
   (b) it can be made easily
   (c) it allows for shrinkage of the hot metal on cooling
   (d) it makes the pattern easier to remove

6. When hand forging if there is a need to increase the thickness of a bar for a short length (i.e. for the manufacture of the head of a bolt) then this process is known as:
   (a) punching
   (b) swaging
   (c) drawing
   (d) upsetting

7. Which of the following forming processes is most suitable for producing plastic bath tubs:
   (a) injection moulding
   (b) vacuum forming
   (c) extrusion
   (d) drawing

8. When using electrical solder a flux or paste may be used, the main purpose of this is:
   (a) to protect the metal from oxidising when heat is applied
   (b) to allow the solder to flow more readily
   (c) to contain the solder to within the fluxed or pasted area
   (d) to make the soldered joint look nice

9. Gas welding requires two cylinders of gas, these are:
   (a) oxygen and nitrogen
   (b) oxygen and acetylene
   (c) acetylene and nitrogen
   (d) oxygen and argon

10. When annealing a mild steel component it is heated and then:
    (a) quenched in oil
    (b) allowed to cool slowly
    (c) quenched in water
    (d) allowed to cool and then re-heated

11. When hardening mild steel the hardness of a component is dependant upon:
    (a) the amount of carbon present in the steel and the rate at which cooling takes place:
    (b) the amount of carbon present in the steel
    (c) the rate of cooling
    (d) the volume of material being hardened

12. Which of the following heat treatment processes is used as the final process for producing a screw driver blade?
    (a) hardening
    (b) normalising
    (c) annealing
    (d) tempering

13. Etching is a process which changes the components features by:
    (a) machining
    (b) corrosion
    (c) erosion
    (d) painting

14. Which of the following processes involves coating mild steel with zinc?
    (a) galvanising
    (b) electrolysis
    (c) anodising
    (d) chrome plating

15. Coating has the prime function:
    (a) to offer surface protection
    (b) to add strength
    (c) to assist in any further manufacturing
    (d) make the surface harder

16. Two of the three basic reasons for employing the grinding process are:
(i) to remove surplus material
(ii) to machine very hard materials

Which of the following is the third?

(a) it is easy to use
(b) the cost is minimal
(c) it obtains a high degree of surface finish
(d) it has a high metal removal rate

17. There are four main categories of portable fire extinguisher, which of the following is colour coded red?

(a) one which use water
(b) one which use a foam
(c) one which use a gas or vapour forming liquid
(d) one which use a dry powder

18. Servicing booklets often contain diagnostic information in the form of a flowchart. These usually comprise of four forms of boxes; which employs the use of a diamond shape?

(a) the process box
(b) the junction box
(c) a termination box
(d) the input/output box

19. When performing a maintenance service the main reason a detailed log is kept is to:

(a) act as a dummy run
(b) list the range of tools to be used
(c) to assist in making future designs
(d) enable steps to be retraced

20. Performing an initial engineering service involves many aspects – how is it best described?

(a) inspection
(b) maintenance
(c) diagnosis
(d) evaluation

Unit 4: Engineering Materials

Introduction

Since the dawn of civilisation, man has used the materials available to him to satisfy his basic human needs. Weapons were fashioned in order that he could hunt for food and defend himself, while other tools, implements and machines enabled him to cultivate the land. Further use of materials for clothing, heating, housing, furniture, transport and entertainment added comforts to his survival.

Although the same needs still form the basis of life in today's society, mass production methods and increased commercial competition has led to the need to maintain product quality, while the conservation of the earth's natural resources becomes an ever increasing and important issue to be addressed.

As a result of these changing needs, the demands placed on the materials engineer or technician, while no less challenging, are vastly different from those of the past. It is, therefore, the engineer's responsibility to consider these many issues when selecting materials for the engineered products that allow us all to enjoy the quality of life that they bring.

In this unit the student will learn how to:

- Characterise materials in terms of their properties.
- Relate materials' characteristics to processing methods.
- Select materials for engineered products.

4.1 Characterise materials in terms of their properties

Topics covered in this element are:

- The properties of materials.
- The structure of materials.
- The relationship between structure and properties of materials.
- The testing of materials.

Properties of materials

A convenient starting point from which a study of engineering materials can be made is to consider the different properties these materials need in order to fulfil the requirements of service in which they will be expected to operate. Properties of materials can be broadly classified into mechanical properties and physical properties.

Mechanical properties

Strength

This is the property of a material which enables it to resist tensile, compressive or shear forces without failure. It is quoted as the applied force per unit area – known as 'stress' – and is measured in newtons per square millimetre (N/mm^2). Alloy 'high-tensile' steels are examples of strong materials.

Toughness

This is the property of a material which enables it to absorb the energy from an applied force, impact or shock load without fracture. An indication of the toughness of a material can be obtained by striking a sample of the material a sudden blow with a hammer above a notch cut in the specimen. The toughness of a material is not in direct proportion to its strength or hardness and is associated with its fibrous nature. For this reason it is usually difficult to obtain a good surface finish on tough materials by machining with cutting tools. Wrought iron, copper and nylon are examples of tough materials.

Elasticity

This is the ability of a material to return to its original shape and size after the removal of a deforming force. All materials possess elasticity to some degree and each has its own elastic limit beyond which permanent deformation and ultimately fracture occurs. At stress levels below the elastic limit the amount of deformation is directly proportional to the applied force which may be tensile (elongation) or compressive (contraction). This relationship is known as 'Hooke's law' after investigation of elasticity was carried out by Robert Hooke (1635–1763).

Plasticity

This is the opposite of elasticity. It is the property which enables a material to be deformed under a force without fracture, the resulting deformation being non-recoverable and non-proportional to the applied force. The deformed material is said to take on a 'permanent set'. The plastic range of a material commences immediately the elastic limit of a material is exceeded and continues until failure occurs. Certain materials, for example lead, have a good plastic range at room temperature and can be cold worked, whereas others like steel must be preheated in order to produce a plastic state suitable for manipulation. Ductility and malleability are particular cases of plasticity.

Ductility

This is the property possessed by a material which allows permanent plastic deformation to take place before fracture when the material is subjected to a tensile force. A good example of a ductile material is mild steel which can be manipulated cold by being 'drawn out' into wire. A measure of the ductility of a material can be made by reference to the percentage elongation of a specimen which has undergone a tensile test.

Malleability

This is the property possessed by a material which allows permanent plastic deformation to take place without fracture when the material is subject to a compressive force. A good example of a malleable material is lead, which can be manipulated cold by being hammered into sheet. Most other materials, notably iron and steel in particular, become malleable only when heated to a high temperature. Typical hot-working processes which make use of this property are rolling, forging and extrusion.

Hardness

This is the ability of a material to resist wear, abrasion, scratching, indentation or penetration by another material. Measurement of this property can only be made relative to other materials and is given in the form of a number only (dimensionless). Extremely hard materials, e.g. diamonds, are taken as reference materials in laboratory-controlled tests to determine hardness. In a practical workshop test, scratching samples with a file will give a rough guide as to their relative hardness.

Brittleness

This is regarded as the opposite of ductility. It is the property possessed by a material which shows little or no plastic deformation after failure resulting from the application of a force. Cast iron is a typical example of a brittle material.

> **Self-assessment task**
>
> Explain what is meant by 'mechanical' properties.

Physical properties

In addition to mechanical properties, the design of an engineered product may require other properties such as electrical or magnetic conductivity or resistance, thermal conductivity or resistance, or temperature or chemical stability. This group of properties of materials is called 'physical properties' in which the melting point and the density of a material are also included.

Electrical conductivity

This is a measure of the ease with which a material will allow the flow of – or 'conduct' – an electric current. Materials which conduct electricity are known as conductors. Metals, in particular, aluminium, copper and iron are good conductors of electricity (Table 4.1).

Table 4.1 Typical electrical conductivity values

Material	Conductivity (Ω^{-1} m)
Metals	
Aluminium	40,000
Copper	64,000
Iron	11,000
Mild steel	6,600
Polymers	
Acrylic	Less than 10^{-14}
Nylon	10^{-9} to 10^{-14}
PVC	10^{-13} to 10^{-14}
Ceramics	
Alumina	10^{-10} to 10^{-13}
Glass	10^{-10} to 10^{-12}

Electrical resistivity

This is a measure of the ability of a material to resist the flow of an electric current and is thus the reciprocal of electric conductivity. Materials which possess high resistivity (resistance) are known as insulators. Examples of insulators are ceramics, plastics, rubber, glass and dry air. A useful British Standard Specification to which reference should be made is BS 5714. In between conductors and insulators lies a range of materials known as semiconductors which can be good or poor conductors of electricity depending on their temperature. Silicon, germanium and certain metal oxides possess semiconducting properties.

Permittivity

In electrostatics, which is the branch of electricity concerned with the study of electrical charges at rest, an electrostatic field accompanies a static charge which is utilised in a capacitor. A dielectric is an insulating medium separating charged surfaces of a capacitor and the ratio of electric flux density in the dielectric to the electric field strength is called permittivity. Permittivity values are dimensionless and include those listed in Table 4.2. The relevant British Standard Specification number is BS 7663.

Table 4.2 Example permittivity values

Material	Permittivity
Air	1
Polythene	2.3
Mica	3–7
Glass	5–10
Ceramics	6–1,000

Magnetic conductivity

This is a measure of the reluctance (resistance) which a material offers to a magnetic field. Good magnetic conductors are the ferromagnetic materials which get their name from the Latin word 'ferrum' meaning iron. Consequently, all metals in which iron is present as the main constituent or 'parent' metal are good magnetic materials.

Magnetic materials are classed as being either 'hard' or 'soft'. Hard magnetic materials are not easily magnetised but do retain much of their magnetism after the magnetising force has been removed. Conversely, soft magnetic materials are easy to magnetise but quickly lose virtually all magnetism when the magnetising force is removed. Soft magnetic materials are said to possess a high permeability.

The original hard ferromagnetic materials, typically quench-hardened high-carbon steel used for permanent magnets have now been replaced by far more powerful alloy materials such as 'columax' which possess over thirty times more magnetisation power than their carbon steel predecessors. In addition, soft magnetic materials like wrought iron and low-carbon steel which were traditionally used for transformer laminations and electromagnet cores have now largely been replaced by new nickel–iron alloys possessing very high permeability (Fig. 4.1) and low hysteresis values. Hysteresis is the 'lagging effect' of flux density whenever there are changes in the strength of a magnetic field. Examples of these materials are mumetal (75 per cent nickel) and permalloy (78.5 per cent nickel).

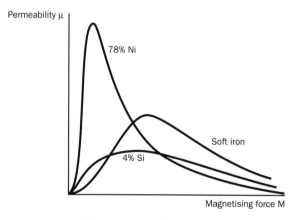

Figure 4.1 Permeability curves for 78 per cent nickel–iron alloy, silicon–iron and soft iron

Thermal conductivity

This is a measure of the rate at which a material will conduct heat. The greater the rate of heat flow, the higher the conductivity of the material. Good heat conductors are normally good electrical conductors, hence metals, in particular, copper and aluminium, are good conductors of heat. Thermal resistivity is the reciprocal of thermal conductivity.

Thermal expansivity

This is the consideration of the expansion of materials due to an increase in heat. The linear expansivity or 'coefficient of linear expansion' of a material is a measure of the amount by which a unit length of the material expands when its temperature is raised by 1 °C (Table 4.3).

Table 4.3 Typical thermal property values

Material	Thermal conductivity (W/m K)	Linear expansivity ($K^{-1} \times 10^{-6}$)	Specific heat capacity (J/kg K)
Metals			
Aluminium	230	24	920
Copper	380	18	385
Mild Steel	54	11	480
Stainless steel	16	16	510
Polymers			
Bakelite (thermoset)	0.23	80	1,600
Nylon 66 (thermoplastic)	0.025	100	2,000
PVC (thermoplastic)	0.0019	700	1,000
Ceramics			
Alumina	2	8	750
Silica, fused	0.1	0.05	800
Glass	1	8	800

Chemical Stability

This is the resistance of a material to the effects of the atmosphere and the environment. There is a marked contrast between the chemical stability of metals and non-metallic materials (see Corrosion, p. 125 and rubber, p. 108).

Types of engineering material

Based on a sound understanding of the properties of materials, suitable selection of material can then be made from the wide range of materials which are available to the engineer. It is convenient at this stage to consider a classification of materials in their different forms in the form of a diagram (Fig. 4.2).

As can be seen from the diagram, engineering materials can be divided into two main types, namely metals and non-metallic materials. The metals can be further divided into ferrous metals which contain iron and non-ferrous metals which contain no iron. Non-metallic materials are split into natural materials and synthetic or man-made materials. Any metal (ferrous or non-ferrous) has a limited number of useful properties and will usually, possess conflicting or incompatible property values. For example, a very hard material is likely to be extremely brittle and have very little tensile strength. In practice, a compromise is made in which a number of metals are combined to give the property values which engineering demands. These metal combinations are known as alloys.

While it would be impossible to attempt to memorise details of the full range of metals which exist, it is useful to know the characteristics and properties of the more common metals and alloys.

Ferrous metals

The classification of ferrous materials is shown in Fig. 4.3.

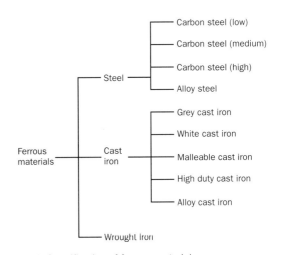

Figure 4.3 Classification of ferrous materials

Steel

Steel is essentially an alloy of carbon contained in iron in a chemically combined form, as distinct from pig or cast iron where carbon exists in a free graphite form freely dispersed as flakes of graphite throughout the metal. The carbon content may be from 0.1 to 1.7 per cent. Where small amounts of silicon, sulphur, manganese and phosphorus are present, and regarded as impurities, the material is known as plain carbon steel.

The properties of plain carbon steel are controlled by the amount of carbon present. As the carbon content is increased, the tensile strength and hardness increase, but the ductility is reduced.

A maximum strength of approximately $960 \, MN/m^2$ is obtained when the carbon content is approximately 0.89 per cent, but the ductility is low.

Mild steel

One of the cheapest and most common of engineering materials, is the range of plain carbon steel containing from 0.1 to 0.3 per cent carbon.

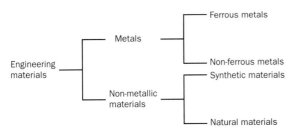

Figure 4.2 Classification of materials

- Mild steel has good tensile strength of approximately 460 MN/m² and is ductile at room temperature, allowing a limited amount of cold work to be performed by rolling or drawing into rods, bars, tubes or sheets with a bright surface finish and good accuracy of shape and size.
- Sheet mild steel may be further cold worked by pressing, spinning, stretching and many other methods of manipulation, however it is not usually cast into shapes as the melting temperature is high (approximately 1,400 °C) and good characteristics are not easily obtained.
- At a temperature of 900 to 1,200 °C, mild steel is extremely plastic and can be easily wrought into basic shapes by rolling, extrusion or forging. Rods, bars, tubes, plates or forged shapes have a black surface finish and form the raw material for structural sections and plates, for buildings and ships, or for further machining to produce common engineering components such as nuts, bolts, shafts and gears.
- Mild steel is easily welded for structural purposes and when machined gives a good surface finish. It cannot be increased in hardness through its section by heat treatment, but some heat-treatment processes impart surface hardness to resist abrasion, while leaving a tough interior to the component.
- Mild Steel is not resistant to corrosion. Damp atmospheres will quickly cause a red oxide commonly known as 'rust' to form, therefore components must be protected by grease, paint or other special surface treatment such as chromium plating.
- Mild steel is magnetic at room temperature.

Medium-carbon and high-carbon steels

Above 0.3 per cent up to 0.7 per cent carbon, there is a range of plain carbon steel called medium-carbon steel, while a further range containing between 0.7 per cent and 1.4 per cent carbon is known as high-carbon steel. Unlike mild steel, the properties of plain carbon steel containing more than 0.7 per cent carbon may be further modified by heating and quenching in water or oil to obtain increased hardness and toughness (see p. 117).

Typical uses for medium-carbon steels are springs, axles, high-tensile tubes and wire, while high-carbon steel is more suitable for cutting tools, e.g. cold chisels, wood chisels, dies, punches, hand files, etc.

Alloy steels

When other elements are deliberately added to steel to improve its properties and to enable it to meet special conditions of service, the resulting steels are known as alloy steels. It is difficult to state precisely the effect of any one alloying element because the effect depends upon the quantity used, the number of other alloying elements used with it and the carbon content of the steel. However, a description of the principal effects of these elements is as follows:

- When **manganese** and **silicon** are present in steel in small amounts a superior grade of carbon steel suitable for use in leaf springs is produced. With 13 per cent manganese content, the alloy steel has very pronounced work-hardening and hence wear-resistant properties making it an extremely useful steel for use in heavy plant and machinery.
- **Chromium steels** are noted for their hardness and resistance to wear and when chromium and nickel are alloyed with steel an extremely useful range of high-tensile steels called nichrome steels is produced. This alloy steel contains 0.7 per cent chromium, 3.5 per cent nickel and 0.3 per cent carbon and is used for highly stressed components such as bolts, shafts and aircraft parts.
- **Stainless steel** is a steel with a chromium percentage between 12 and 18 and a smaller percentage of nickel. Stainless steels possess high corrosion resistance and are used in fittings subjected to high temperatures.
- **Nickel steels** are tough, ductile and possess high strength and when alloyed with smaller amounts of manganese, silicon and chromium, these steels are used in such items as engine crankshafts and armour plating.
- **Vanadium** is alloyed with chromium to produce alloy steels which possess high tensile strength and are very resistant to repeated stresses. These steels are particularly suitable for spanners, springs, engine components, etc.
- 14–18 per cent **tungsten** is used as the main alloying element in high-speed steel together with 3 to 5 per cent chromium and small percentages of vanadium and molybdenum. High-speed steels are air hardening and have the ability to retain their hardness at red heat. Consequently their use as a metal-cutting tool material for machining hard materials is common practice.

Cast iron

This is an alloy of iron and carbon which is very close in composition to the crude pig iron produced in a blast furnace. To produce components from cast iron, pig iron is remelted, impurities are removed where possible and extra elements are added to obtain particular properties. The metal is then poured into an impression or mould.

Carbon is present in cast iron in amounts varying between 2.5 and 4.5 per cent. While a small amount of the carbon combines chemically with the iron, most exists in a free state as graphite. The free graphite forms flakes of various sizes, this effect being influenced by other elements present for example silicon, sulphur, manganese and phosphorus which in turn vary the properties obtained.

The total amount of carbon, silicon and impurities is usually in the region of 6 per cent.

The graphite form is also influenced by the rate of cooling of the component, and this may be varied in order to obtain various properties and types of cast iron. Unless specially alloyed, cast irons are magnetic at room temperature and are prone to corrosion in ordinary atmospheric conditions.

Grey cast iron

This is the most common form, containing approximately 3.5 per cent carbon. When fractured it shows a crystalline structure with grey, flake graphite giving it the characteristic of brittleness with low tensile and low impact strength and restricting its use where high strength is important.

Grey cast iron is brittle at all temperatures and cannot be wrought. Other disadvantages with cast iron are mainly concerned with the cost and rate of production as moulds have to be broken up after casting each component, necessitating a new mould for each casting.

Its assets make grey cast iron a useful material for the following reasons:

- It has high compressive strength.
- It is a cheap material and has a low melting temperature (approximately 1,130 to 1,250 °C).
- It is highly fluid when molten making it possible to manufacture large components or intricate shapes by simply pouring the metal into an impression moulded in sand. This allows the component to be designed with varying thicknesses, giving strength where required and saving metal or subsequent machining work by casting holes, recesses or pockets where required.
- The free graphite permits an easy, if somewhat dirty, machining of the metal by cutting tools.
- The provision of good sliding and bearing surfaces on a component are possible as the graphite gives a degree of self-lubrication and helps to retain an oil film.
- Its crystalline nature absorbs vibration and makes it an ideal material for machinery frames and beds.
- The ready flow of the metal into any shape or form gives a component a pleasing appearance, a point which should be considered in any component design.

White cast iron

This is a type of iron which has little free graphite, the carbon being in chemical combination with the iron. This is caused by reducing the total carbon content to approximately 2.5 per cent, or cooling the cast component very quickly. White cast iron is a hard, white, brittle material which is easily broken.

Components made from this material are usually subject to a heat-treatment process called the 'Whiteheart Process' which gives a tough, high-strength material known as malleable cast iron, the components being better able to withstand shock loading. The name Whiteheart refers to the colour of the fracture of castings processed in this way.

High duty cast irons

These, and **alloy cast irons**, are the results of special alloying or special methods of cooling. These alloys have been developed to withstand high loads, corrosion, high temperatures and high abrasion.

Wrought iron

This is the nearest commercial material to pure iron, containing 0.03 per cent carbon with up to 1.8 per cent of other impurities. Wrought iron is the result of further refinement of pig iron in a furnace, causing the impurities to be oxidised or formed into a slag.

The melting temperature is approximately 1,500 °C, but before the metal finally solidifies during cooling, it forms a pasty state which can be wrought into shapes by hammering, pressing or rolling processes. Bars can readily be joined together when the metal is in this condition, forming a true weld. The slag inclusion causes the material to exhibit a fibrous character, giving extreme toughness, malleability and ductility when cold.

When under load, wrought iron will show a considerable amount of elongation before fracturing making it a safe and useful material for hooks, chains and lifting equipment. Wrought iron is a relatively weak material having an ultimate tensile strength of approximately $310 \, \text{MN/m}^2$ and for this reason it is limited in application.

Non-ferrous metals and alloys

The classification of non-ferrous materials is given in Fig. 4.4.

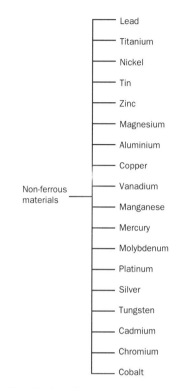

Figure 4.4 Classification of non-ferrous materials

Aluminium

This is a very widely used metal, having only one third the specific gravity of steel, which allows its use for lightweight components. The metal is the most abundant in the earth's crust, but the process of refinement into a commercial metal is costly.

Aluminium of 99.99 per cent purity can be obtained, possessing high electrical and thermal conductivity, but with a very low tensile strength of $93 \, \text{MN/m}^2$. To make use of aluminium in this condition it is used together with other materials for example, steel in the manufacture of electrical cables where the aluminium conductor is built around a steel core. It is soft and ductile and will tear when machined with cutting tools, making a good surface finish difficult to achieve.

Aluminium of 99.5 per cent purity is more usual, having slightly decreased conductivity but 50 per cent increased ultimate tensile strength. Cold-working processes can increase this still further on, for example, rolled strip or drawn wire. Its low melting temperature of 660 °C makes it economical to use in casting processes.

An important property of aluminium is its high corrosion resistance due to the formation of a dense surface oxide film. This is very thin but if scratched will form again. The oxide film will not resist a high corrosive environment, as, for example, that encountered in sea water, but it can be artificially thickened to give greater protection by a process known as anodising.

Aluminium alloys

The low strength of pure aluminium makes it unsuitable for many uses, but with small amounts of alloying elements, such as copper, silicon, manganese or magnesium, it forms the basis of a wide range of important lightweight, high-strength engineering alloys. e.g. duralumin, hiduminium and 'Y' alloy. Aluminium alloys can be classified as either cast alloys or wrought alloys. Cast alloys are grouped separately according to their response to heat treatment into 'as-cast' alloys, which derive no benefit from heat treatment and heat-treatable alloys. Wrought alloys are grouped into work-hardening and heat-treatable types. Aluminium–copper alloys can be heat treated, resulting in an increase in strength and hardness. These alloys possess good strength at elevated temperatures making them suitable for engine pistons for example. Duralumin, however, is unable to retain its strength at temperatures much above 200 °C and finds suitable applications where these temperatures are not encountered, e.g. in motor vehicle, bicycle and domestic goods manufacturing.

Copper

When refined to a metal of 99.9 per cent purity, copper has very high electrical and thermal conductivity and high resistance to corrosion, making it suitable for many uses in the electrical, chemical, heating and building industries.

It will melt at approximately 1,100 °C but when cast it has a low tensile strength, approximately 240 MN/m^2.

It has high ductility and malleability which allow cold working, drawing or rolling. This has the effect of increasing the strength which in turn reduces the ductility and increases the hardness. An annealing process restores the ductility but again lowers the strength. By controlling the amount of cold working after annealing various states of hardness and strength can be achieved and when selecting copper for a particular component this should be taken into consideration.

Pure copper is expensive and for certain components not requiring high conductivity, some impurities, giving a cheaper and sometimes stronger material, are permissible. The commercial grades of copper divide into three main groups:

- *High conductivity copper* – Being 99.9 per cent pure, and of the highest electrical and thermal conductivity, this is the best quality and has many uses in the electrical industry.
- *Best select copper* – This is cheaper, having a purity of approximately 99.75 per cent but lower conductivity. This quality is useful for many general purposes, e.g. pipework, boiler tubes, etc.
- *Arsenical copper* – Containing up to 0.5 per cent arsenic, this copper has a higher strength than pure copper at elevated temperatures, as it raises the softening temperature from approximately 190 °C for the pure metal, to approximately 550 °C. The material has a high thermal conductivity and is useful for boiler tubes and domestic plumbing.

In addition to cold working, copper may be hot worked, by rolling, forging or extrusion, having its highest plasticity between 800 and 900 °C. Copper machines easily with cutting tools, allowing a good surface finish and close dimensional accuracy to be obtained. Joints are easily made in copper by soldering and brazing. Welding is possible, but the material must be oxygen free or brittleness will result. Copper can be highly polished and gives a good surface for various plating processes, e.g. chromium plating.

An important use of copper in industry is as the main element in two large groups of alloys, namely brass (copper and zinc), and tin–bronze (copper and tin).

Zinc

This is a bluish-white metal which is brittle and weak at room temperature. It melts at 419 °C, but between 100 and 150 °C it is very plastic and easy to manipulate by rolling. It is a difficult material to join by soldering, brazing or welding.

Zinc has a high corrosion resistance and is used in sheet form on buildings, also as air- and water-tight packaging containers and in electrical cells. A more economical use of zinc is as a protective coating on sheet steel by a process known as galvanising. Galvanised sheet has many uses, such as water tanks, corrugated roofing sheets, etc.

Zinc is a safer coating material than tin on sheet steel, as scratching of the surface will cause the zinc to corrode before the steel. In certain applications, advantage is taken of the fact that zinc corrodes more quickly in the presence of iron than will the iron itself. Such cases are often referred to as 'sacrificial coatings'.

High-grade zinc (99.9 per cent pure) with small additions of aluminium and copper (up to 6 per cent) form a group of important low-melting temperature alloys which have aided the development of the mass production technique of die-casting. Known as zinc-based die-casting alloys, they are extensively used for lightly stressed, lightweight, thin-sectioned components for domestic appliances, door handles, office equipment, instrument casings, carburettor bodies and small toys.

Zinc is alloyed with copper in varying amounts up to 50 per cent and forms the basis of an extensive and important range of alloys known as the brasses.

Brass

Brass is an alloy of copper and up to 50 per cent zinc. It may also contain small quantities of tin, manganese, lead, aluminium and silicon.

Brasses with up to 37 per cent zinc are called alpha [α] brasses. Brasses in the range 37–45 per cent are called alpha/beta [α/β] brasses and brasses with more than 45 per cent zinc are termed beta [β] brasses. The Greek prefix describes the 'phase' or condition of the microstructure of the alloy.

Important α brasses are (1) *cartridge brass*, which contains about 30 per cent zinc and is used for ammunition cases and (2) *admiralty brass*, in which an addition of 10 per cent tin to cartridge brass improves the corrosion resistance of the resulting alloy. α brasses are suitable for cold-working after pre-annealing and post-stress-relieving treatments.

Important α/β brasses include (i) *'muntz' metal*, containing 60 per cent copper and 40 per cent zinc, (ii) *leaded brass*, in which 0.5 to 3.5 per cent lead is added to muntz metal to improve its machineability, (iii) *naval brass*, in which approximately 10 per cent tin is added to muntz metal to improve its corrosion resistance and (iv) *high tensile brass*, in which up to approximately 1.7 per cent (total) of iron nickel, tin, aluminium and manganese are added to muntz metal to give improved mechanical properties.

Tin

This is a soft, white metal having a low melting temperature of approximately 232 °C. At room temperature its structure consists of large crystals, making it weak and brittle. In its pure state it is therefore of little practical use.

With lead it forms a range of solders, having many uses, from rapid solidifying solders for joints in electrical wiring, to pasty, slower solidifying solders for plumbing joints.

Tin forms a part of many anti-friction alloys, from the soft white metals of tin and antimony to the hard, complex, heavy-duty bearing alloys of copper and tin, which form tin–bronzes and gun-metals.

The most useful property of tin is its high resistance to acid corrosion, making it useful as a plating material in the manufacture of mild steel containers for food and fruit juices. Care must be taken when choosing this material, however, as scratches penetrating the surface layer of tin will cause a more rapid corrosive action than if the steel had been left in an uncoated state.

Bronze or 'tin-bronze'

This is an alloy of copper and tin with small quantities of other elements such as phosphorus, lead, nickel or aluminium added to modify its properties. Tin–bronze containing zinc is called gun-metal.

Bronzes are used where ease of casting, good corrosion resistance and high resistance to wear are required. Depending on their microstructure, bronzes can be classified as either wrought bronzes (α bronzes) or cast bronzes (β bronzes).

Magnesium

This is one of the lightest engineering metals and is found in the ores magnesite and dolomite and in sea water as magnesium chloride from which it can be extracted by electrolysis. It possesses similar characteristics to aluminium, being slightly less dense and having a slightly lower melting point. Like aluminium, it combines readily with oxygen to form an oxide which acts as a skin to prevent corrosion. However, it must be protected by paint or lacquer if the air is humid or contains traces of salt.

Pure magnesium is soft and has a low modulus of elasticity. Its tensile strength is improved significantly by working; however, its hexagonal lattice structure limits its elongation by cold working to about 5 per cent as this arrangement of crystals restricts the amount of 'slip' which can take place. Magnesium is more easily hot worked.

Magnesium alloys

Magnesium is not strong enough to be used structurally without alloying. Elements that form strengthening compounds with magnesium or cause it to respond to strengthening heat-treatment processes include manganese, zinc and zirconium. Like aluminium alloys and copper–tin alloys, magnesium-based alloys can be classed as wrought alloys or cast alloys, some of which will respond to heat treatment.

Titanium

The high-strength/low-density properties of titanium make it a very useful structural metal. Unfortunately, the great affinity or attraction for oxygen which this metal possesses has made its production difficult, necessitating the use of special extraction and processing techniques.

Titanium can be formed by rolling, drawing, forging and extrusion and can be machined with no special problems as long as the workpiece is held rigidly and low cutting speeds and coarse feeds are used. Titanium can be welded and brazed successfully but it is necessary to use a shielded gas when fusion welding to avoid scaling of the surface. Any scaling of titanium or titanium-based alloys by heating in air can be removed by pickling or by grit blasting.

Titanium-based alloys

Aluminium, copper, manganese, molybdenum tin and vanadium can be alloyed with titanium to improve the properties of the resulting alloy.

> **Self-assessment tasks**
>
> 1. Explain why most metallic materials used in the manufacture of engineered products are alloys.
> 2. List TEN non-ferrous metals giving ONE use for each material selected. Try to include metals not already covered in the text.

Non-metallic materials

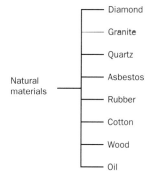

Figure 4.5 Classification of natural materials

Like pure metals, the non-metallic materials, found naturally as minerals or organic matter, usually have a limited range of properties having extreme values. In the past, these materials have only been used for engineering purposes where the extreme value of a particular property was required, as for example the flexibility of rubber or the hardness of diamonds.

From mineral sources asbestos, granite, ceramics, diamonds and oil are converted into various forms for many different purposes. Wood, rubber and fibrous materials, such as cotton, silk, etc., are organic materials having various engineering applications. Many organic materials can now be changed by chemical processes to give a wide range of synthetic or 'man-made' materials, generally known as plastics materials, or simply plastics.

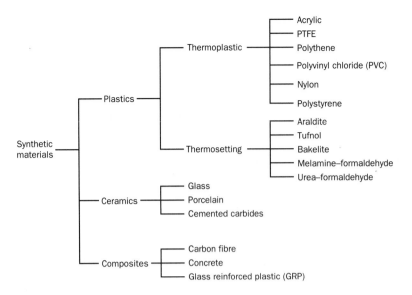

Figure 4.6 Classification of synthetic materials

The history of plastics

The history of the plastics industry goes back over 150 years to 1862 when it was discovered that nitrocellulose could be mixed with camphor and alcohol to form what became known as 'Celluloid'. It was not until after the turn of the century that the second major plastic emerged when phenol–formaldehyde resin known as 'Bakelite' was produced commercially. Unfortunately, Bakelite articles could only be made in dark colours, e.g. brown and black; however, it was found that if phenol was replaced by another substance called 'urea' producing urea–formaldehyde, this difficulty could be overcome. In the 1930s, it was discovered that another substance, melamine, used with formaldehyde gave a product melamine–formaldehyde (melamine) which absorbed far less water and stained far less readily than urea–formaldehyde.

As mass production continued to rise, the real potential of the manufacture of different plastics materials in which long chains of molecules (polymers) were produced from single molecules (monomers) began to be seen, and with it the concept of a future large-scale industry producing goods by chemical synthesis. Although initially only produced in small quantities, this period produced the first of the plastics we know so well today, such as polystyrene, polyvinylchloride (PVC) and polymethylmethacrylate (Perspex). However, it was not until after 1945 that the public was to be made aware of the rapidly expanding plastics industry.

Plastics

The mechanical properties obtainable in different plastics vary between rubbery flexibility, glass-like rigidity, brittleness or fibrous toughness. When the good strength characteristics of the majority of plastics are combined with their low densities, the designer is able to design a strong product without incurring the penalty of excessive weight. In addition, other useful properties like electrical resistance, atmospheric and chemical corrosion resistance and low coefficient of friction are combined with lightness in the manufacture of plastic products.

Unfortunately, plastics will not normally resist high temperatures or flames and their limitations become clear if we consider their behaviour when subjected to heat. One of the great virtues of most plastics is the fact that they start life as almost colourless substances which can be given any colour simply by the addition of suitable dyes and pigments. Moreover, the colour goes right through the material in contrast with metal or wood which have to be painted.

Classification of plastics

It is possible to classify most plastics, technically 'polymers', into one of two groups by reference to the method of manipulation or process required to form the component. These groups are: (1) thermoplastic materials; (2) thermosetting materials.

- *Thermoplastic materials* – These can be softened and manipulated into shape by the application of heat and pressure, the shape being retained after the removal of the heat. Further heating will cause re-softening and the material can be heated and caused to flow an indefinite number of times, provided that the temperature reached is not sufficient to decompose the material completely.
- *Thermosetting materials* – These can be manipulated into shape by heat and pressure, but above a certain temperature they undergo a chemical change which prevents further manipulation. The material cannot then be re-softened and will withstand further heat up to a particular limit, when decomposition or charring occurs.

A further practical difference between the two (with a few exceptions) is that most thermoplastics are soluble in specific organic solvents (e.g. alcohol, acetone, benzene and chloroform), whereas thermosetting materials, once hardened, cannot be dissolved in any solvent without breaking down, i.e. decomposing. Some common examples of thermoplastic and thermosetting type materials, their properties, uses and forms of supply are described in Table 4.4; however, more and more are continually being developed to meet various industrial requirements.

Table 4.4 Properties, uses and forms of supply of some typical plastics

Proprietory name	Properties of finished components	Typical uses	Forms of supply
Thermoplastics			
Celluloid xylonite	Tough and water resistant, highly inflammable, transparent	Films and packaging	
Polythene alkathene	Stable and inert to most liquids, tough and fairly elastic, translucent and can be coloured, flexible dependent upon thickness	Packaging films, thin-walled tubing; rigid mouldings, bottles and containers made by 'blowing'	
Polyvinylchloride (PVC)	Tough and rubbery almost non-inflammable, can be welded by heat; high electrical resistance, resistant to liquids and chemicals	Electric cable insulation; chemical plant piping, moulded shapes, floor covering; sound-recording tapes	
Perspex	Rigid in sheet form but easily manipulated when heated, very high light transmission	Mainly used in sheet form instead of glass	Moulding powders, sheets, filmstrips, rods, tubes, some as highly fluid resins, paints or varnishes
Polystyrene	Similar to Perspex but tends to be brittle, recognisable by metallic ringing sound when lightly struck; takes delicate colours and can become opaque	Hygenic storage boxes, drawing instruments, e.g. set-squares, curves and protractors	
Nylon	Wax-like moulding material with high chemical and liquid resistance; very low coefficient of friction, tough and wear resistant; components need little lubrication	Small mouldings, bearings, gears, rollers, etc.; can be machined from rods or tubes; spun fibres to make high strength ropes	
Polytetrafluoroethylene (PTFE)	Non-inflammable, high chemical resistance, will not absorb water, very low coefficient of friction, being likened to ice on wet ice; high softening point (320 °C)	Small mouldings for bearings, gears rollers, etc.; coatings sprayed on to metal rollers and sliding components in chemical plant and food handling machines; electrical insulators and tapes	
Terylene (polyester resin)	High strength, will not absorb water, melts quickly to form a very fluid resin (often classified with the thermosetting plastics)	Spun into fibres, and thin sheets; resin impregnated into glass fibres, etc. to make low pressure laminated components; foundry patterns, press tools, jigs, and fixtures, car bodies and aircraft components	
Epoxy resins	Similar to polyester resin		
Thermosets			
Bakelite	Tough if impregnated into cotton or other fibres, heat resistant if with asbestos fibres; dark colours due to action of light	Moulded forms, sheets and bars for electrically insulated components, e.g. terminal bars and control panels.	Moulding powders and resins
Urea & Melamine	Hard and of high strength, heat, water and detergent resistant, lightweight, resembling porcelain, takes delicate colours	Moulded forms with thin walls, e.g. cups, tumblers, etc.	

Additives

Plastics materials for use in plastics moulding processes are normally in the form of powders or small chips known as granules, or as preforms. In their pure unmodified state, these polymer materials are in most cases unsuitable for processing into finished articles. Before being moulded into shape by the application of heat and pressure, they have to be mixed with other ingredients known as 'additives', in order to modify or eliminate undesirable properties and to develop their useful characteristics.

Additives can be incorporated during the polymerisation reaction or alternatively with the polymer itself. Some additives modify the properties of the finished product by physical means, while others achieve their effect by chemical reactions and are used not only to influence the properties of the finished product, but also to improve processing characteristics. Some of the more common additives are as follows:

- *Fillers* – added to improve physical properties and in some cases to produce a cheaper product by acting as extenders. Fillers used include wood, cork dust, asbestos, carbon black, chalk and chopped glass fibre. A filler may also take the form of a gas producing expanded or foamed plastics like expanded polystyrene and polyurethane foam.
- *Stabilisers* – added to protect polymers from the adverse affect of heat or exposure to ultra-violet radiation, the main source of which is sunlight. e.g. carbon black.
- *Plasticisers* – added particularly to PVC to give it greater flexibility and make it easier to form.
- *Lubricants* – widely used to facilitate the processing of a variety of polymers, by reducing the forces between molecules and by reducing adhesion of the polymer to hot metal surfaces during processing.
- *Antioxidants* – added to prevent oxidative degradation, i.e. gradual breakdown in the presence of oxygen, which most polymers are subject to at the elevated temperatures necessary for processing and at atmospheric temperatures over a period of time.
- *Flame retardants* – added as most polymers are flammable to a greater or lesser extent.
- *Colourants* – may be added as dyes or pigments and are available in a vast range of colours. They are added for the following reasons:
  - to give greater product appeal and make the product more saleable

- as a means of identification, e.g. cable insulation
- to make the product more readily visible, e.g. garments for roadworkers and motor cyclists
- to simulate a natural or traditional product, e.g. leather luggage

Rubber

This is produced from the sap or latex of a tree called *Hervea Brasiliensis*; however, the term rubber comprises a broad spectrum of synthetic materials in addition to the natural product.

Natural rubber has little value as an engineering material in its crude state because it quickly becomes tacky at elevated temperatures and has low tensile strength at room temperature. When compounded with other substances and vulcanised however, it changes to become a useful and versatile material. The reinforcing filler material used in rubber production is carbon black. The raw material and filler are mixed with small amounts of chemicals which speed up the vulcanising process and extend the working life of the product by improving its ageing characteristics.

Large amounts of reclaimed rubber are used either alone or mixed with raw rubber compounds, in much the same way as thermoplastic materials are reused. Typical articles produced are tyres, inner tubes, footwear, face masks and rubber gloves.

Semiconductor materials

Semiconductor materials or semiconductors are materials which act as insulators when very cold, conduct an electric current slightly at normal room temperature and show rapidly increasing conductivity with further temperature rises. Semiconductor materials are capable of having their conduction properties changed during manufacture. Examples of semiconductor materials are silicon, germanium and certain metal oxides and these are used extensively in the electronics industry in the manufacture of a wide range of solid-state devices. Typical examples of devices incorporating semiconductor materials are diodes, thermistors and transistors. Transistors are specially treated crystals of semiconductor materials and their importance lies in their ability to link two circuits in such a way that the current through one controls the current in the other. Integrated circuits are built up from transistors and are incorporated into control systems for electronic appliances like computers, televisions and domestic goods.

Ceramic materials (Ceramics)

Ceramics is the name given to a wide range of materials which include brick, concrete, stone and clay. Due to the high refractive index of these materials they find application in industry where high thermal resistance is required, e.g. firebrick linings in furnaces, chimney flues and crucibles.

The same materials also possess equally high electrical resistance properties and porcelain (from clay) in particular is used extensively as an insulating material in power transmission systems and sparking plugs.

More recently metal-cutting tool material has been developed, successfully utilising the exceptional hardness properties possessed by a range of commercially produced ceramics. These materials comprise metallic oxides, carbides and nitrides, e.g. aluminium oxide 'alumina' and tungsten carbide. Unfortunately the extreme hardness properties of ceramic materials means that they are also very brittle. This necessitates modification by bonding with other materials in order to render their structure and properties more suitable to meet the kind of conditions under which they are expected to operate (see 'composite materials' below).

Glass

The majority of glass produced commercially is made from the raw materials silica, lime and soda. Other ingredients are used to produce particular effects or to produce special physical properties.

Glass is a hard-wearing, abrasion- and corrosion-resistant material possessing excellent weathering properties. It is a good electrical insulator but is an extremely brittle material with no ductility, giving no warning of failure when a load is applied.

Composite materials (composites)

A material which is made from or 'composed' of different materials is called a composite material or composite. In a composite, one material in the form of rods, strands or fibres is bonded together with the other main matrix material(s). The fibres themselves have some of the highest moduli and greatest strengths available; however, they do possess little resistance to compressive or bending forces. The solution is to embed the fibres in a suitable matrix material in order to take advantage of their outstanding properties.

Naturally occurring composite materials include examples such as wood, bone and horn which are based on fibres of cellulose, collagen and keratin respectively. The advantages of deliberately combining materials in order to obtain improved or modified properties was appreciated by ancient civilisations. An example of this is the making of air-dried bricks by mixing clay with straw in an attempt to reduce the effects of shrinkage stresses. Nowadays, especially with the growth of the plastics industry and developments in fibres, a vast range of combinations of materials is available for use in composites.

Glass reinforced plastic (GRP)

This important composite material is produced when a plastic material, usually polyester resin, is reinforced with glass fibres in strand or mat form. The resin is used to provide shape, colour and finish, while the glass fibres laid in all directions give mechanical strength. When correctly laid-up or 'laminated', the thick, viscous liquid resin surrounds the strands of the glass fibre mat and sets to give a hard, rigid structure. Resin is normally supplied pre-accelerated to shorten setting time but if a much shorter setting time is required, a catalyst, in the form of a liquid or paste, can be used which causes the liquid resin to set shortly after it is mixed in. In order to improve appearance by giving a smoother surface finish, a surfacing tissue made up from more closely spaced and finer strands of mat is often used on the rough side, i.e. non-mould side of a lay-up (Fig. 4.7).

Glass reinforced plastic is used extensively where products require a high strength-to-weight ratio and good corrosion resistance such as boat building and performance car

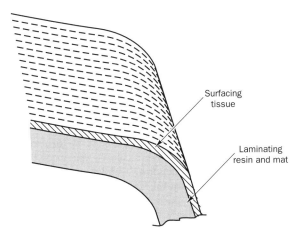

Figure 4.7 The composition of a GRP 'lay-up'

manufacture. It is particularly suitable for the construction of complex shapes which would be difficult to form in sheet metal or wood.

Cermets

Certain metal oxides and carbides bonded together 'sintered' in a metal powder matrix form important composite materials called 'cermets' which are used for metal cutting tool tips.

The properties of the resulting composite material enable extremely hard metals to be machined at production rates (cutting speeds) and finer quality far greater than was previously possible.

Reinforced concrete

Probably the most common example of a composite material in civil engineering is reinforced concrete, in which steel reinforcing rods are embedded in the concrete (Fig. 4.8(a)). Concrete itself is also a composite material, being made by bonding aggregate (stone chippings) in a mortar (sand, cement and water) matrix (Fig. 4.8(b)).

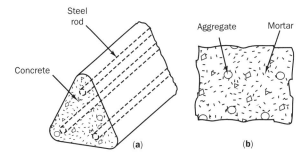

Figure 4.8 Reinforced concrete: (a) steel rods in a concrete matrix; (b) aggregate (stone chippings) in a matrix of sand and cement (mortar)

Self-assessment task

What is meant by the term *composite*? Give THREE examples of composite materials, stating the materials involved.

The structure of materials

Metals

Despite the continued advance and development of synthetic materials, metals still remain the most important materials used in engineering. Just like salt or sugar they are in fact crystalline in form and the form in which a particular metal exists has a direct influence on the properties which that metal possesses.

Metal crystals form when the molten metal cools and solidifies, whereas crystals of other substances, for example, copper sulphate and sodium chloride (common salt; Fig. 4.9) form when a saturated solution of the compound evaporates causing the solid to crystallise out. Closer examination shows that the crystals themselves can each be considered as being identical 'unit cells' (Fig. 4.10) which fit together to form a geometrical network or lattice structure (Fig. 4.11).

In aluminium, copper, nickel, lead, silver and gold and several other metals, the atoms are spaced evenly in rows at right angles to each other in three dimensions (i.e. atoms are arranged at each corner of millions of small cubes), while other atoms occupy positions at the centre of each of the cube faces. This particular atomic lattice pattern is known as *face-centred cubic* and is illustrated in Fig. 4.11(a). In iron at room temperature, and in several other metals such as vanadium, tungsten, molybdenum and sodium, the atoms are disposed in a different cubic pattern. In this type, there is again an atom at the corner of each imaginary cube but instead of other atoms occurring in the middle of the cube faces, a single atom is located at the centre of every cube. This structure is known as *body-centred cubic* and is illustrated in Fig. 4.11(b). Iron is a particularly interesting case, for at room temperature

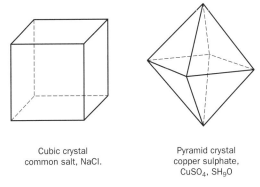

Cubic crystal
common salt, NaCl.

Pyramid crystal
copper sulphate,
$CuSO_4, 5H_2O$

Figure 4.9 Common crystal shapes

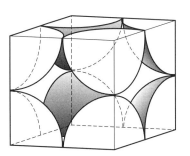

Figure 4.10 Simple unit cell

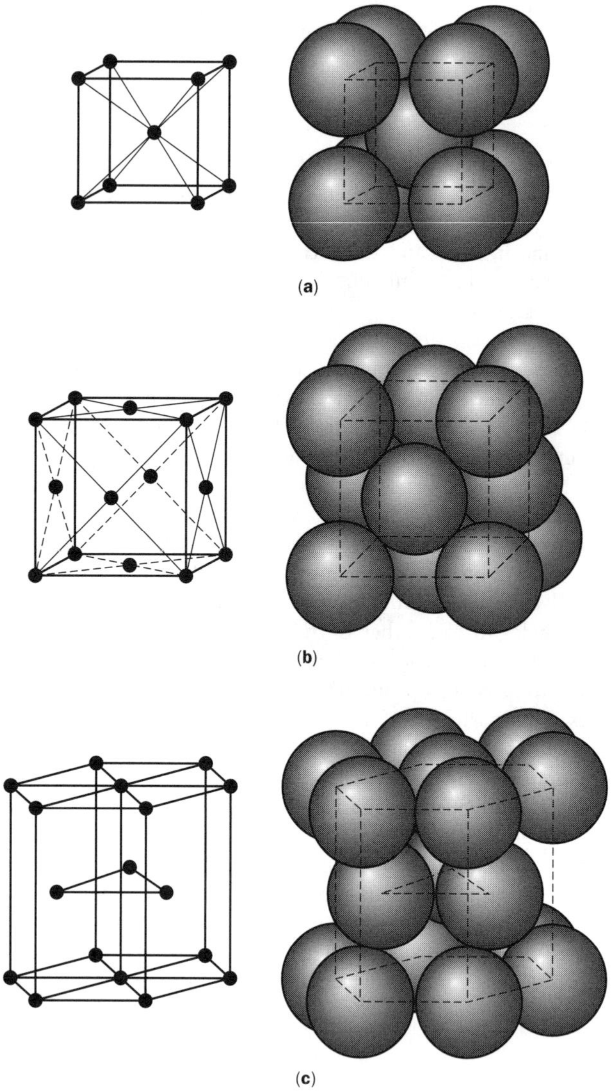

Figure 4.11 Arrangement of atoms in: (a) body-centred cubic (b.c.c.) structure; (b) face-centred cubic (f.c.c.) structure; and (c) close-packed hexagonal (c.p.h.) structure

its atoms are arranged in the body-centred cubic form, however at 910 °C the atoms rearrange themselves into the face-centred cubic pattern, while at a still higher temperature of about 1,400 °C the iron atoms change back to a body-centred cubic lattice. (The structure of plain carbon steels is discussed on p. 106.) The atoms in zinc, magnesium and cadmium are arranged in a *close-packed hexagonal* pattern as illustrated in Fig. 4.11(c). While other metals have still more complex lattices, most of the common metals exist in face-centred cubic, body-centred cubic or close-packed hexagonal form.

Self-assessment task

The common metals crystallise into one of three main types of structure. With the aid of suitably labelled diagrams describe these different forms. Give one example of each type.

Metal solidification

Observation of a pure metal cooling from the liquid state shows that it does so in a particularly well-defined manner. As soon as the freezing point is reached, nucleii begin to form at random throughout the cooling liquid, perhaps 'touched off' by specks of dust or a sudden shock or vibration. As soon as a nucleus has initiated the formation of a crystal a number of branches spread out rapidly in three dimensions. Soon these begin to grow secondary branches which, in turn, produce their own side-shoots until eventually all the space between the original branches or 'dendrites' is filled in, giving rise to a solid mass or crystal of metal. This phenomenon is repeated many times throughout the solidifying metal from numerous dendrite systems until they touch and no more liquid is left (Fig. 4.12).

In an alloy where 'phases' with different freezing points exist (see phase diagrams, p. 105) the high-freezing point phase always crystallises first and the low-freezing point phase solidifies last at the grain boundaries between the dendrite systems.

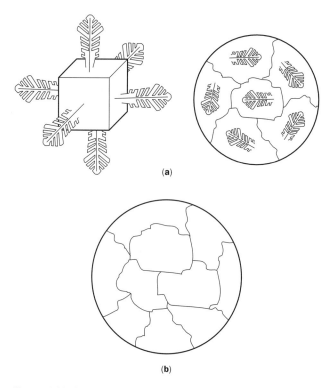

Figure 4.12 Crystals: (a) dendritic growth; (b) structure

Microscopy

Although the arrangements of atoms in a crystal of a metal or alloy can only be determined to any real degree of precision by the use of X-ray techniques, the arrangement of the crystals and the phases which are present in an alloy can be identified in the laboratory by ordinary microscope.

Specimens for micro-examination are prepared by grinding a flat face on the specimen followed by careful treatment with successively finer grades of waterproof abrasive paper. Final polishing is carried out on a horizontally revolving disc to which a diamond paste or an aqueous suspension of powdered alumina or magnesia is applied. The result will be a specimen polished to a mirror-like finish which may then be examined microscopically to reveal any physical defects or flaws in the structure, for example blowholes, hairline cracks, etc. Subsequent etching of the polished surface of the specimen with a suitable reagent for a few seconds followed by washing in water and then in alcohol and finally drying in a stream of hot air enables a closer examination of the structure of the metal or alloy to be made. The etchant is a chemical reagent chosen to attack the constituents in the sample to varying degrees, one constituent appearing dark in colour (etched) while the other appears light (unetched). Examination of etched specimens enables closer attention to be made to the size and shape of crystals and the conditions at grain boundaries.

A rather crude but nevertheless useful method of examination may be carried out without the microscope using instead a hand-held low-power magnifier. This method of inspection called 'macro-examination' can reveal details of methods of manufacture, the presence of physical defects such as blowholes in castings, evidence of non-uniform heat treatment and the inclusion of non-metallic impurities, for example slag, etc. Some skill in the interpretation of these results is, however, necessary with this method of examination, as is careful specimen preparation.

Self-assessment tasks

1. Describe how you could determine the method of manufacture from the examination of the microstructure of a carefully prepared specimen.
2. With regard to the face to be examined, why do you think it is necessary to consider the position on the component from which the specimen is taken?

[*Hint*: You will need to refer to Element 4.2 in order to answer these questions.]

Equilibrium (phase) diagrams

Whereas pure metals possess a single and sharply defined melting point, alloys, except in special circumstances, melt and freeze over a range of temperature, the extent of which is governed by their composition. The point at which an alloy just becomes entirely molten on heating or begins to solidify on cooling is known as the 'liquidus' while that at which it becomes completely solid on cooling or starts to melt on heating is termed the 'solidus'. Between the solidus and the liquidus the condition of the alloy is pasty. The situation is best illustrated in the form of an equilibrium or phase diagram, the simplest case being provided by copper–nickel alloys which display complete miscibility (ability to mix).

All the arrest points which indicate changes that take place for different percentages of the alloy are plotted in Fig. 4.13. From the diagram it can be seen that the molten alloy consisting, say, of equal amounts of the two metals will start freezing at 1,312 °C and will thereafter remain in a pasty state until the temperature has fallen to 1,248 °C, at which stage solidification is complete.

It is, however, often the case that the constituents of an alloy display neither complete miscibility nor complete immiscibility but occupy an intermediate position up to a certain percentage of one metal dissolving in the other, and vice versa. Consequently, more than one solid solution exists and in such circumstances the diagram adopts a somewhat more complicated form, as will be evident from that representative of tin–lead alloys (Fig. 4.14).

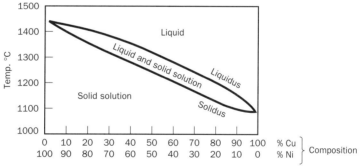

Figure 4.13 Equilibrium 'phase' diagram for copper–nickel alloys

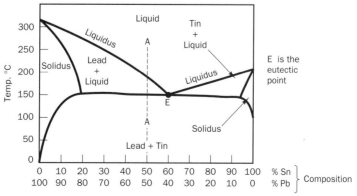

Figure 4.14 Equilibrium 'phase' diagram for tin–lead alloys

The solidification of tin–lead alloys

In order to demonstrate the solidification of a 50/50 alloy, consider the line AA. Solidification commences at about 210 °C. Crystals of lead are precipitated until about 180 °C. When the remaining liquid has a proportion of 63 per cent tin/37 per cent lead, the remaining liquid solidifies, the final solid consisting of crystals of lead mixed with lead–tin eutectic (see also behaviour of tin–lead solders, p. 119).

The structure of plain carbon steels

Before a study of the thermal equilibrium diagram for steel can be made it is essential to understand that structural changes take place during the heating and cooling of iron itself. Pure iron exhibits allotropy which may be defined as the ability of an element to exist in more than one physical form.

At ordinary temperatures iron is a soft, grey, ductile and magnetic element called 'ferrite'. However, if it is heated to a temperature above 910 °C, called the critical temperature, it undergoes an atomic rearrangement which causes it to lose its magnetism. In this condition it is called 'austenite'. Subsequent slow cooling causes the iron to change back into its original state and its magnetism returns. We know what this rearrangement is. At ordinary temperatures the atoms forming a crystal of iron are arranged in a body-centred cubic structure (Fig. 4.15(a)), while at temperatures above the critical point the atoms rearrange themselves into a face-centred cubic structure (Fig. 4.15(b)) with atoms at the centre of the faces of each cube. These two types of iron are called by the Greek symbols α (b.c.c.) and γ (f.c.c.) iron. The change from α iron to γ iron is also accompanied by a slight increase in volume of the mass of metal.

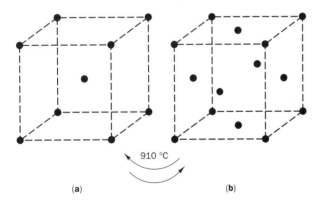

Figure 4.15 The allotropy of iron showing the two atomic structures: (a) body-centred cubic (alpha) iron or 'ferrite'; (b) face-centred cubic (gamma) iron or 'austenite'

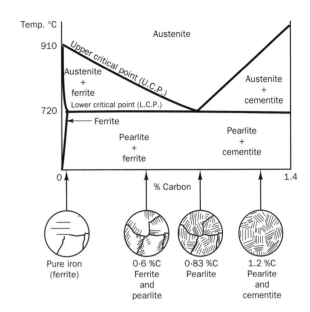

Figure 4.16 The steel portion of the iron–carbon equilibrium diagram

The various microconstituents present in plain carbon steels may be defined as follows:

- **Ferrite** This is a solid solution having a maximum of 0.04 per cent carbon dissolved in body-centred cubic (α)iron. Ferrite can be regarded as almost pure iron and is soft, ductile and easily worked.
- **Pearlite** This structure exists at the 'eutectoid' and consists of alternate layers of ferrite and cementite. It contains 0.83 per cent carbon and is formed by the breakdown of austenite at temperatures below the lower critical point (720 °C).
- **Cementite** This is a hard, brittle compound of iron and carbon and may exist in the free state usually at the grain boundary or as a constituent of the eutectoid pearlite.
- **Austenite** This is a solid solution of carbon in face-centred cubic (γ)iron and contains a maximum of 1.7 per cent carbon at 1,130 °C. Austenite only exists in plain carbon steels above the upper critical point and is a soft, non-magnetic compound.

Self-assessment task

Sketch and label the steel section of the iron-carbon equilibrium diagram using the terms ferrite, pearlite, austenite and cementite.

The iron–carbon equilibrium diagram

This diagram (Fig. 4.16) is important as it provides the essential information required for the heat treatment of all carbon steels. The various heat-treatment processes are described in Element 4.2. It is used in the same way as the lead–tin diagram except that the steel remains in the solid state while the structural 'allotropic' changes take place. The diagram shows the various microconstituents and the temperature change points, 'critical points', which occur in the steel part of the iron–carbon diagram.

The mechanical properties of plain carbon steels.

The mechanical properties of slowly cooled plain carbon steels will depend upon the proportions of each of the microconstituents present. They vary linearly from 0 per cent carbon content (100 per cent ferrite) to 0.83 per cent carbon content (100 per cent pearlite); however, beyond 0.83 per cent carbon content, free cementite appears in the microstructure and the linear relationship ceases to exist.

Using a range of test values, it is a simple matter to construct a graph to show how the microconstituents of carbon steel affect its properties (Fig. 4.17). It should be appreciated that the properties obtained in normalising, which involves air cooling, vary according to the thickness of the section and also, fully annealed (furnace cooled) steels will give a softer and more ductile steel. A further limitation of the graph is that it does not take into account the effect of variation of other elements present in plain carbon steels, e.g. manganese, silicon, sulphur and phosphorus.

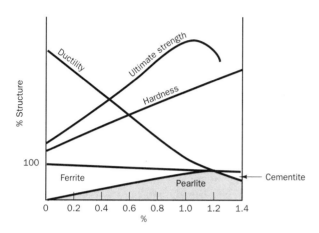

Figure 4.17 The effect of carbon on the microstructure and properties of plain carbon steels (normalised)

The composition and structure of non-metallic materials

Plastics

Most man-made plastics, technically known as 'polymers', are produced from the by-products of crude oil and coal, both of which are organic compounds based on carbon. In addition to carbon, most plastics contain hydrogen, many contain oxygen and a few contain nitrogen, chlorine or flourine.

Whereas metals are generally obtained reasonably easily from their ores, most plastics are produced by a more complicated process called polymerisation which involves the building up of long chains of molecules (polymers) from single molecules (monomers). Figure 4.18 shows the basic form of the polymeric material polyethylene and ethylene, the monomer from which it is derived.

Note: For many polymers their names can be determined by adding the prefix (poly) to the name of the monomer, e.g. (poly)ethylene, (poly)styrene.

The ways in which different types of polymers behave can be explained by the manner in which the long molecular chains are arranged inside the material. Figure 4.19 shows the three forms the chains can take, viz. linear, branched and cross-linked. Linear chains can move readily past each other as they have no side branches or cross-links to impede movement. However, relative movement in the branched chain is restricted and in the cross-linked chain no movement at all is possible.

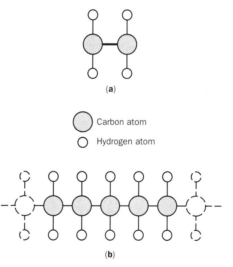

Figure 4.18 Polymers: (a) single monomer; (b) polymer chain

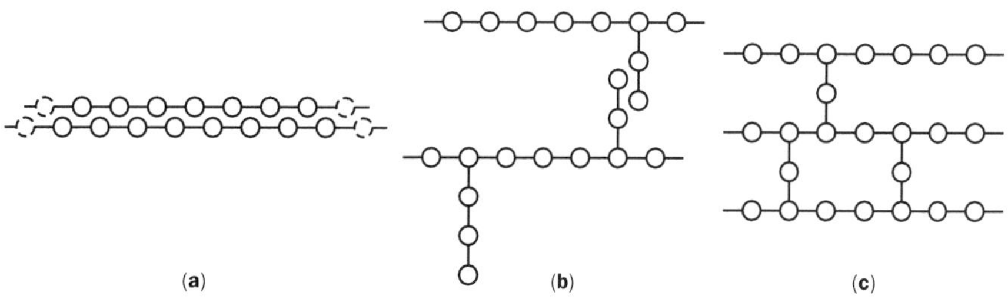

Figure 4.19 Polymer chains: (a) linear; (b) branched; (c) cross-linked

High-density polyethylene (polythene) is a thermoplastic material which has linear molecular chains. Consequently, polythene is easily stretched and is not rigid because the chains are independent and can easily slide past each other. The energy in the form of heat needed to break bonds between chains in this case is low, resulting in a material which has a low melting point. The absence of strong bonds between chains also means that when the material is heated, the subsequent removal of heat allows the material to revert to its initial state.

Other thermoplastic materials, for example, polypropylene, have molecular chain arrangements with side branches which produce a harder and more rigid material. Another consequence of a material having branched chains is that they are unable to pack together as tightly as material composed of linear chains, resulting in the material having a lower density than linear chain type polymers.

Thermosetting polymers like bakelite possess cross-linked chains resulting in extremely hard, rigid materials in which energy is needed to break bonds before flow can occur. They have higher melting points than thermoplastic materials with linear or branched chains and, in addition, the effect of heat causes the cross-linked bonds to break producing an irreversible change which ultimately results in the breakdown of the structure of the material.

Unlike the orderly crystalline arrangement of the molecules in most compounds, the arrangement of the molecular chains in polymers is completely random and is said to be amorphous (non-crystalline). Linear chain type polymers can be amorphous and Fig. 4.20(a) shows the type of structure that would occur in this case, the chains being all entangled with one another. Linear polymer chains can, however, assume a more orderly arrangement and Fig. 4.20(b) shows a typical arrangement of polymer chains which can occur where the chains fold backwards and forwards on themselves in certain areas of the structure. These areas are called *crystallites* and the arrangement is said to be *crystalline*. Polymers with branched chains show less tendency to crystallise while cross-linked polymers have zero crystallinity.

Crystallinity in polymers affects the properties of the polymers. Linear polyethylene with 95 per cent crystallinity has a relative density of about 0.95 and a melting point of 135 °C, whereas branched polyethylene with 50 per cent crystallinity has a relative density of 0.92 and a melting point of 115 °C. The greater density results from the more closely packed molecules in the crystalline type of polymer. The two forms of polyethylene are known as high-density polyethylene (linear) and low-density polyethylene (branched). The more closely the molecules are packed, the more energy (heat) is required to break the bonds between them and melt the polymer. In addition, crystallinity in polymeric materials leads to stiffer, stronger materials, i.e. tensile strength and tensile modulus.

Elastomers

An elastomeric material, or rubber as it is more commonly known, is defined as any material which after vulcanisation is capable of being extended to several times its original length and of rapid recovery to or very near to its original length when the extending force is removed, e.g. rubber band.

The term 'rubber' comprises a broad spectrum of synthetic materials, in addition to the natural polymer product. Both natural and synthetic rubbers are combined with additives such as plasticisers, fillers and antioxidants to give the required product.

Similar to plastics, rubber products consist of linear chain molecules with some cross-linking between chains (Fig. 4.21). The cross-linking ensures that the material will be elastic. The introduction of sulphur during vulcanisation controls the degree of cross-linking and hence the degree of flexibility of the rubber. Fully vulcanised rubber is the strong insulating material known as ebonite.

Of the natural rubbers, butyl rubber is produced from the gases isobutylene and isoprene; nitrile rubber is made from butadiene and acrylonitrile; and neoprene rubber is made from chloroprene.

Synthetic rubbers are more resistant to sunlight, temperature changes, ozone, ageing and many organic and inorganic chemicals than natural rubber products. Table 4.5 gives some of the properties of typical rubbers.

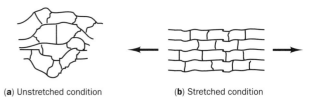

(a) Unstretched condition   (b) Stretched condition

Figure 4.21 Cross-linked elastomer chains: (a) unstretched condition; (b) stretched condition

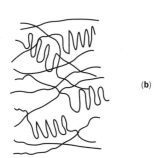

Figure 4.20 Linear polymers: (a) an amorphous polymer; (b) a folded crystalline polymer

Table 4.5 Properties of some typical rubber

Material	Tensile strength (MN/m² or MPa)	Maximum elongation (%)
Natural rubber	30	700
Neoprene	28	600
Nitrile	28	550

Self-assessment task

With the aid of suitably labelled diagrams, compare and contrast the structures of (a) polymeric materials and (b) elastomeric materials.

Semiconductor materials

In semiconductor materials, the conduction of electricity is achieved by means of the movement of thermally generated electrons and is controlled solely by the temperature of the pure 'intrinsic' semiconductor material. In order to make practical semiconductor devices, the composition and structure of pure semiconductor materials is changed by adding small amounts of other substances or 'impurities' to the material, which have a dramatic effect on its conducting properties. When a pure semiconductor is 'doped' in this way it is designated as either an 'n-type' or a 'p-type' semiconductor depending on the kind of substance added. Silicon doped with phosphorus is an example of an n-type material. Alternatively, antimony or arsenic can be used as impurity elements. The presence of these impurity atoms increase the number of negative electrons which are free to move through the material (Fig. 4.22(a)). Conduction occurring as a result of the movement of these spare electrons is called 'extrinsic' conduction. Silicon doped with indium is an example of a p-type material; alternatively, gallium or boron can be used as the impurity elements. Here the indium atoms create gaps or positive holes in the electron structure and electrons are then able to move through the material by passing from one hole to another (Fig. 4.22(b)).

Materials testing

Data obtained from tests performed on engineering materials enable the design engineer to select a suitable material from which an engineered product will be made. Tests on both metallic and non-metallic materials fall into two categories: destructive tests and non-destructive tests.

In general, the destructive tests give information about the mechanical properties of the material while the non-destructive tests indicate whether the material is free from defects. Destructive tests involve the use of a test specimen which is either produced from a representative sample of the raw material or is specially formed on the article itself. The two standard and most frequently used tests in engineering are the *tensile test* and the *impact test*. In both cases the material is stressed until it breaks. A third test, the hardness test compliments the tensile and impact tests; however, this type of test is not strictly speaking destructive as only a small mark remains on the test specimen after the test.

Tensile test

In the tensile test (Fig. 4.23) the test piece is either in rod or strip form with a central section reduced in diameter for a suitable length which is known as the 'gauge length'. In the test, the testpiece is gripped at its ends in a machine which is capable of exerting a tensile force (pull) until fracture occurs and simultaneously measuring the load applied. An extensometer, which is normally fitted to the gauge length, enables the extension to be recorded as the load is increased. From the results, a load–extension curve for the testpiece can be plotted (Fig. 4.24).

The first part of the graph shows that a linear relationship exists between incremental increases in load and extension of the testpiece. This continues through the elastic range up to the limit of proportionality (A). During this period the material is said to be elastic and if the load were to be removed at any point up to the limit of proportionality, the material would return to its original shape and size. Beyond the limit of proportionality, the graph follows a curve which

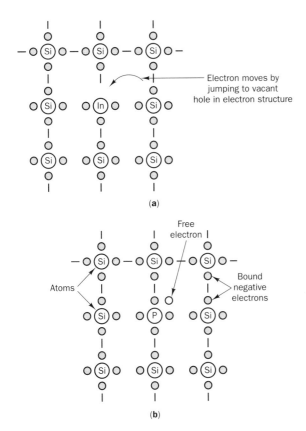

Figure 4.22 Semiconductors: (a) p-type semiconductor – silicon (Si) doped with indium (In); (b) n-type semiconductor – silicon (Si) doped with phosphorus (P)

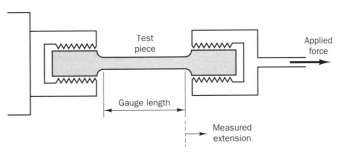

Figure 4.23 Principle of tensile testing

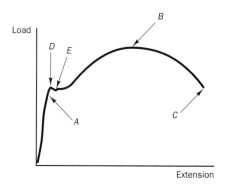

Figure 4.24 Load–extension diagram for low carbon steel

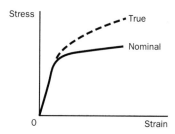

Figure 4.25 Effect of different definitions of stress

reaches a peak at the maxiumum load (B) at which point the ultimate tensile strength can be calculated. From this point onwards there is a steady fall until point (C) where fracture occurs. This is known as the *breaking load* and at this point the stess to produce fracture can be calculated. Points D and E are known as the upper and lower yield points. After the testpiece has been broken, the two halves are removed from the machine and fitted together. The resulting extended gauge length is now measured from which the percentage elongation can be calculated.

If the material under test is ductile, the fracture will have occurred in the middle of a well-defined 'waist'. Measurement of the diameter at this point enables the percentage reduction in area to be calculated. Both the percentage elongation and the percentage reduction in area are indications of the ductility of the material under test. Table 4.6 shows how a range of metals and plastics perform during tensile tests. Further valuable information relating to tensile tests performed on metals can be found by referring to BS EN10002.

Table 4.6 Tensile strengths of a range of materials

Material	Tensile strength (MN/m^2 or MPa)
Metals	
Mild steel	400
Cast iron	140–300
Brass	120–400
Aluminium alloy	146–600
Thermoplastics	
Polyvinylchloride (PVC)	35–60
Acrylonitrile-butadiene-styrene (ABS)	25–50
Polycarbonate (PC)	55–65
Polytetrafluoroethylene (PTFE)	15–35
Cellulose acetate (CA)	13–62
Thermosets	
Phenol formaldehyde (PH)	50–55
Polyester (unsaturated) (UP)	40–90

Stress–strain graphs

In addition to load–extension curves, the behaviour of materials subject to tensile and compressive forces can also be shown in graphical form in terms of their stress–strain relationships, where stress is the applied force per unit area and strain is the extension or contraction per unit length of the particular material under test. The area referred to in the stress definition is taken to be the area before any forces are applied, i.e. no account is made of the reduction in area that occurs when the material is stretched or of the increase in area if it is compressed. The true stress at all points beyond the limit of proportionality is therefore different to the nominal stress defined in this way (Fig. 4.25). On the part of the graph where the strain is directly proportional to the stress, a modulus of elasticity, or 'Young's modulus' as it is often called, is defined as:

$$\text{Modulus of elasticity } (E) = \text{Stress/Strain}$$

Young's modulus is a measure of 'stiffness' or the ease with which a material can be stretched or compressed. The higher the value of the modulus, the more stress is needed to produce a given extension. Some typical values are given for metals and plastics in Table 4.7.

Table 4.7 Modulus of elasticity of a range of materials

Material	Modulus of elasticity (GN/m^2 or GPa)
Metals	
Mild steel	220
Cast iron	150
Brass	120
Aluminium alloy	70
Thermoplastics	
Polyvinylchloride (PVC)	2.5–4.0
Acrylonitrile-butadiene-styrene (ABS)	2.0–3.0
Polycarbonate (PC)	2.0–3.0
Polytetrafluoroethylene (PTFE)	0.3–0.6
Cellulose acetate (CA)	0.5–2.8
Thermosets	
Phenol formaldehyde (PH)	5.2–6.0
Polyester (unsaturated) (UP)	2.0–4.4

Impact test

An impact test (Fig. 4.26) is carried out in order to determine the ability of a material to resist a sudden 'shock' load. In the test, a standard size 'V' notch is machined across a testpiece which is produced from a length of round or square section bar. The purpose of the notch is to concentrate the stress to which the testpiece is subjected when it is struck by a blow from a weight at the end of a pendulum, while it is held in a vice. The energy in newton-metres (Nm) absorbed in fracturing the testpiece is related to the distance which the pendulum swings after impact and is known as the impact value. The most common impact tests are the Charpy test and the Izod test. The tests are very similar in principle, the main difference between the two being the way in which the

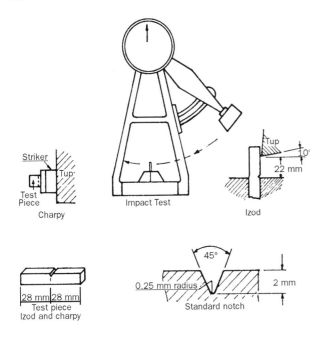

Figure 4.26 Izod and Charpy tests

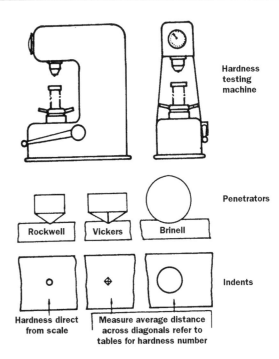

Figure 4.27 Hardness testing

specimen is held. Table 4.8 gives typical impact strength values for some common metals and plastics materials (polymers) while more comprehensive data can be obtained by referring to BS EN10045.

Table 4.8 Impact strengths of a range of materials

Metal	Impact strength (J)*
Aluminium, commercial pure, annealed	30
Aluminium–1.5% Mn alloy, annealed	80
Aluminium–1.5% Mn alloy, hard	34
Copper, oxygen free HC, annealed	70
Cartridge brass (70% C, 30% Zn), annealed	88
Cartridge brass (70% Cu, 30% Zn), $\frac{3}{4}$ hard	21
Cupronickel (70% Cu, 30% Ni), annealed	157
Magnesium–3% Al, 1% Zn alloy, annealed	8
Nickel alloy, Monel, annealed	290
Titanium–5% Al, 2.5% Sn, annealed	24
Grey cast iron	3
Malleable cast iron, Blackheart, annealed	15
Austenitic stainless steel, annealed	217
Carbon steel, 0.2% carbon, as rolled	50
Plastic	**Impact strength (kJ/m^2)†**
Polythene, high-density	30
Nylon 6.6	5
PVC, unplasticised	3
Polystyrene	2
ABS	25

*Metals tested by Charpy V
†Plastics tested by Izod (notch-tip radius 0.25 mm, depth 2.75 mm)

The hardness test

The various hardness tests (Fig. 4.27) are concerned with different forms of material deformation as an indication of hardness. There are several methods of determining the hardness of a metal but, of these, the indentation method is the most popular. Three examples which are similar in principle but different in detail are the Brinell, Vickers and Rockwell type tests. In the Brinell test, a heavy load (3,000 kg) forces a hard steel ball (dia. 10 mm) into the metal sample under test. The diameter of the resulting impression is measured and transferred into hardness units from standard tables. The Vickers test uses a pyramidal diamond penetrator and, in this method, the average length of the diagonals of the impression are converted into hardness units. The Rockwell test uses either a ball or diamond penetrator but, in each case, the depth of penetration is measured after the application and measurement of an initial light load. The resulting hardness number can be read directly from the scale on the front of the instrument. Although perhaps less well known, a fourth type of test known as Shores scleroscope (Fig. 4.28) merits inclusion in a consideration of hardness tests on materials. In this test a small diamond-pointed hammer weighing 2.4 g is allowed to fall inside a graduated glass tube from a height of

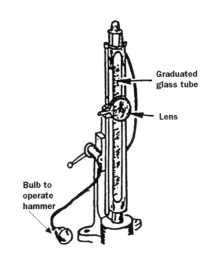

Figure 4.28 Shore scleroscope

250 mm onto the material under test. After striking the specimen, the hammer bounces up from the surface and the height of the first rebound is observed and used as the index of hardness. The harder the material the less energy it will absorb when it is struck by the hammer and thus the greater will be the height of the rebound.

While, strictly speaking, this test uses the indentation principle, the resulting mark on the testpiece is so small as to have little or no detrimental effect on the specimen.

Hardness tests on plastics

The Vickers, Brinell and Rockwell hardness tests can also be used on plastic materials. The Rockwell test, with its measurement of penetration depth rather than surface area of indentation, is more widely used. The Shore durometer is another form of test. This involves an indenter in the form of a truncated cone with either a flat or a spherical end.

Table 4.9 gives typical hardness values for metals and plastics at room temperature and it is important that reference is made to the following specifications for more detailed information, viz. Plastics testing BS 278, Vickers hardness BS EN233878, Brinell hardness BS 240 and Rockwell hardness BS 4175.

Table 4.9 Hardness (Brinell) of a range of materials

Material	Hardness (HV)
Cast iron	140–240
Stainless steel	170
Brass	100–160
ABS	6–8
Polycarbonate	10–20
PVC	9

Self-assessment tasks

1. What is meant by the property of hardness, and how can this be useful in engineering?
2. Explain briefly a simple method of determining the hardness of two apparently similar pieces of steel.
3. Why is ductility an important property of a metal?
4. How could a metal be simply tested to find its degree of ductility?
5. What is meant by the strength of a material, and how could this be ascertained (a) accurately under controlled laboratory conditions and (b) approximately in the workshop?
6. What is the elastic limit and yield point of a material?
7. What results will a tensile test reveal about a material?
8. What is the purpose of impact testing?
9. Explain briefly how a material can be tested by the Izod method. Describe also the testpiece.
10. What is the purpose of the Brinell test?
11. What is a Shore's scleroscope? How does it operate?
12. Write a few simple notes on the Vickers diamond test.
13. What is the Charpy impact test and how is it carried out?
14. What is a Brinell hardness number and how is it determined?
15. Explain the value of a load–extension graph of a material under test.

Conductivity test

An indication of the relative thermal conductivity of different materials can be given by passing rods of different materials through rubber bungs inserted in holes in the side of a container (Fig. 4.29). Each rod is first dipped into molten paraffin wax, withdrawn and left to allow the coating of wax to solidify. Boiling water is then poured into the container so that the ends of the rods are all heated to the same temperature. After some minutes have elapsed it is noticed that wax along the rods has melted to different distances, indicating the differences in the thermal conductivities of the materials.

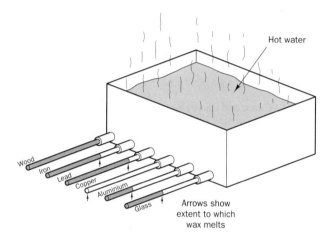

Figure 4.29 Comparisons of thermal conductivities

Tests for environmental stability

Corrosion

A simple test perfomed to monitor corrosion is shown in Fig. 4.30. In the test, a length of mild steel angle iron is supported in a tank of water over a long period of time, perhaps months and periodically, during this time, evidence of corrosion is recorded.

Different materials or galvanised or painted testpieces can be used and the results compared. In addition, the effects of different environmental conditions, e.g. salt-laden atmosphere, humid conditions, altitude, can be studied and their effect noted.

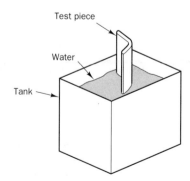

Figure 4.30 Corrosion test

Ultraviolet degradation

A simple test to investigate the degradation of plastic materials is illustrated in Fig. 4.31, in which similar plastic containers are exposed to different environmental conditions over a long period of time. In this way, the ultraviolet effects from sunlight, the bio-degrading effects from bacteria in the ground and acidic or alkaline 'chemical' effects of water and the atmosphere can be studied in order to evaluate their effects.

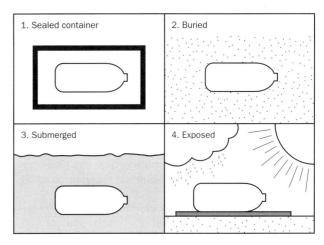

Figure 4.31 Degradation of plastic bottle

4.2 Relate materials' characteristics to processing methods

Topics covered in this element are:

- The identification of appropriate processing methods which take account of materials' structure and properties.
- The effects of processing methods on materials' structure and properties.
- The implications of different property values.

The properties of materials have implications for how they are best used and processed. For example, materials that are hard find application where resistance to wear is a requirement, while materials possessing high electrical conductivity are suited to processes such as arc welding.

In this element, a range of appropriate processing methods are described and can be considered as being those methods which are carried out when the material is in: (a) the liquid state or (b) the solid state.

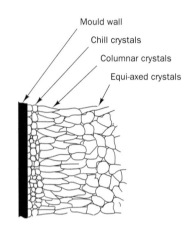

Figure 4.32 The grain structure of a casting

Material processing in the liquid state

Material, which can be either metallic or non-metallic, is processed in the liquid state by pouring or 'casting' into a mould.

Metallic materials

In the case of metallic materials, the grain structure within a cast product is determined by the rate of cooling of the metal. The metal which is in contact with the mould surface will cool faster than that in the centre of the casting giving rise to small 'chill' crystals. Nearer the centre, however, the cooling rate is much less resulting in the formation of large, elongated 'columnar' crystals at right angles to the mould walls. In the centre of the mould, the cooling rate is slowest and, in this region, medium-sized crystals of uniform size and shape called 'equi-axed' crystals finally develop (Fig. 4.32). A casting with a structure entirely of equi-axed crystals gives better strength properties and this type of structure can be promoted by cooling the casting more rapidly.

Castings themselves are nearly always composed of masses of metal having various shapes, volumes and thicknesses and as these sections will cool and contract at different rates, very complex stresses are often set up. The effect of this cooling is shown in Fig. 4.33(a) and (b). At Fig. 4.33(a) a large boss of metal is joined by arms to a much thinner section. When cooling from the molten state, the thin sections will have solidified and be well cooled down before the large volume of metal at the boss has cooled very much. Ultimately, this will cool down and contract, but as the arms have already cooled and become rigid, the force exerted by the contraction of the large boss cannot be relieved by any movement on the part of the two thin sections. This means that a permanent internal stress is left in the casting between the large boss and the outer frame, and even if the casting appears to be sound, a sudden blow or shock could cause it to fracture at one of the sections shown. Machining the hard and rigid skin from the casting on its outer frame would relieve some of the stress and probably allow a certain amount of spring to take place. Figure 4.33(b) shows how the grain structure pattern, which develops at sharp corners of castings on solidification, results in a plane of weakness and Fig. 4.33(c) shows how a simple design modification would eliminate this fault.

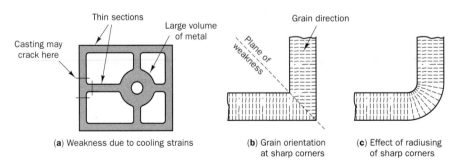

Figure 4.33 Cooling effects on castings: (a) weakness due to cooling strains; (b) grain orientation of sharp corners; (c) effect of radiusing sharp corners

These two simple cases serve to show how internal stresses and weaknesses can exist in a casting, and although the pattern-maker endeavours to minimise these effects by studying the method of casting, they cannot be prevented altogether. Castings with internal stresses will often warp when these stresses have been relieved by machining off the hard and rigid outer skin. When it is important for the shape of the casting to remain stable (e.g. for a gauge, large worm-wheel, etc.), it should be rough machined and 'weathered' or 'seasoned' by being placed in the open for a few months. This allows the dissipation of internal stresses to take place and minimises the possibility of distortion after the casting has been finished. For small castings a quick method of seasoning is to place a number of castings in a tumbling barrel for about half an hour. The vibration and light blows which the castings receive while being tumbled round in the barrel form an effective method of relieving internal stresses.

Cooling contraction

When a metal or alloy cools from liquid state to solid state, it contracts in three distinct stages: (i) *liquid contraction*, which takes place when the alloy cools from the pouring temperature to the solidification temperature; (ii) *solidification contraction*, which occurs when the metal passes through the stage of solidification; and (iii) *solid contraction*, which spans the period when it cools from solidification temperature to room temperature (Fig. 4.34). Of these three contractions, the first two, i.e. liquid and solidification contraction, have no effect on the material's structure and properties as they are taken care of by a proper risering system – the void created by these two shrinkages in the casting being filled up by the molten metal supplied by the riser. However, solid contraction of the casting, which is the most significant of the three and which is accounted for by making the pattern dimensions larger than those of the casting by the amount of contraction, may cause warping or hot tearing of castings as well as producing dimensional errors if not properly controlled.

Traditionally, cast iron has always been used as industry's main casting metal, hence its name. It has a relatively low melting point and possesses good fluidity which is essential for a casting material. Cast iron is used in sand casting where large, heavy castings with high compression strength are required. Die casting, in which metal moulds are used, has a much faster rate of cooling because the mould, being metal, has a high thermal conductivity. Zinc is preferred to cast iron in the die-casting process where lightweight castings are more important and high strength is not such a priority.

Unlike metals processed in the solid state, for example by forging, drawing, etc., cast metals do not show directionality of properties, the properties being the same in all directions. They do, however, have the additional problems of blowholes and slag inclusions, caused by working from liquid metal, which occur during solidification.

Non-metallic materials

Like metals, certain non-metallic materials, for example thermosetting-type plastics (thermosets), can be cast successfully. The solidification of the plastic is due to the chemical reaction that takes place between the liquid resin and hardener which are mixed together before pouring (see p. 102). An accelerator may also be used to speed up the curing process by acting as a catalyst. Thermoplastic-type materials can also be cast; however, due to their different structure and properties, they have to be in the form of powder or granules which fuse into a solid block in the heated mould (*sintering* process). The casting of plastics has been adopted in the electrical and electronic industry to encapsulate very small parts such as capacitors, coils or diodes, or larger items like cable joints, transformer windings, etc.

The properties of the encapsulating material enable it to:

- Exclude moisture from the parts.
- Exclude dust and foreign matter from the parts.
- Prevent damage to delicate parts by vibration in use.
- Prevent damage by exertion of undue force, e.g. blows.
- Improve insulation at a point where failure or partial failure has occurred or may occur in operation.

Material processing in the solid state (metallic)

In the case of metallic materials, the actual processing of the metal is closely monitored in order to control the effect the manipulation is having on the internal structure of the metal. In particular, overstressing during processing and heat-treatment processes carried out on the metal before or after manipulation all have an effect on grain size, grain growth and orientation of the crystal structure of the metal.

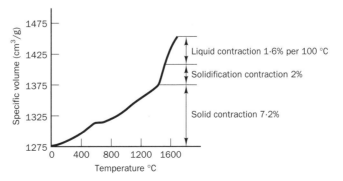

Figure 4.34 Change in specific volume of steel with temperature

Structural changes due to hot working

Hot working is carried out when metal is manipulated just above its recrystallisation temperature. In hot working, deformation and recrystallisation occur simultaneously allowing some grain growth to take place leaving the product in a relatively soft condition. (Fig. 4.35(a)). It is important to note that if the finishing temperature is too high, further grain growth takes place while the metal is cooling above its recrystallisation temperature or, alternatively, if the finishing temperature is too low, work hardening will result.

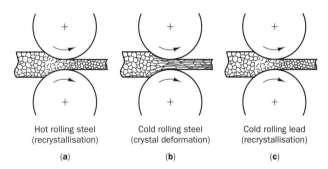

Figure 4.35 Actions of rolling metal: (a) hot rolling steel (recrystallisation); (b) cold rolling steel (crystal deformation); (c) cold rolling lead (recrystallisation)

Hot-working processes like rolling, forging and extrusion increase strength, ductility and toughness by introducing directional properties in the product. The resulting improvement being greatest at right angles to the direction of working. In forged products the fibres will follow the shape of the section resulting in a product possessing superior mechanical and magnetic properties than similar shapes which have been machined from hot-rolled material (Fig. 4.36).

Since the metal is softer, more plastic and easier to shape, hot-working processes require less power and are therefore more economical than cold-working processes.

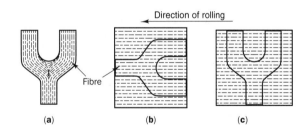

Figure 4.36 Grain orientation in forged and machined products: (a) forged; (b) machined (longitudinal); (c) machined (transverse)

Structural changes due to cold working

Cold working, as the name suggests, is carried out when the metal is manipulated 'cold' by processes such as rolling, drawing and pressing. It is usually carried out on previously hot-worked metals and alloys as the finishing stage in production.

The effect of cold working is to break down the crystal structure of the metal by subjecting it to a deforming force (Fig. 4.35(b)). Cold work destroys the lattice structure of the metal with its regular crystal planes along which deformation can occur. Much greater pressures are necessary for cold working than for hot working as the metal can only be permanently deformed when the elastic limit is exceeded. Since for most metals there can be no recrystallisation of grains in the cold-working range, there is no recovery from grain distortion and fragmentation. As grain deformation proceeds, greater resistance to this action builds up, resulting in increased strength and hardness of the metal. The amount of cold work that a metal will withstand depends upon its type and composition, i.e. the more ductile a metal, the more it can be cold worked. Hardening due to cold working is known as strain hardening and it is the only possible method of increasing the hardness and tensile strength of pure metals. When metal is deformed by cold work, severe stresses known as 'residual' stresses are set up in the metal. These stresses may be undesirable and, where this is the case, they are removed by a method called *process annealing*, which involves reheating the metal to a temperature below its recrystallisation temperature. Within this range of temperature, the stresses are rendered ineffective without any appreciable change in the physical properties or grain structure of the metal. Further heating into the recrystalline range eliminates the effect of cold working altogether, restoring the metal to its original condition. As increasing annealing time displaces recrystallisation to a lower temperature, it is important that annealing temperature and time are carefully controlled in order to maintain the final quality of the product. The effect of heating on the structure and mechanical properties of cold-worked metal are shown in Fig. 4.37. Certain metals, for example lead, tin and zinc, will not work harden since, due to their recrystallisation temperature being below room temperature, recrystallisation immediately follows deformation (Fig. 4.35(c) and Table 4.10).

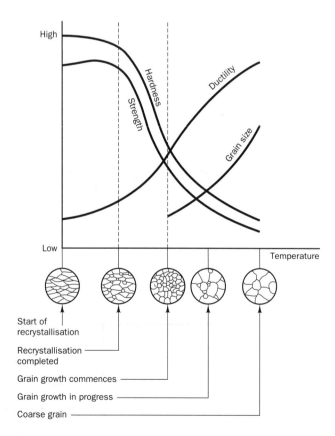

Figure 4.37 The effect of heating on the structure and mechanical properties of cold-worked metal

Table 4.10 Temperature properties of a range of materials

Material	Recrystallisation temperature (°C)	Melting point (°C)
Tungsten	1,200	3,410
Molybdenum	900	2,620
Nickel	600	1,458
Iron	450	1,535
Brasses	400	900–1,050
Bronzes	400	900–1,050
Copper	200	1,083
Silver	200	960
Aluminium	150	660
Magnesium	150	651
Zinc	70	419
Lead	20	327
Tin	20	232

Cold-worked products possess a characteristic clean, bright finish and accurate dimensional control of size is possible. In industry, mild steel has traditionally been used and today is still accepted as the favourite material for processing by hot- or cold-working techniques. Its properties, particularly its ductility and malleability, make it extremely suitable for processing by all the recognised forms of hot and cold working.

Material processing in the solid state (non-metallic)

Solid-state processing of plastics material cannot be carried out on thermosetting plastics because of their nature, but thermoplastic-type plastics can be formed in much the same way as their metallic counterparts albeit at much reduced processing temperatures. The most common thermoplastic materials likely to require forming are the acrylics (Perspex) which become pliable at 120 °C and are ready for forming at about 160 °C. (*Note*: above 160 °C the material starts to decompose.) The main solid-state plastic-forming processes are *vacuum forming*, which is the term used to describe a thermoplastic sheet-shaping technique employing reduced air pressure in order to impart the final detail to a moulded article, and *blow moulding*, which is similar in principle to vacuum forming except that in this process the heated plastic material is blown onto the mould surface by air pressure instead of being sucked onto it by drawing a vacuum. Two methods of blow moulding exist: *extrusion blow moulding* and *injection moulding*.

How heat treatment influences structure and properties

The metallurgy of steel largely depends on the fact that when the temperature of iron is raised above a critical level it exists in γ form and will hold carbon in solid solution, while below this level it reverts to α form, i.e. *ferrite*, and throws this carbon out to form either *pearlite* or *cementite*. It might be thought, therefore, that at any given carbon content the same structure would always appear. However, this is not the case as by altering the rate at which the temperature is lowered from above the upper critical point, we can produce a wide variety of different structures in steel, even though the carbon content remains the same.

The reason for this apparent ability to disobey the basic rules of physical chemistry is that the system requires time to reach equilibrium (balance), and for a system to do this fully, time is required for the various constituents to diffuse throughout the mass so that the equilibrium phases can develop. Normally, as processed steel cools, the rate of diffusion of the different constituents into a system of solid phases is inevitably slow. However, it is quite possible to speed up the rate of cooling so that by the time the metal is quite cold, and therefore diffusion is complete, the system has been 'frozen' in a state of disequilibrium.

There are three principal single-stage methods of heat treatment by which the structure of carbon steel can be altered by varying the rate of cooling. They are called *annealing*, *normalising* and *quenching*, and while they are in fact just methods of cooling the steel more and more rapidly, a more detailed knowledge of these important processes and the results which they have on the structure of steel is essential.

Annealing

As has already been mentioned, one form of annealing 'process annealing' is carried out where steel or other metal is heated to a comparatively low temperature after cold work in order to cause recrystallisation. In the process annealing of steel the metal is not generally heated to a temperature above the upper critical point, as this is not generally necessary. In the other form of annealing the steel is heated to a temperature just above the upper critical point in order that all the carbon is absorbed into solid solution in α iron forming what is known as *austenite*. Care must be taken, however, not to overheat the steel at this stage or the crystals of austenite will start to grow and this unfavourable crystalline structure will be repeated in the cold metal. When the carbon has all been absorbed into austenite, the steel is allowed to cool slowly in the furnace, i.e. conditions are so arranged that the steel is brought as near as possible to equilibrium conditions.

As the metal cools through the upper critical point, the δ iron will become unstable and start to change into the α form. Let us assume that this is happening to a piece of 0.50 per cent carbon steel (refer to the equilibrium diagram illustrated in Fig. 4.16). The diagram shows that initially, ferrite is produced as the temperature slowly drops to below the upper critical point and new crystals of this carbon-free metal will start to appear around the original crystal boundaries. As more carbon-free, ferrite metal comes out it will form a series of large new crystals. Meanwhile the carbon diffuses into the centre of the remaining austenite crystals until they contain 0.85 per cent carbon. When the temperature passes through the lower critical point all the remaining austenite immediately changes to pearlite.

In annealed steel, therefore, the ferrite and pearlite are divided up into separate areas. The slow cooling ensures that the carbon is allowed time to diffuse into the austenite as the ferrite appears along the crystal boundaries, and a comparatively coarse crystal structure results. By keeping the rate of cooling slow, we ensure that the separation into ferrite and pearlite is complete. Since we want annealed steels to be soft, we ensure that the areas of soft ferrite are as large as possible and normally an annealing operation of this sort is only carried out on low carbon steels.

Normalising

Normalising is a similar operation to annealing but conducted rather faster, so that the carbon does not have time to diffuse so completely from the ferrite areas into austenite. Normalising is conducted by pulling the hot steel out of the furnace and allowing it to cool in still air. Draughts may cause it to cool too quickly.

Once again the steel is heated to just above the upper critical point and therefore completely converted into austenite. However, in this process, the rate of cooling is so fast that the carbon does not have time to diffuse through the metal, and thus initial small crystals of ferrite first appear both around the crystal boundaries and in the bodies of the original crystals.

These crystals are not able to join together and form large areas and the separation into ferrite and pearlite is incomplete. Instead of the final ferrite and pearlite areas being large and well separated in a coarse crystal structure, they are intimately mixed together in a much finer crystal structure as much of the ferrite has come out, not before, but with the pearlite.

The actual quantity of pearlite in a normalised steel is not greater than the quantity of pearlite in an annealed steel of the same carbon content, but by spreading it over the mass of soft ferrite and by refining the crystal structure, its effect on hardness and toughness of the steel is much increased.

Lower carbon steels such as bars, beams and sections are generally normalised.

Quenching

In quenching steel, the temperature is made to fall so fast that not only does the carbon fail to diffuse through the mass and form large concentrations of ferrite, but the entire precipitation of carbon is suppressed altogether and instead the metal is 'frozen' in an intermediate and slightly unstable condition in which carbon atoms are held in solid solution in a distorted body-centred-cubic (b.c.c.) lattice which, largely for this reason, is intensely hard and brittle. Under the microscope this type of steel, which is called *Martensite*, appears in the form of needles, known as an 'acicular' structure.

With normal carbon steels, a satisfactory method of quenching is to plunge the steel at a temperature just above the upper critical point into cold water in order to cause a drastic and rapid drop in temperature. However, quenched steel is so intensely hard and brittle that, in general, there are not a great many uses to which it can be put. The principal value of quenching in the treatment of steel is as the first operation in the two-stage treatment known as quenching (hardening) and tempering.

Quenching and tempering

Quenched steel, in which the carbon is held in solid solution in a severely distorted b.c.c. lattice, is in a very unstable condition and consequently a relatively low temperature is sufficient to 'loosen' the atomic structure and allow the carbon to be precipitated into a more stable condition. When quenched steel is heated to some temperature below the lower critical point, some of the carbon comes out of solution, not in the form of pearlite but in a very finely divided form known as *troostite*, or *sorbite*. This gentle reheating of the steel after quenching is known as *tempering* and its effect is to make the steel less hard but very much tougher.

Sometimes these two operations are combined in one, by cooling the steel rapidly to its tempering temperature and then holding at this temperature for some time to allow the precipitation of carbon, as this lessens the danger of cracking during the volume expansion which accompanies the δ to α change. The carbon then appears in a rather different finely divided form which is referred to as *bainite*.

Very high stressed items such as gun-barrels, armour plate, boiler drums, turbine rotors and similar products will be made of alloy steels and then given an elaborate quenching and tempering treatment. This ensures both high tensile strength and maximum toughness.

Surface-hardening treatments

The previous heat-treatment processes would all have the effect of producing a uniform structure throughout a component; however, there are obvious advantages to be gained by increasing the hardness of the surface layers only, thereby converting the surface into a wear-resistant 'case'.

The methods involved are often termed *case-hardening* processes and the oldest, in which the carbon content of the surface layers of a component are increased by heat treatment in a carbon-rich environment, is known as *carburising*. The high carbon case can subsequently be heat treated to a higher hardness level than that of the core.

Another way of altering the surface chemistry of a steel component is to diffuse nitrogen into the surface layers, forming intensely hard nitride particles in the structure. A combination of both carburising and nitriding is sometimes employed, i.e. by carbo-nitriding in a cyanide salt bath.

Other processes are based on localised rapid hardening by flame or induction heating followed by quenching and light tempering to raise the hardness of the affected zone to a level substantially above the core of unaffected material. In greater detail, a description of the processes is as follows:

Carburising

There are three methods of carburising using different physical carburising media:

- 'Box' or 'pack' carburising employs a solid medium, charcoal, plus an activator, and, as the name implies, the component to be treated is packed in charcoal inside a sealed box. The box is heated to about 900 °C for several hours during which time carbon diffuses into the surface layers. The carbon content and depth of the case depend on the time and temperature of the heating.
- Salt baths employ molten cyanide salts, the carbon diffusing from the liquid medium.
- Parts can also be carburised by soaking in controlled atmosphere furnaces in which uniform case depths are obtained by the circulation of hydrocarbon gases.

Having obtained a high carbon case on a component, it is then necessary to carry out further heat treatment in order to obtain the best combination of a hard skin on a soft, tough core. Various combinations of heat treatment are applied, all aiming to produce the maximum surface hardness together with a fine grain, and a tough and shock-resistant core.

Special low carbon (less than 0.2 per cent carbon) and alloy steels are suitable for carburising, and are made to give a wide range of core properties with easily controlled heat treatment.

Nitriding

Steels can be case hardened in a similar manner to that of carburising by using nitrogen. In this process, the parts are enclosed in a gas-tight box heated to about 500 °C and a nitrogenous atmosphere is maintained within the box during the heating, soaking and cooling cycle, by the continual introduction of ammonia, which partially dissociates to nitrogen and hydrogen. The nitrogen diffuses into the surface layers of the steel components and forms iron nitride in a fine dispersion. An extremely hard case is produced, which is superior in hardness and wear resistance to any other form of treatment. The case depths obtained in nitriding are somewhat shallower than carburised cases; however, nitrided cases will outlast carburised surfaces in some applications, i.e. high-speed gears and bushes.

Special nitriding steels have been developed which contain alloying elements such as chromium and aluminium and the alloy nitrides formed in the cases of these steels impart enhanced hardness and wear resistance and a higher fatigue resistance.

An advantage of the nitriding process is that a component can be heat treated and finished to size before nitriding, and because of the relatively low treatment temperature, distortion is minimised and many complicated shapes are ready for immediate use.

Carbo-nitriding/cyaniding

Carbo-nitriding or cyaniding in a bath of molten cyanide salts works on the principle that the breakdown of cyanide salts produces carbon monoxide in a form suitable for carburisation and, in addition, also produces nitrogen, some of which is absorbed by the surface layers of the steel. By careful control of the molten bath composition and by the addition of activating compounds, it is possible to produce cased parts in which the carburisation is supplemented by nitriding. This method is usually applied to small parts where thin cases of high hardness are required.

Flame and induction hardening

Many steels can be flame or induction hardened to produce a hard surface on a reasonably tough core. In this process, the surface is rapidly heated either by a flame or by a high-frequency induction coil and the surface is promptly quenched by water sprays. The resulting surface layer of martensite is not as hard as a carburised or nitrided case, but the method is a useful and cheap way of improving strength and wear resistance. This method is widely employed in the manufacture of gears where, because the depth of the surface zone can be substantially deeper than that obtained by case hardening, quite large components can be treated. The faces of the teeth of very large gears can be individually hardened satisfactorily, with little distortion, using relatively cheap equipment.

How joining processes affect structure and properties

Metallic materials

Bolted, screwed or riveted connections are used when a semi-permanent joint is required. However, these joining techniques do tend to set up regions of high stress where the connections are made.

Processes considered in which both metals and non-metals in the solid state are permanently joined and where local stress concentrations are avoided are:

- soldering (hard and soft)
- welding
- joining by adhesives.

In the case of metals, the condition of the parent metal at the joint itself, i.e. liquid or solid, depends upon the particular joining method selected.

Soldering (hard and soft)

Soldering involves the joining of metal by the addition of a molten filler metal of substantially different composition to, and at a temperature well below, the melting point of the metals to be joined. In this process, melting or fusion of the parent metal does not take place, instead the molten filler metal is drawn by capillary action into the space between the parts to be joined. Of the common metals, copper and mild steel solder well, whereas brass does not 'whet' so readily due to its zinc content. The particular temperature at which the filler metal or 'solder' melts determines the type of joining process. Up to a melting temperature of approximately 400 °C soft soldering is carried out, while hard soldering, which covers silver soldering and brazing, uses filler materials which melt in the range between 550 and 1,200 °C.

The basis of almost all soft solders used for joining ferrous and non-ferrous materials, other than aluminium, are the alloys of tin and lead and it is useful at this stage to consider in more detail the behaviour of tin–lead alloys during the soldering process.

Tin–lead alloys (solders)

These alloys are characterised by the fact that they all become completely solid at the common temperature of 183 °C, known as the *solidus*. The *eutectic* alloy of 63 per cent tin and 37 per cent lead changes from completely solid to completely liquid at this temperature. All other compositions have an excess of either tin or lead which remains solid within the molten eutectic as the temperature is raised above 183 °C and finally melts completely at a higher temperature known as the *liquidus*. The liquidus temperature varies depending upon the exact composition of the alloy resulting in all compositions having a 'pasty' stage between 183 °C and their particular liquidus, except the eutectic (see Fig. 4.14).

In general, solders near to the eutectic composition are used for fine electrical work and soldering operations where rapid solidifying of solder is important. Medium tin solders are used for general work and the low tin solders, which have a very wide 'pasty' range, are eminently suitable for 'wiped' joints as used in plumbing. Some lower temperature 'fusible' solders are available for special applications where ordinary soldering temperature would damage the work or any adjacent soldered joint. Another important application of low-temperature solders is in the manufacture of fuses and fusible plugs for use in boilers. It should be remembered that these solders are of lower strength and suit lower surface temperatures than do the tin–lead solders.

Soldering fluxes

In the soldering process, a flux is necessary to remove existing oxides from the work and the solder and to prevent reoxidation during heating. In addition, the flux must be capable of being displaced from the joint area by the molten solder. Fluxes range from extremely high strength or 'active' ones to those which are very mild in action. Unfortunately, in general, the more active the flux the more corrosive is its residue and although there are wide differences within each group, it is convenient to try and group most fluxes as:

- active fluxes
- non-corrosive fluxes

Active fluxes
Zinc chloride 'killed spirits' forms the active base of many proprietory fluxes. Originally made by 'killing' spirts of salts (hydrochloric acid) with an excess of zinc scrap, it is now commercially prepared by dissolving zinc chloride in water. Fluxing is thorough and rapid but the residues are very corrosive and the whole component should be thoroughly washed after soldering has taken place. Acids, in particular hydrochloric acid and phosphoric acid, also serve as active fluxes.

Non-corrosive fluxes
Resin or tallow while inactive at room temperature have a mild action when heated and constitute true non-corrosive or 'safe' fluxes.

Fluxes are frequently mixed in paste form with petroleum jelly (vaseline) or similar ingredient which does not drain off the work as readily as a liquid flux. Flux paste is also carried inside hollow solder wire to form cored solder from which both flux and solder are applied simultaneously to the work. Either safe or active fluxes may be made up in any of these forms.

Brazing

From the metallurgical point of view, brazing is similar to soft soldering. In the brazing process, higher melting point brass alloys are used, resulting in stronger joints which are capable of withstanding higher service temperatures.

Effects of welding on materials' structure and properties

The metallurgical aspects of welding are particularly interesting since, in a welded joint, examples of cast, wrought and heat-treated structures exist together. The weld deposit will possess a typical cast structure with all its inherent defects. The heat-affected zone of the parent metal will exhibit the effects of heat treatment, while unaffected regions will reveal a typical wrought structure. The effects of welding as a processing method may therefore be studied under the following headings:

- The weld-metal deposit.
- The heat-affected zone of the parent metal.

The weld-metal deposit

The weld metal is in effect a miniature casting which has cooled rapidly from an extremely high temperature. Long columnar crystals may therefore be formed giving rise to a relatively weak structure (Fig. 4.38(a)). In a multi-run weld each deposit 'normalises' the preceding run and considerable grain refinement is obtained with consequent improvement in mechanical properties. In this case, only the top run exhibits a coarse 'cast' structure and this can largely be removed after welding if necessary (Fig. 4.38(b)). The effect of welding temperature on the structure of a spot weld is shown in Figs 4.39(a) and (b). If the welding temperature is too high, the columnar crystals will meet at the centre forming a plain of weakness which may lead to intercrystalline cracking (Fig. 4.39(b)) whereas if the temperature is correct equi-axed grains will form at the centre before the columnar crystals can meet (Fig. 4.39(a)). The importance of correct control of current and time in spot welding is therefore apparent. Other possible defects in the structure of the weld metal include non-metallic inclusions, gas porosity and cracking.

Non-metallic inclusions
The formation of oxide and nitride inclusions due to atmospheric contamination is usually avoided by the use of a flux. Modern flux-coated electrodes usually provide good quality weld deposits substantially free from harmful inclusions. In the argon-arc welding process, the metal is deposited under a shroud of the inert gas argon, which prevents oxidation and no flux is necessary. Slag inclusions can be avoided in multi-run welds by effective removal of the slag after each deposit.

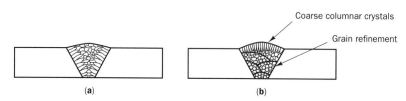

Figure 4.38 Diagrammatic representation of structure of weld metal and in single (a) and multi-run weld (b) deposits

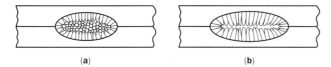

Figure 4.39 Structure of spot welds: (a) welding temperature correct; (b) welding temperature too high

Gas porosity
The chief cause of gas porosity is the presence of hydrogen in the weld metal or the formation of steam from the reaction of hydrogen with any oxide present in the melted parent metal. There are numerous sources of hydrogen in welding, the chief ones being the welding flame in gas welding or the flux coating in metallic arc welding.

Weld-metal cracking
Welded joints which are prepared under restraint are liable to intercrystalline cracking in the weld deposit due to contractional strains set up during the cooling of the metal. Such cracking, usually known as 'hot cracking', is largely related to the grain size and the presence of grain boundary impurities. At high temperatures, the grain boundaries are more able to accommodate shrinkage strains than the grains themselves. A coarse grain deposit with large columnar crystals possesses a relatively small grain boundary area and is therefore more susceptible to hot cracking.

Heat-affected zone of the parent metal

It is not easy to generalise when considering the effect of welding heat on the structure and properties of the parent metal. The extent of any structural change will depend upon the time at temperature and consequently such factors as the thermal conductivity, specific heat and dimensions of the plate, together with the speed and method of welding are important. Welded joints prepared from metals possessing high thermal conductivity, such as copper and aluminium, possess wider heat-affected zones than those prepared from nickel or steel. Metallic arc welding produces a more concentrated heating effect than gas welding. An increase in the welding speed also reduces the width of the heat-affected zone.

The heat-affected zone in mild steel plate can exhibit various structures ranging from an overheated structure for those parts heated to well above the upper critical range, to an under-annealed structure for those parts heated to within the critical range. There is a marked increase in ductility and elasticity in the heat-affected zones of steels while, in contrast to most non-ferrous alloys, an increase in hardness will occur under these conditions. The degree of hardening increases with increasing carbon content of the steel. In general, with non-ferrous metals and alloys it would be expected that softening of work-hardened or age-hardened metal would occur in the heat-affected zone. In addition, grain growth usually occurs in non-ferrous alloys but this is usually insufficient to have any great effect on the strength of the welded joint. The structure of the heat-affected zone of a typical weld preparation in hard-rolled aluminium sheet is shown in Fig. 4.40.

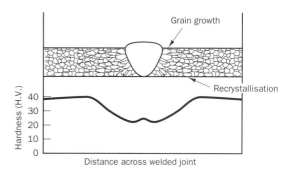

Figure 4.40 Effect of welding on the structure and hardness of hard-rolled aluminium sheet

Welding (non-metallic)
Welding can successfully be used to join thermoplastic-type material. The technique of welding sheet plastic is similar to that of welding steel by the oxyacetylene method insofar as a heat source and filler rod are used. However, the low thermal conductivity and softening temperature of thermoplastics mean that much reduced welding temperatures are needed. Unlike metals, which have a sharply defined melting point, thermoplastics usually have a wider range of temperature between which they soften and eventually degrade. During welding, some degradation inevitably occurs, resulting in the strength of the joint being slightly below that of the surrounding material. Unlike metal welding, where the metal filler rod is melted, the plastic filler rod is only softened and the resulting bead left standing above the surface of the sheet as reinforcement can increase the joint strength by as much as 20 per cent.

Adhesives

An adhesive is a material that sticks (adheres) two similar or dissimilar materials together. The bonding material is called the *adhesive*, the materials being stuck together the *adherends*, and the area of the material adhered *the joint*. The strength of an adhesive bond depends upon two factors:

- *Adhesion* – This is the ability of the adhesive to stick to the adherend faces. It can occur in two ways:
  - Mechanical bonding, which occurs when the adhesive penetrates the pores of the adherends and produces a physical cementing.
  - Specific bonding, which occurs when the adhesive reacts chemically with the adherends bonding them through intermolecular attraction.
- *Cohesion* – This is the ability of the adhesive to withstand forces within itself without failure.

Surface preparation

A bonded joint is of little or no use if the joint surfaces are not properly prepared before bonding is carried out. The best joints are made when the surfaces are absolutely clean and have good affinity for the adhesive (whetting capacity). Surfaces are prepared by one of the following pretreatment processes (listed in order of effectiveness):

- Degrease only.
- Degrease, abrade and remove loose particles.
- Degrease and chemically pretreat.

Care must be taken to avoid contaminating the surfaces during or after pretreatment. Whatever pretreatment process is used, it is good practice to bond the surfaces as soon as possible after completion of the pretreatment.

The design of bonded joints

No matter how effective the adhesive is and how carefully it is applied, the joint itself will be a failure if it is not correctly designed and executed. Bonded joints may be subjected to tensile, compressive, shear or peel stresses often in combination. They perform relatively poorly under peel and cleavage loading while they are strongest in compression, tension and shear.

A bonded joint needs to be designed so that the loading stresses are uniformly distributed over the full joint areas and not concentrated into a single region of high stress. Thus; for maximum strength, cleavage and peel stresses should be designed out of joints wherever possible. Further efficiency and joints of greater strength result from close consideration of joint design (Fig. 4.41).

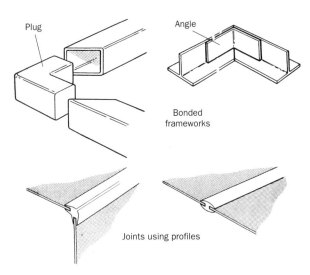

Figure 4.41 Joint construction and reinforcement

Powder metallurgy

This is the study of processing methods which produce solid metal articles from metal in powder form. The process joins the particles of powder together more or less mechanically although it is recognised that some superficial welding possibly takes place. The compact or 'biscuit' as it is sometimes called has very little strength and a much lower density than that of the finished article due to the presence of numerous cavities within its structure. Sintering is carried out by heating the compact in a furnace to a temperature near to but below the melting point of the metal in an inert gas (oxygen-free) atmosphere. The grains of powder diffuse together so that the cavities disappear and the density rises.

Like pure metals, alloys can also be successfully sintered as long as the constituent metal powders are thoroughly mixed prior to pressing and sintering. On close examination, after sintering has been carried out, the metals are found to have combined so thoroughly that instead of a mixture of individual grains of the separate metals, grains of a homogeneous alloy appear. This is one of the reasons why the metal powders must be in an extremely fine state.

An interesting example of the powder metallurgy process is the manufacture of self-lubricating bearings in which an alloy of copper and tin is pressed and sintered but the compact is intentionally left porous so that oil can be forced into the cavities or 'pores'. Sufficient oil remains in the bearing to enable it to run a lifetime without recharging.

Hot pressing and sintering can be carried out simultaneously, as in the manufacture of high-speed tool tips from a mixture of cobalt and tungsten carbide. In this process only the lower melting point metal melts so that it cements together the grains of the higher melting point metal. Hot pressing is, however, more expensive than cold pressing due to the difficulty of heating the dies and the resulting shorter die life.

Despite the initial cost of producing metals in powder form, powder metallurgy has found numerous applications where the more traditional processing methods can no longer be easily or economically used (see BMW Case Study in Element 4.3).

Self-assessment task

List TEN common engineering materials and, considering a range of processing methods, describe the characteristics which make each chosen material suitable or unsuitable for the processes selected.

4.3 Select materials for engineered products

Topics covered in this element are:

- The identification of criteria for material selection to meet design specifications for engineered products.
- The use of data sources to determine material selection to meet design specifications for engineered products.
- The selection of materials for engineered products which meet given design specifications.

Criteria for selection of materials for engineered products

The selection of materials from which engineered products are to be made involves a careful consideration of the following criteria:

- property requirements
- processing requirements
- economic requirements

Ultimately, the final choice will involve a compromise.

Property requirements

In order for a component to be successful in service it must possess suitable properties, i.e. weight, strength, hardness, toughness, elasticity, rigidity etc. In addition, certain other properties such as electrical, magnetic or thermal properties and fatigue or creep resistance may be required.

Corrosion resistance must usually be as high as possible or, alternatively, the material must respond to corrosion-resistance treatment.

Processing requirements

Materials, in particular metals, can be classified as either casting materials or wrought materials and the suitability with which each can be cast or worked can be very important. The ease with which a material can be machined and the quality of the finish obtained are usually very important, as is its ability to be joined by welding, brazing or soldering. If heat-treatment processes are a requirement of its manufacturing process, the metal selected must respond to such a treatment.

In addition, it is necessary to consider both the duty and the shape of the component and the quantity required in order to select a suitable material, with the required processing characteristics whether it be metallic or non-metallic.

Economic requirements

Economic considerations are important and should take into account the total cost which comprises the cost of the raw material plus the cost of its manipulation, machining, joining and finishing. The overall costs of a number of alternative materials, both metallic and non-metallic, should be compared when production rates are to be met. The availability of a particular material is important (Fig. 4.42) and will influence its cost. (*Note*: a material is not necessarily inexpensive because the raw material from which it is derived is plentiful, as the material cost is also associated with the method used to extract it and to prepare it for use.) Adverse geographical conditions or political unrest in a country will also affect raw material costs.

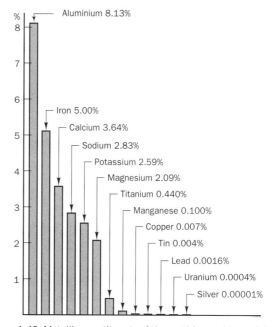

Figure 4.42 Metallic constituents of the earth's crust by weight

---

Self-assessment task

Selection of a material for a particular component is controlled by three factors. State these factors and use a simple example to illustrate the importance of considering all three at the design stage.

The failure of materials in service

It is essential that engineered products are designed in such a way that any stresses which are encountered in service are insufficient to cause failure. Consequently, component design based on material yield strength figures is clearly unsatisfactory, necessitating the application of a suitable factor of safety in order to arrive at a maximum allowable stress.

Unfortunately however, despite such measures, many components still fail in service and designers now recognise that operating conditions which produce brittleness, fatigue, creep or environmental attack all influence design and if ignored will ultimately lead to failure.

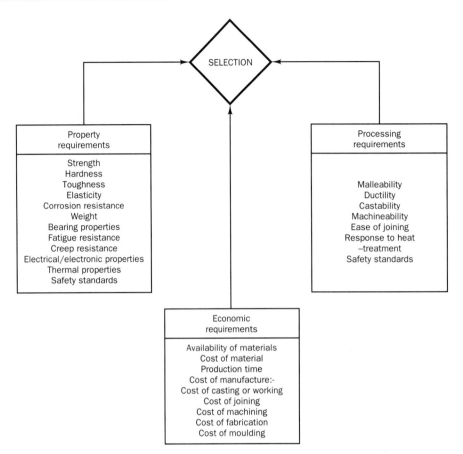

Figure 4.43 Selection of material

Fracture

Fracture can be classified as either 'brittle' or 'ductile' and depends on the stress at which it occurs in relation to the elastic/plastic properties of the material (Fig. 4.44).

In brittle fracture, failure occurs before any appreciable plastic deformation takes place and follows the paths between adjacent crystal planes. These are generally materials with the weakest atomic bonding, typically cast iron, glass and concrete, and all metals with body-centred cubic and close-packed hexagonal structures. Brittle fractures often start at some defect in the material or component which acts as an area of stress concentration, although this is not always the case; corrosion, for example, can also lead to 'stress raisers' in the form of pits or pockets in the metal.

Ductile fracture takes place at some stress figure above the yield point of the material so that some plastic flow precedes failure. The resultant fracture is of the 'cup and cone' variety and occurs in face-centred cubic metals like mild steel.

The type of fracture produced is dependent mainly on the nature of the material and in particular on its lattice structure. It is, however, affected by other factors such as: (i) the rate of application of stress, (ii) its environmental and temperature conditions, (iii) the amount of previous cold work the material may have received and (iv) the quality of its surface finish.

Fatigue

Failure of a component due to fatigue can be seen to be a particular case of brittle failure. Fatigue failure occurs under the actions of changing stresses, i.e. repeated or fluctuating, which are frequently below the yield strength of the material.

Aircraft wings and rotating shafts are examples of members subjected to alternating stresses, and where similar stresses are to be encountered careful consideration at the design stage and during subsequent maintenance is extremely important if fatigue failure is to be avoided.

The nature of a fatigue fracture is generally easy to identify as the crack-growth region is burnished by the mating surfaces

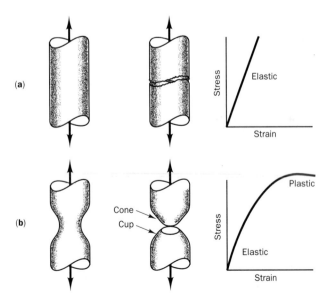

Figure 4.44 Types of fracture: (a) brittle; (b) ductile

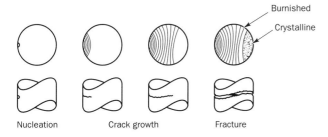

Figure 4.45 The stages of fatigue failure

rubbing together as the stresses alternate. The ultimate fracture is crystalline (Fig. 4.45).

The German engineer Wohler carried out a great deal of investigation into the phenomenon of fatigue and developed the useful fatigue-testing machine which bears his name.

Creep

Like fatigue failure, failure due to creep occurs below the tensile strength of the material. However, whereas fatigue failure results from the application of fluctuating stresses, creep failure occurs due to constantly applied forces resulting in the slow extension of the material. Elevated operating temperatures for example in steam turbines, boilers, etc., increase creep rate, as movement or 'slip' in the material structure becomes easier. Lead, with its pronounced plastic range and low melting point is a good example of a material which offers poor resistance to creep, as are thermoplastic type polymer materials.

Materials which offer improved creep resistance are those possessing closed-packed crystal structures (face-centred cubic and close-packed hexagonal) and their resistance to creep can be improved still further by the addition of suitable alloying elements, e.g. nickel in Nimonic alloy steels. Here the finely dispersed particles form effective barriers against creep. Additionally, suitable heat treatment carried out in order to increase grain size in a material improves creep resistance properties.

Corrosion

Due to the costly destructive effect of corrosion, a study of its causes and methods of prevention are of vital importance to the materials engineer. The rate of corrosion is affected by:

- Factors relating to the metal, viz.
  - the position of the metal in the electrochemical series (Table 4.11)
  - contact with dissimilar metals
  - the presence of impurities or a second constituent
  - the presence of internal stress.
- Factors relating to the environment, viz.
  - relative humidity
  - the presence of impurities in the atmosphere
  - rate of supply and distribution of oxygen
  - the acidity or alkalinity of any liquid present
  - the rate of flow of any liquid present
  - the presence of external stress.

When it is appreciated that normally more than one of the above occur simultaneously, the complex nature of the corrosion problem can begin to be appreciated.

In general, two principal forms of corrosion can be distinguished:

- direct chemical corrosion
- electrochemical corrosion

Direct chemical corrosion

This covers corrosion where there is direct chemical reaction between the metal and dry gas (oxygen) usually at high temperature. The nature of the oxide film produced has an important effect on the extent of subsequent corrosion. If the film is hard and adherent, as in the case of heat resisting nickel–chromium alloys, then the metal will be protected and corrosion will eventually cease. On the other hand, the corrosion of iron and steel under the same conditions is progressive due to the formation of a loose, black oxide film called *millscale*.

Electrochemical corrosion

This covers all forms of corrosion which occur where the metal is in contact with a liquid or moist atmosphere.

Electrochemical theory assumes that all metals have a tendency to dissolve or corrode. The corrosion resistance of metals is determined by their position in the electrochemical series in which metals are arranged according to their electrode potentials (Table 4.11). The value of these potentials is very small and they are expressed relative to hydrogen which is taken as zero. The greater the negative potential, the greater is the tendency of the metal to dissolve or corrode. The tendency of an individual metal to corrode is relatively small; however, it is greatly increased when it is brought into contact with a dissimilar metal in the presence of a conducting liquid (*electrolyte*). Since the metals are at different potentials, a current will flow between the two, resulting in corrosion of the metal which is higher in the table (*anode*) while the metal which is lower in the table (*cathode*) is protected. One factor which governs the rate of corrosion is the relative areas of the anode and cathode.

Table 4.11 The electrochemical series of metals

Metal	Electrode potential (V)
Sodium	−2.71
Magnesium	−2.40
Aluminium	−1.70
Zinc	−0.76
Chromium	−0.56
Iron	−0.44
Cadmium	−0.40
Nickel	−0.23
Tin	−0.14
Lead	−0.12
Hydrogen	0.00
Copper	+0.35
Silver	+0.80
Platinum	+1.20
Gold	+1.50

The effect of design on corrosion

In order to keep corrosion to a minimum the importance of good design cannot be overemphasised. Direct contact with the corrosive medium can be reduced by considering requirements like drainage and ventilation at the design stage. For example, in order to prevent the build-up of moisture at pockets, crevices and joints in fabricated components, butt-welded joints may be preferred to bolted brackets or riveted joints.

Ultimately the best solution is to design out of the product wherever possible any areas which may be vulnerable to attack.

Protection of metals and alloys from corrosion

Although alloys like stainless steel, monel and inconel have been produced to resist corrosion, their use cannot be economically justified except for special applications. It is therefore necessary to adopt other methods to protect the cheaper metals and alloys from corrosion. The methods used may be broadly classified as follows:

- *Metallic coatings* – These coatings may be applied by (a) dipping, (b) electro-deposition, (c) cladding, (d) spraying and (e) cementation methods, e.g. sherardising, calorising.
- *Non-metallic coatings* – These include oxide, ceramic and phosphate films and protection by various paints, varnishes and lacquers.

British Standard Specifications

The British Standards Institution is the approved body for the preparation and issue of the British National Standards. It issues standards which cover minimum standards of quality and performance for a multiplicity of materials, products and procedures. British Standards are identified by the prefix 'BS' followed by the reference number of the specification and the date of issue. A selection of British Standards for materials and test procedures is listed below, while several others have been referred to previously in the text of this unit.

Aluminium alloys	
Wrought	BS 1470–1475
Cast	BS 1490
Brass	
Muntz metal	
Hot rolled	BS 1541
Rods and sections	BS 1949
Die cast	BS 1400–B4–C
Cartridge brass	BS 267, 378, 885
Basis brass	BS 265
Naval brass	BS 251 and 252
Die cast	BS 1400–B5–C
Bronze	
Admiralty gunmetal	BS 1400–G1–C
Leaded gunmetal	BS 1400–LG2–C
Phosphor bronze	BS 1400–PB3–C
Cast irons	
Grey iron castings	BS 1452
Whiteheart malleable iron castings	BS 309
Blackheart malleable iron castings	BS 310
Pearlitic malleable iron castings	BS 3333
Spheroidal or nodular graphite	BS 2789
Austenitic cast iron	BS 3468
Nickel silvers: Wrought	BS 1824
Steel	
Wrought	BS 970
Cast	BS 3100
Magnesium alloys	
Wrought	BS 3370, 3372, 3373
Cast	BS 2970
Zinc and zinc alloys:	
For die casting	BS 1004
Handbook: Mechanical tests for metals	BS Handbook No. 13
Tensile testing of metals	BS 18
Brinell hardness testing	BS 240
Diamond pyramid hardness numbers	BS 427
Direct reading hardness testing:	
Rockwell Principle	BS 891
Impact test: test pieces	BS 131
Ductility tests: simple bend	BS 1639
Creep tests	BS 3500
Fatigue tests	BS 3518

The BSI represents the United Kingdom in the International Standards Organisation (ISO) and, apart from being actively engaged in the preparation of international standards, it is at the same time ensuring compatibility where this is possible between international and British standards.

> **Self-assessment task**
>
> Visit your college/public library and confirm the reference numbers of SIX BS specifications referred to in this unit. State also the latest date of issue of each specification.

Case studies

The following case studies cover in some detail the design implications and material selection criteria of this section and it is anticipated that they will prove to be informative and helpful to the student during assignment work on material selection based on a specification. A selection of student-based assignments are included at the end of this element.

Material selection in the civil engineering industry: spanning the Firth of Forth

The Forth Bridge was a necessity. No effective route along the east coast of Scotland was possible until those two great parallel indentations of the coastline, the Firth of Forth and the Firth of Tay had been bridged. The Tay may appear to have been the greater obstacle because of its width, but this was not the case since the Firth of Tay is comparatively shallow.

It was possible to sink foundations at regular intervals across the whole width of the Firth of Tay and to build an ordinary girder viaduct. This was of exceptional length certainly, but presented no particular difficulty from the point of view of design or construction. Not so the Forth. Immediately north of Edinburgh this waterway has a depth of 50 metres and more and whatever the type of bridge thrown across its waters it was impossible to avoid spans of unprecedented length. On the site chosen for the Forth Bridge, a little north-east of Edinburgh, the Firth of Forth narrowed somewhat though it still had a width of 1,600 metres (over 1 mile). Here there was also a valuable asset in the mainstream islet of Inchgarvie to serve as a foundation for the centre and most massive cantilever of the bridge. But to the north and south of the island the sea bed shelved rapidly, forming channels from 30 to 50 metres in depth and from 500 to 525 metres wide. Over these two waterways, therefore, the bridge had to stride without any intermediate supports for it was almost impossible at that time to sink foundations for other bridge piers through such a depth of water.

The total length from the beginning of the first cantilever to the end of the third is 1,600 metres. There are also fifteen high-approach viaducts which lead to the bridge proper and a number of smaller masonry arches at the two extreme ends. The sum of these spans makes up a total length of more than 2,500 metres (1.5 miles).

With a vast structure of this type, the effect of expansion, particularly in the spring and autumn between a cold night and a warm day are considerable. The general allowance for expansion all over the bridge is 1 in 1,200. Certain joints in this immense latticework of steel act as hinges to take up the expansion and sliding joints in the rails allowed for the necessary longitudinal movement of the track without breaking the continuity of the running surfaces. The total expansion allowance in the entire length of the three cantilevers is 2.5 metres. The bedplates which transmit the weight of the cantilevers to the three clusters of stone piers are provided with elliptical holes for holding-down bolts to allow for the slight movement that may be necessary, and this is facilitated by the highly polished and greased condition to which the bedplates are maintained.

Another matter of considerable importance is the resistance of the entire structure to the effects of corrosion. At the time when the Forth Bridge was built, no modern corrosion-resisting steels such as those containing small percentages of chromium were available on a commercial scale. Ordinary mild steel was therefore used for all steel plates and sections, though special care was taken to ensure that only steel of the highest quality, which had been subject to tests and inspection considerably more rigorous than those normally devoted to structural steel, were used in the fabrication.

But exposure to salt-laden air hastens oxidation of other than special alloy steels and they can be protected only by suitable and regular painting. To cover the Forth Bridge with a coat of paint takes a gang of some 45 men three years, requires approximately 54 tonnes of paint and necessitates the thorough scraping and painting of some 135 acres of steel work.

Thousands of cubic metres of masonry and cement, together with over 54,000 tonnes of steel and 6.5 million rivets were used in the building of the bridge. The last rivet was driven by King Edward VII, then Prince of Wales, at the official opening on 4 March 1890, seven years after work had begun. The total amount of money spent on the construction of the Forth Bridge was about £2.5 million and about £500,000 was spent on the railway approaches. Even to this day, the Forth Bridge remains one of the Wonders of Engineering.

Lessons of a disaster

The Firth of Tay at Dundee is wider than the Firth of Forth at Queensferry, yet it was the Tay which was first spanned by a giant bridge. The first ill-fated bridge was opened on 31 May 1878. In due course, Queen Victoria travelled over it and knighted its designer. The bridge was undeniably one of the wonders of the year and excited a great deal of admiration. However, when old pictures of it are studied one cannot help observing that the whole superstructure looked strangely slender for so great a work and that the visible changeover from the brick piers to the light cast-iron columns after the fifteenth pier was not reassuring. The bridge was still young when a number of weaknesses became only too apparent. After the iron columns had been erected they were filled with Portland cement concrete. As this set it swelled, causing certain of the columns to split and crack. These cracks were remedied with wrought-iron bands by the engineers in charge of the maintenance of the bridge. The cross-bracing between the cast-iron columns also tended to slacken and there is no doubt that it was insufficient for the strains which it had to undergo from lateral wind pressure on the superstructure.

The remainder of the bridge's history is brief and tragic and on the evening of Sunday, 29 December 1879, during a fierce storm, gale-force winds struck. The high girder spans and a train passing over the bridge offered a solid broadside resistance to the gale. The tilting girders snapped the frail cast-iron columns and the train and 73 passengers were hurled to their deaths into the waters of the Tay.

The terrible accident produced a number of valuable lessons. The official findings that the bridge was badly designed, badly constructed and badly maintained, sounded damning enough but they showed that not nearly enough allowance had been made for the greatest strain which a bridge can be called upon to bear – that of lateral wind pressure. After this nobody was likely to use cast-iron columns filled with Portland cement again for the support of a giant viaduct.

Self-assessment tasks

1. With regard to the use of materials, contrast the success and failure of the first bridges to span the Firth of Forth and the Firth of Tay.
2. Nowadays, significant changes have been made in the choice of materials and the way in which they are used to build bridge structures. Describe the changes you would make in material selection giving your reasons for making these changes.

Material selection in the aviation industry (airframes)

The design and manufacture of any structure as complicated as an aircraft involves a series of compromises. For example, the need for great size, which may involve a heavy structure, may conflict with an equally pressing need for lightness in the interests of economical operation. Alternatively, the need for great range, which will require allocating a lot a space and weight-carrying capacity for fuel, may shrink the payload beyond acceptable limits. These problems were as difficult to solve in the early days of aviation as they are today. In addition, the pioneers were greatly restricted by the materials at their disposal and by the techniques available for cutting, joining and forming them.

In the very first airframes, wood was selected as the primary structural material. This was braced with wire and covered with fabric. Fabric-covered wooden structures were not, however, the way ahead and, during World War I, Junkers employed wings made of a series of tubes running spanwise, braced internally with metal and covered with a corrugated aluminium skin which proved to be much more durable than other aircraft of this era. The Dutch engineer Anthony Fokker (1890–1939) was another leading designer of the period who built military aircraft for the Germans in World War I. Fokker, however, favoured a fuselage made of welded steel tubes covered with fabric and wings made of wood but with a load-bearing plywood skin.

In spite of Fokker's success with wooden frames and skins, metal was eventually to become the dominating material in the construction of airframes. It is almost impossible to determine who first used light aluminium alloys for this purpose. In any case, it is clear that quite a number of designers envisaged the use of aluminium alloy from aviation's very infancy (as early as 1910). However, in 1926, William B. Stout designed the all-metal Ford Tri-Motor employing a metal structure with a corrugated skin based on Junkers' philosophy.

At the beginning of World War II most front-line fighter aircraft, including the Hurricane, Spitfire and Short Sterling heavy bomber, were still not of true 'stressed-skin' construction since all had areas which were fabric covered. The end of the war, however, saw the introduction of a new generation of all-metal aircraft using light alloy materials. There were, of course, exceptions to the general rule, notably the De Haviland Mosquito made of wood and the Vickers Wellington made in a lattice of light alloy covered with fabric.

The first jet aircraft which flew in the late 1940s relied entirely on load-bearing, light alloy fuselage design; however, as speeds approaching the speed of sound were reached in the 1950s, evidence of fatigue failure meant that certain alloys were no longer used in load-bearing structures.

More recently, supersonic flight posed the problem of kinetic heating due to friction. One of the reasons why Mach 2 (twice the speed of sound) was chosen as the cruise speed for Concorde was that at higher speeds the level of kinetic heating, which varies considerably from one part of the airframe to another, would have ruled out the use of structures made primarily of light aluminium alloy and this would have added greatly to the cost of the aircraft. On Concorde, for instance, temperatures at cruise range from about 160 °C at the nose to just over 110 °C at the tail, and the designers of the aircraft had to make provision to allow for the safe expansion and contraction of the structural components at different rates and by different amounts. The Mach 3 supersonic transport (SST) planned by the Americans but abandoned in 1971 would have used titanium and stainless steel in many critical areas. Although light alloys are still found in many different types of aircraft, designers are making ever-increasing use of new plastics and fibre-reinforced composite materials in airframes. Composites based on carbon or graphite fibre have nearly two-thirds the tensile strength of steel at about one-sixth the weight, and three times the strength of aluminium alloy at half the weight. Their properties are, however, highly directional and their strength in compression and at right angles to the orientation of the fibres is low. This means that they are very suitable for structures subjected to tension in only one direction, such as rotors and spars, but that they have to be built up from laminations rather like plywood if loads are likely to vary in magnitude and direction.

Lessons of a disaster

The De Haviland Comet which first flew in July 1949 was significant as the world's first jet-powered passenger aircraft. Unfortunately, the early years of the Comet, which might have given Britain a decisive lead in jet-powered transport, were blighted by two mysterious disasters in 1953 and 1954 in which the planes seem to have blown up in mid-air. At least one of these disasters, it was later discovered, was due to explosive disintegration of the fuselage. Like other modern airliners the Comet had a pressurised cabin and it emerged that, after a period of continuous service, the repeated application of pressurisation loads had caused metal fatigue in the corner of a small but poorly designed aperture at one point in the structure. From this point failure spread quickly throughout the fuselage structure.

The Comet disasters led to some concentrated research into metal fatigue and the findings taught aircraft designers some important lessons. Certain alloys, for instance, are now no longer used in load-bearing structures, the permitted overall stress levels have been reduced and much greater attention is now given to avoiding design details that could lead to local concentrations of stress. Fatigue, of course, cannot be entirely eliminated, so all aircraft structures are now designed to one of two fatigue categories, 'fail-safe' and 'safe-life'. A fail-safe structure will not fail even if a crack develops in one part of it. In a safe-life structure, the designer specifies a fixed life for each component after which it must be replaced even though it may show little or no evidence of fatigue.

Self-assessment tasks

1. Give TWO examples of: (a) wood and fabric covered aircraft; (b) aluminium alloy and fabric covered aircraft; and (c) aluminium alloy and alloy skin aircraft.
2. Give the year of manufacture and the number of years in service of each named aircraft. (*Note*: You will need to visit the college/public library in order to answer this question.)
3. Write an account tracing the use of materials in aircraft from the early days of powered flight to the present supersonic age. Identify the reasons behind the changes made in material selection in the aircraft industry.

Material selection in the motor industry; the ultimate driving machine.

The conrods in an eight cylinder BMW engine can move up and down 108 times a second. They accelerate with the force of a bullet leaving a gun, then change direction, bearing two and a half tons of weight as they do so.

With such massive forces applied to them, any imperfections of assembly or manufacture will be greatly exaggerated. This, in turn, will make the engine run less smoothly. So BMW have developed a process by which the conrods are made stronger, lighter and exceptionally well balanced.

The conrods are made from powdered steel sintered under high pressure at 1,100 °C. A process that achieves an unprecedented ±0.2 per cent tolerance of target weight. They are then literally snapped along a predetermined fault line by a purpose-built machine.

Consequently when the two halves are reassembled around the crankshaft, the result is always a perfect fit. So the engine runs more smoothly. Broken parts and all.

4.4 Unit test

Test yourself on this unit by answering the following multiple-choice questions.

1. Tensile testing of materials is carried out in order to determine the properties of:
   (a) ductility and malleability
   (b) malleability and elasticity
   (c) ductility and strength
   (d) fatigue and creep

2. Four different materials tested to destruction gave the resultant load–extension graphs shown in Fig. 4.T1. Which material was the most brittle?
   (a) i
   (b) ii
   (c) iii
   (d) iv

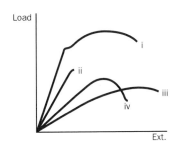

Figure 4.T1

3. Which of the materials in Fig. 4.T1 was the most ductile?
   (a) i
   (b) ii
   (c) iii
   (d) iv

4. Which of the following tests is a non-destructive test?
   (a) tensile
   (b) X-ray
   (c) Izod
   (d) Charpy

5. Which of the following metals has the greatest tensile strength?
   (a) mild steel
   (b) bronze
   (c) duralumin
   (d) cast iron

6. Which of the following metals has the highest melting point?
   (a) tin
   (b) zinc
   (c) magnesium
   (d) aluminium

7. The percentage carbon content of mild steel is from
   (a) 0.1–0.3%
   (b) 0.3–0.6%
   (c) 0.7–1.7%
   (d) 2.5–3.5%

8. Indicate the correct heat-treatment procedure carried out to prepare a cold chisel for use:
   (a) hardening and tempering
   (b) hardening and quenching
   (c) annealing and hardening
   (d) tampering and annealing

9. Surface scaling during the hot-rolling of steel is caused by a gas combining with the iron in the steel. This gas is:
   (a) nitrogen
   (b) carbon dioxide
   (c) hydrogen
   (d) oxygen

10. Directional properties will be introduced into a component which has been:
    (a) cold rolled
    (b) hot forged
    (c) sand cast
    (d) die cast

11. When a molten metal cools, solidification commences at:
    (a) the grain boundary
    (b) the crystal
    (c) the dendrite
    (d) the nucleus

12. Bronze is an alloy of
    (a) tin and zinc
    (b) tin and copper
    (c) copper and zinc
    (d) copper and lead

13. When alloyed with steel, 13 per cent of one of the following metals gives stainless steel. Which one is it?
    (a) cobalt
    (b) chromium
    (c) manganese
    (d) molybdenum

14. The best material suitable for use as a permanent magnet alloy is:
    (a) carbon steel (hardened)
    (b) permalloy
    (c) mumetal
    (d) columax

15. The failure of materials in service which are subjected to repeated stress over a long period of time is due to:

    (a) creep
    (b) fatigue
    (c) temperature
    (d) thermal conductivity

16. Which of the following is a composite material?

    (a) aluminium alloy
    (b) steel reinforcing rods
    (c) reinforced concrete
    (d) polyethylene

17. Which material is used in order to vulcanise natural rubber?

    (a) carbon
    (b) sulphur
    (c) carbon dioxide
    (d) hydrogen

18. The process of polymerisation involves:

    (a) the return of the material to its original condition
    (b) breaking down the structure of the material
    (c) building up long chains of molecules
    (d) injecting the hot plastic into the mould

19. Which of the following materials is a thermosetting plastic?

    (a) polystyrene
    (b) PVC
    (c) nylon
    (d) bakelite

20. The approved body for the preparation and issue of British National Standards is:

    (a) ISO
    (b) BSI
    (c) DIN
    (d) BSF

Unit 5: Design Development

The fundamentals of the design process enable products or processes to be successfully developed by exploring such themes as: problem identification, research investigation, manufacturing processes and testing. A design has to encompass all of these themes either independently or in various combinations with an evaluation as to its overall suitability to serve today's society. The process is a complex affair and utilises design brief, design solutions and technical drawings.

In this unit the student will learn how to:

- Produce design briefs for an engineered product and an engineering service.
- Produce and evaluate design solutions for an electromechanical engineered product or an engineering service.
- Use technical drawings to communicate designs for engineered products and engineering services.

After reading this unit the student will be able to:

- Identify customer requirements and prioritise essential elements.
- Identify the relevant standards and legislation as applied to the product or process.
- Identify operational or technical constraints.
- Produce design briefs for given products or services.
- Identify requirements of products and processes from design briefs.
- Obtain technical information relevant to the design brief.
- Generate feasible design solutions.
- Evaluate design solutions and select a final design solution.
- Communicate through graphical methods incorporating standards and conventions a final design for a product or process.
- Produce parts or component lists.

5.1 Produce design briefs for an engineered product or an engineering service

Topics covered in this element are:

- The identification and prioratisation of customer requirements.
- The identification of standards and legislation.
- Identifying technical and operational constraints.
- The production of design briefs for specific engineered products or services.

As language is an accepted form of communication it is important that an established form of communicating technical information is also maintained. This language must have rigorous standards which are acceptable to designers, craftsmen, customers, technicians and salesmen alike. It must:

- Be easy to understand.
- Be unambiguous so that it can only be interpreted one way.
- Have enough detail that further information is not required.
- Be unrepetitive.

Identifying customer requirements

This is the crucial aspect of the design process, for it sets the scene for the project as a whole. If there is any misunderstanding between the designer's interpretation and the customer's requirements at this stage, there will undoubtedly be major problems (Fig. 5.1).

Identifying the customers needs will vary from customer to customer; however, a technical specification may be drawn by clarifying the task.

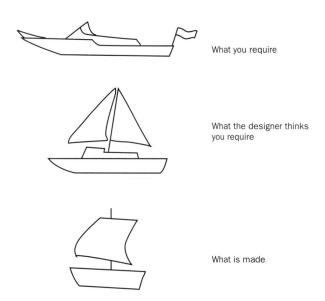

Figure 5.1 Identifying customer requirements

It is important to evaluate what are demands and what are wishes.

- Demands are requirements that have to be met (i.e. max length 50 mm, min width 25 mm, max weight 1.2 kg, etc).
- Wishes are the requirements that the customer would like to have, and should be considered whenever possible (e.g. it would be nice if the colour could be blue but I will accept red).

Try to classify, in a rank order of importance, the wishes listed in Table 5.1. In order to arrive at a technical specification, there are many aspects to consider before arriving at an acceptable solution. (The examples listed in Table 5.1 may form the basis of a check-list.)

Table 5.1 Customer requirements

Criterion	Requirements (needs and wishes)
Physical geometry	size, width, height, length, weight
Production	methods, quality, number, tolerances
Ergonomics	type of operation, lighting, shape, operating height, layout
Operation	wear, marketing, quietness, working environment
Maintenance	servicing, inspection, repair, cleaning, painting
Safety	protection systems, operational and environmental safety
Energy	output, efficiency, ventilation, pressure, temperature, storage, capacity, supply
Signals	input, output, display, control

Self-assessment task

Using Table 5.1, compile TEN questions in the form of your own check-list you would ask someone about designing a security system for them. This is an exercise normally carried out in order that it may offer the opportunity to turn a design brief into a technical specification.

Ergonomics

This term is used to describe how human aspects have to be considered by the designer. It needs to involve things like ease of use, comfort, positioning of dials, handles and seats. The human form differs a great deal and every configuration has to be considered. The following shows some of these aspects:

- What are the size of hands or fingers (for control knobs)?
- The colour and shape of parts. (Can it be seen easily?)
- The type of noise produced. (Can it be heard?, is it annoying?)
- How does it feel? (Is it too smooth to grip?)

An important element in the ergonomics of design is that with the presence of a human interacting with the product, feedback loops are required to provide information for control purposes.

For example, visual displays remain the main method of initiating feedback and are found in four main categories:

- moving pointer (analogue, fixed scale pointer moves)
- moving scale (analogue, pointer fixed, moving scale)
- digital (number form)
- lights (simple colour code, on or off)

The designs of visual indicators should be made with the following in mind:

- The display should be made visible to the operator and not tucked away in some obscure place.
- The display should be illuminated.
- The scales need to be clear and simple.
- There should be no requirement to perform conversion of scale readings.
- If precise reading are required, digital readouts are preferred.

If the visual display also includes control mechanisms, then there are several control layout considerations to take into account:

- They must be positioned so that they are within easy reach, and those used most frequently are closest to the operator.
- The displays should show a degree of consistency, i.e. the zero points on all dials are in the same relative position.
- Any sound warning is distinguishable from another.
- If a control element moves an item, the control device (i.e. handwheel) should operate in the same direction.
- The information displayed should be kept to a minimum.

Aesthetics

The way a product physically looks may be just as important to a customer as its function. Its appearance is often known as its aesthetics, which may be defined as the appreciation of beauty or good taste. This relates almost entirely to the appearance of the product and sometimes is the major factor to be considered. Shape, colour, texture and decoration are all elements of appearance, but so is the situation within which the product operates. These differing environments play an important part in the consideration of aesthetics; for example, manufacturing a modern style colourful three-piece suite. Certain countries have taboos about the colour red, while others prefer sharp squarish shapes rather than soft rounder ones. The following give some examples of the types of environment considerations required:

- light, dark
- hot, cold
- dirty, clean
- public place, private place
- acid, alkali
- wet, dry

Self-assessment task

Discuss the requirements of controls and displays of the following:

(a) The capacity of a petrol tank for a family size car.
(b) A thermostatically controlled refrigeration unit.
(c) A digital telephone.

Quality

The customer must decide upon the level of reliability, the type of materials to be used and the expected life span of the product. The designer may advise on certain issues, and any discrepancy must be resolved early on in the design stage. This enables designing for production to begin at the earliest stage, while giving due consideration to quality control.

Cost

Generally the attempts to reduce cost centre around production techniques and simplifying designs. The functions of the product need to be examined carefully, for they too may be simplified or eliminated altogether. Therefore, concentrating on the price of components or components used in vast numbers may add up to a substantial saving in costs or time.

Ask the customer such questions as:

- Can a cheaper material be used?
- What is the expected purchase price?
- Has the product to be self-maintaining?
- What amount of maintenance is required?
- Are there to be any future variations to take into account?
- Can parts be standardised?

A summary of cost reduction with regard to value, increases or maintains the value of a product to the purchaser while reducing the cost to the producer. This may be achieved by trying to:

- eliminate
- simplify
- standardise
- reduce
- modify

Quantity

This links very closely with cost, for a designer there has to be a direct relationship between the cost of production and the quality of the product. There is little point in designing and producing knives and forks, in vast quantities, in a low grade steel to make them as cheap as possible, only to find that they soon rust and are unusable.

Size

The physical size constraints of the product have to be determined. This is especially true if the product forms part of a range of other products. Aesthetically they have to look pleasing and be proportional to the others within the range. In today's microchip world no longer is the phrase 'you get more for your money' apt, for smaller is not a requirement for costing less. The following advantages may derive from size ranges:

- The design work is done once and is then transferred though scaling.
- Production of selective sizes may be repeated in batches and thus become more cost-effective.
- It is easier to establish higher quality levels.
- Shorter delivery times are possible.
- Replacement parts are often held in stock.

If manufacturing is based on mass production, then the tolerance on any one given component will vary depending upon the production method.

The fit that the component has to make with a mating part will also vary, due to the mating part having a tolerance as well. It is possible in mass production to utilise a technique known as selective assembly, which simply selects components to be mated from either the top, middle or bottom end of their respective tolerance bands.

Thus a pair of components with individual sizes of:

A: 10.00 mm to 10.03 mm; tolerance 0.03 mm
B: 10.03 mm to 10.06 mm; tolerance 0.03 mm

could have a fit of

A = 10.00 mm and B = 10.06 mm, tolerance 0.06 mm

or

A = 10.03 mm and B = 10.03 mm tolerance 0.00 mm

Selective assembly would take 10 per cent of the largest A sizes and match these with the largest 10 per cent of B sizes. Then the next 10 per cent size range would follow a similar approach, giving much tighter tolerances and improving quality. It may also be used as a method of widening the individual tolerance bands of components in order to offer a speedier production method.

> **Self-assessment task**
>
> Obtain the co-operation of a friend, parent or ideally someone with an industrial background, and ask if they have a need for something to be made or serviced. Then produce a list of customer requirements for the manufacture or service of the product. (*Note*: it need not be a new product but a modification of an existing one.)
>
> In order for you to achieve this, remember to include elements such as: ergonomics, aesthetics, quality, cost, quantity, etc.
>
> Once you have compiled this list of customer requirements, have it signed by the customer as a true account of his or her wishes, as this may help to avoid any later misunderstandings.

Identifying standards and legislation

Standards form the foundation of all design work. They give a base from which to work. Their origins may come from various organisations and have different levels, i.e.:

- National standards
  - BSI (British Standards Institution)
  - DIN (Deutsche Institüt für Normalisation)
- European standards
  - CEN (Comité Européen de Normalisation)
- Universal standards
  - ISO (International Standards Organisation)

The following list, however, gives some of the variations in the content of these standards:

- classification standards
- communication standards
- dimensional standards
- material standards
- operational standards
- planning standards
- procedural standards
- quality standards
- safety standards
- type standards

The level of the standard is developed from its breadth, depth and range of application.

Company codes of practice

Company standards and regulations may be derived from a range of general standards, only if they are useful and economical.

Any standards used must not conflict with the law of that particular country (i.e. monopoly restrictions or safety regulations). They must be stated in clear and concise terms so that they are readily understood by all. If a new standard is being proposed it must be examined by a working party and subjected to trials.

> **Self-assessment task**
>
> With regards to standards, determine what the following abbreviations mean:
>
> (a) ANSI
> (b) COSHH
> (c) RoSPA
> (d) HSE
> (e) CENELEC

Health and Safety Legislation

One of the most important pieces of legislation that a design engineer has to be aware of, is that of The Health and Safety at Work, etc., Act 1974. It places specific responsibilities on employers, employees, designers, manufacturers, importers, suppliers, substances used at work and occupiers of premises. Inspectors will be given powers of entry at any reasonable time. They may inspect, under section 6 of the Act, that articles designed and constructed should be safe and without risk to health at all times when it is being set, cleaned, used or maintained by a person at work. Steps should be taken to ensure that any person supplied with a product has adequate information about the use for which it was designed.

Ergonomic designs involving layout, function and aesthetics play an important role in everyday products, and often fall under the generalised standards category of company standards.

However, in today's consumer-led market, many recommendations are being strongly urged, none more so than the design of computer workstations. It has been found that poorly designed layouts have led to operational stress, i.e. fatigue, discomfort, annoyance and RSI (repetitive strain injury). This has developed into a recommendation that the VDU workstation shown in Fig. 5.2 be adopted.

The Control of Substances Hazardous to Health Regulations 1988 (COSHH) demand that reference to the classification, packaging and labelling of dangerous substances should be made when hazardous substances are used in a design.

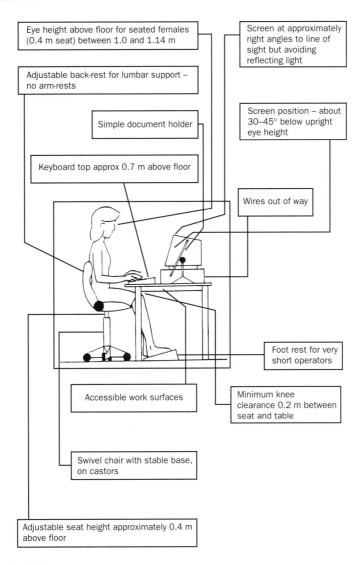

Figure 5.2 A recommended VDU workshop

Designers should also be aware of BS 5304: 1988 'Safeguarding of machinery', the principle aim of which is to eliminate hazards associated with machinery.

Thus, designers must consider the following:

- Avoid the possibility of coming into contact with any rotating parts.
- Eliminate the chance of becoming entangled or being struck by machinery.
- Stop any possibility of being struck by any material ejected from the machinery.

The legal requirements of the safe use of electricity is covered in the Electricity at Work Regulations 1989, and is supported by the Institute of Electrical Engineers (IEE) 'Wiring Regulations'. A summary of this is given below:

- Any system should be made so that it prevents as far as is reasonably practical, danger.
- Any system shall be maintained to prevent such danger.
- Any work activity, use or maintenance, should be carried out in a safe and reasonable manner.
- Any equipment provided for the purpose of protecting people at work, should be suitable for that use and maintained as such.

Mechanical handling

Whenever possible, handling of goods should involve mechanical rather than manual systems.

In general, there are three main types of mechanical handling system:

- Elevators, use for lifting or lowering on a vertical plane.
- Conveyors, used for moving goods horizontally or at inclined angles.
- Mobile equipment, such as fork-lift trucks which transfer goods from one point to another.

Elevators, should have fixed guards at both ends, where there are usually chains and sprockets. Conveyors, have problems with 'nips' or traps between moving parts (i.e. belt and drive or slats and guide frames), therefore fixed guards totally enclosing the trap should be the norm. If the conveyor requires frequent lubrication, interlocking guards should be employed to automatically stop the conveyor when it is lifted.

Further design points to consider with conveyors are:

- Minimise all bends to prevent items jamming or falling off.
- Rails have to be fitted if the conveyor rises more than 1 metre above floor level.
- An emergency stop trip should be fitted if the conveyor is longer than 20 metres.

Mobile handling equipment, including fork-lift trucks and pedestrian-operated stacking trucks, should adhere to the following rules:

- Never exceed the maximum rated load capacity, which should be clearly labelled.
- Passengers should not be carried on such equipment.
- Unauthorised or untrained personnel should never operate such equipment.
- When not in use, the forks should be lowered.
- A formal maintenance programme should be enforced.
- Trucks must comply with the Road Traffic Acts if they operate on a public highway.

A safe working environment

To maintain safety standards within the workplace, the layout of the environment becomes significant. An example of this is in the prevention of overcrowding, which is detailed in the Factories Act 1961 and in the Offices, Shops and Railway Premises Act 1963. The general requirement of 3.7 square metres of floor space should be provided to each person, or 11 cubic metres when the ceiling is less than 3.3 metres high.

Exposure to extremes of temperature is also covered by the Act and offers the recommendations given in Table 5.2 according to the type of work being done.

Table 5.2 Environmental extremes

Type of work	Temperature range (°C)
Office work	19.4–22.8
Light work	15.5–20.0
Heavy work	12.8–15.6

Humidity should be kept to within 30 and 70 per cent.

Lighting standards are outlined in HSE Guidance Notes HS(G) 38 'Lighting at work'. This draws a relationship between the average illuminance and the degree of detail which needs to be seen in a particular situation or task (both are measured in lux (lx); see Table 5.3).

Table 5.3 Illuminance standards

Locations/type of work	Activity	Illuminance (lx)	
		Average	Minimum
Car parks Corridors	Movement of people machines and vehicles	20	5
Construction site Excavation of soils Loading bays Bottling plants	As above but in hazardous areas not requiring detail	50	20
Kitchens Factories Assembling large components	As above but requiring limited detail	100	50
Offices Sheet metal work Drawing office Electronic assembly plants	Work requiring detail Work requiring fine detail	200 500	100 200

Ventilation of the workspace is covered by section 4 of the Factories Act 1961 and section 7 of the Offices, Shops and Railway Premises Act 1963. This requires that there must be sufficient circulation of fresh air to maintain comfortable conditions and a healthy working environment. In the event of areas where there is emission of dust, fumes, gases or forms of airborne pollutants from processes, then additional effective systems of ventilation must be provided.

Noise is a form of pollutant and people exposed to noise in excess of 90 dBA (the decibel) stand the risk of going deaf. This means that designers should be aware of the threshold levels and of the types of soundproofing available. Table 5.4 shows the maximum recommended exposure to noise levels.

(*Note*: the decibel scale is a logarithmic scale and the increase in pressure of 3 dBA represents a doubling of the sound.)

Table 5.4 Maximum noise exposure levels

Noise level	Maximum duration
90 dBA	8 hours
93 dBA	4 hours
96 dBA	2 hours
99 dBA	1 hour
102 dBA	30 minutes
105 dBA	15 minutes

Self-assessment task

Identify the following British Standards. Choose one and describe the influence it has on the design brief of a product or service.

(a) BS 9000
(b) BS 308
(c) BS 4870
(d) BS 3939

Identifying technical and operational constraints

To achieve a final product from an initial idea will have taken many design decisions. These decisions would have been taken with all of the relevant information known at that time. In many cases as the design process (or even the drafting of a design brief) progresses, other constraints appear, which restrict potential solutions to the problem. This continues until only one dominant design remains (which is often decided upon by a combination of its function and on economic grounds).

There are many forms of constraint, ranging from the technologies available, resources used, the type of environment and the amount of money spent. Although these form the everyday types of constraint to producing a design brief, there are others which have an obscure yet potent influence.

Consider the following two constraints. Firstly, a designer's knowledge of a particular process constrains the solution. Since every designer has different experiences and knowledge, there is a tendency for different designers to produce different solutions. The second constraint is in the result of a design decision. When a designer says, 'I will use rivets to join these two pieces of sheet metal,' then a constraint has been placed which may affect other design decisions.

The next section will address the four main areas that constrain designers.

- technology
- resources
- the environment
- cost

Technology constraints

Before identifying the technology constraints, it would be wise to have an awareness of the variety of properties which make demands upon these technological choices. The choice of material is a major factor in any design process, and the following represent some of the considerations a designer must address:

- Mechanical properties, such as strength, stiffness, fatigue, toughness, hardness and the influence of temperature.
- Corrosion resistance; coefficient of friction.
- Wear resistance.
- Special properties, such as optical, magnetic, electrical conductivity, damping capacity, permeability, etc.

Structures

There are two main forms of structures that offer a constraint to a designer: *natural* structures and *man-made* structures.

Natural structures have evolved over centuries and may be seen in mountains, caves, trees, rivers and wild life. The designer often takes a leaf out of nature's book when designing – i.e. a bird's wing and an aeroplane wing must have similar properties, they must both be light, strong and aerodynamic. Man-made structures have often evolved out of necessity and have led to such items as aircraft, tables, chairs, bridges, houses, etc.

When designing a structure there are many aspects to consider before and during the designing process. The following are but a few of these considerations:

- cost
- life span
- size
- safety factors
- materials and method of construction
- strength required
- maintenance requirements

Forces within a structure play an important part in the design process. Consideration has to be given to whether the forces acting on the structure are static (not moving) or dynamic (moving).

These forces can also be applied in differing forms, for example:

- Tension is the force of pulling or stretching and structures often have ties built into them to resist this tensile force (Fig. 5.3(a)).
- Compression is the opposite of tension, it is the force of squashing and a structure may have struts built into it to resist the compressive force (Fig. 5.3(b)).
- Shear forces act on two or more parts of a structure in a sliding or cutting action, similar to the way in which scissors cut (Fig. 5.3(c)).
- Torsion is the force that involves a twisting motion (Fig. 5.3(d)).

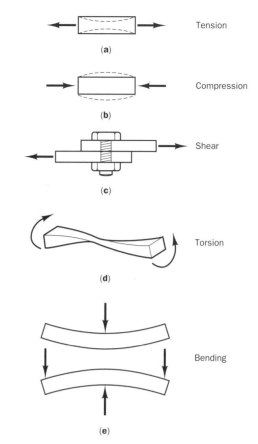

Figure 5.3 Various forces acting on structures

- Bending, unlike the other forces mentioned, usually has an exerting force acting perpendicularly to the axis of the beam, trying to deflect it. The orientation and shape of beams can greatly enhance its bending properties (Fig. 5.3(e)).

Self-assessment task

Make sketches of two bridges: one must be an arched bridge and one must be a cantilever bridge. Compare the structures of the two bridges, indicating where you think the ties and struts are within each structure.

Mechanisms

Mechanisms come in many forms such as gears, cams, threads, pulleys, ratchets, etc., and are used in everyday machines like cars, bicycles, lawnmowers, door closers, etc.

The advantages offered by mechanisms are:

- If designed correctly, they are usually very efficient.
- They can perform a variety of tasks.

The disadvantages of mechanisms are:

- The moving parts often need lubricating.
- Moving parts require guarding.
- A high degree of accuracy in manufacture is required for mating parts.

Mechanisms have to cater for different kinds of movement such as linear (moving in a straight line), rotary (turning) or reciprocating (moving back and forth). They may also make use of levers and resulting forces (moments), which in turn are often used to form linkages.

A common form of mechanism is one in which rotary motion is turned into a linear motion (i.e. that of the lead screw of a lathe driving the saddle). Gears transmit rotary motion and forces and can give a variety of ratios that have positive drive and no slippage, as in belt drives which occurs through lack of friction, giving a direct drive. There are several forms of gear drive mechanisms, and each offers the choice to increase or decrease the rate of rotation by the use of gear ratios (velocity ratio). This involves the following simple calculation:

$$\text{Gear ratio} = \frac{\text{No. of teeth on drive gear (input)}}{\text{No. of teeth on driven gear (output)}}$$

Other forms of mechanisms useful to the designer are pulley drives, belt and chain drives, screwthreads, clutches and brakes (Fig. 5.4).

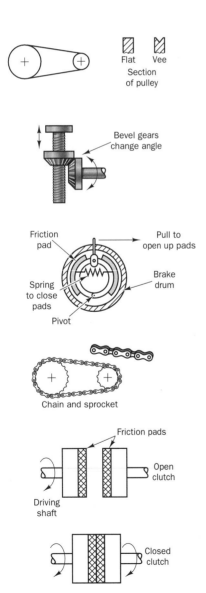

Figure 5.4 Various types of mechanism

Energy

Defined as force, vigour and activity, energy is required at some stage in order to make items work. (Almost all the energy we use originates from the sun in one form or another, with the exceptions of nuclear power and heat from the centre of the earth.)

Energy can be neither created nor destroyed, but it can change its form. There are several types of energy and the following gives examples of the main forms:

- *Chemical energy* is stored in food, coal, oil, petrol and batteries.
- *Electrical energy* offers a clean convenient form of energy as electrons move thorough an electrical conductor.
- *Electromagnetic energy* includes difficult to define items such as radio waves, microwaves, X-rays, light, etc.
- *Heat energy* involves the burning or heating of material in which the molecules expand and move faster.
- *Sound energy* is carried through the atmosphere by vibrating the air, i.e. as in a loudspeaker.
- *Nuclear energy* is created by the splitting of a uranium or plutonium atom by nuclear fission. (1 kg of uranium can produce about 1 million times the energy of 1 kg of coal.) This massive increase in energy has its downside in the form of dangerous radiation, which is invisible.

Electronics

The majority of technological advances in the last 40 years or so has been in the electronics field. In the 1950s very few people had television sets or even radios; now there are computers, electronic mail, infrared remote controls, etc., all with a tendency to be made smaller, cheaper and more reliable. The development of the various components that go to make up an electronic circuit has aided the overall rise in electronic consumer products. The following list shows some of the components that are most often used within an electronic circuit:

batteries	bulbs
capacitors	diodes
integrated circuits	operational amplifiers
potential dividers	power supplies
relays	resistors
switches	transistors

Resources

There are many forms of resources, but for the production of a design brief these can be broken down into three major categories:

- labour
- materials
- plant

Labour is the term given to the workforce. It may force several indirect constraints on a design, such as the availability of trained people, flexibility of these people towards working patterns, the amount they cost per hour and do they possess transferable skills. This means the designer must take into account who the design is intended for and establish if the current workforce is capable of producing the design through to production. It is no use designing specialist racing cars requiring skilled production techniques if the present workforce are only assemblers.

Materials have to be chosen with care, not only because of rising costs but because they are becoming scarce. Efforts to employ reuseable resources are in their infancy at present and are not yet totally cost-effective. Other considerations must today be given to the use of environmentally friendly and aesthetically pleasing materials. Designers must be aware of developing a commonality of parts that includes using similar materials for different jobs, as this will cut down on storage, waste and eventually costs. An example of this would be the standardisation of fixing screws used in the assembly of television sets. A single appropriate screw should be able to fasten the majority of components, thus eliminating coded storage bins, different sizes, different tools and wasted time in selecting the right fastener.

Plant resources refer to the technology hardware that goes into the manufacturing process, therefore consideration as to what is available and how it is used is essential to the designer.

The development of the microchip has led to a new revolution in the machine tool industry, by controling digitally the working actions of the manufacturing process. New, quicker and more accurate methods have evolved offering the designer a wider choice of processes with which to work. A major impact has been the development of computer-aided engineering packages allowing the designer to design, develop and simulate a working model of the design prior to production. This allows detailed examination of the design and modifications to be made, especially towards any safety features, at a minimal cost.

A further major development in this area in recent years has been in the effective control of plant, using computer-aided product planning (CAPP) systems such as Materials Requirement Planning (MRP) and Manufacturing Resource Planning (MRPII). These enable plant to be more efficiently controlled and monitored utilising it to its full potential, offering greater flexibility.

The environment

A designer will take a problem from the environment and return a solution to the environment. Therefore, the design must not be considered in isolation but with the environment in mind, and should be treated as a dynamic process, one in which the design offers information feedback so that an optimum solution may be found.

The environment throws up some interesting contradictions and challenges for the designer. For example, cars and lorries play an important part in our daily infrastructure, yet they also build up dangerous levels of carbon monoxide.

Another similar problem is that of the sulphur dioxide emitted from coal-burning power stations which, when combined with the air and rain in the atmosphere, produces a dilute form of sulphuric acid, which kills trees. Nitrates used by farmers as a fertiliser to increase yields is now finding its way into water supplies. And nuclear waste has to be stored safely away from human habitation for thousands of years.

As our standard of living has increased, then so has our demand for the amount of energy we use. Oil and coal play a big part in our consumption of energy, even with the side-effects mentioned above. Other cleaner power sources are being investigated like wind power, solar energy and wave power, but none has as yet reached the economic efficiency levels of oil or coal.

Cost

Most attempts at reducing costs concentrate on simplifying designs, manufacture or assembly. However, there are strategies to target cost reduction, which involve concentrating on high-cost components with the view to substituting them with a low-cost alternative. They may involve reviewing components used in large numbers so that any small saving for a single component would add up to a large saving overall.

In general the following may offer a checklist of cost reducing a product:

- **Reduce**   Reduce the number by combining components if possible (the assembly of which costs more).
- **Eliminate** Are there any components/functions that are redundant?
- **Simplify**   Look for alternatives, i.e. is there an easier method of assembly?
- **Standardize** Duplicate components wherever possible.
- **Modify**   Can the method be improved? Can a different, cheaper material be used?

Self-assessment task

Choose any one of the following environmental changes that have occurred in recent years and comment on the technical or operational constraints it or they had to go through:

(a) Obtaining information from television.
(b) Plastic money (as in a cashless society).
(c) Entertainment.
(d) Hypermarket shopping.
(e) Noise pollution due to traffic.
(f) The centralisation of information (computer files).
(g) Package holidays.
(h) Education.

Producing design briefs

The design brief starts with the identification of the problem itself. An actual full design brief is made by writing a specification of the problem and may take the form of only two or three sentences or statements (however, it is more likely to be pages in length). These sentences or statements combine to form detailed specifications of the main functions and limitations of what is required, enabling the designer to work within a prescribed boundary and to focus on key points.

The following is a simple example of such a design brief.

The problem:
I have a garden shed that now stores expensive gardening equipment.
It needs protecting.

- *Brief*
  – Design and make a product that will deter a thief from stealing my gardening equipment.

- *Detailed specification*
  - Must be simple to use.
  - Must prevent a thief from stealing equipment.
  - Must act as a deterrent.
  - Must be strong.
  - Must not rust.

- *Limitations*
  - It has to be made within a simple college/school workshop.
  - It has to be made within 6 weeks.
  - It must not cost more than £15.

Self-assessment task

Produce a design brief specification for any one of the following:

(a) A baby alarm.
(b) A bike security device.
(c) A doorbell for a deaf person.
(d) A liquid lever indicator for a blind person.
(e) A rainfall recorder.
(f) A fire door security alarm.

Please note: you do not have to manufacture the chosen item; just produce a design brief specification.

5.2 Produce and evaluate design solutions for an electromechanical engineered product and an engineering service

Topics covered in this element are:

- Identify the requirements of a product or service from a design brief.
- Technical information relevant to the design brief.
- The generation of feasible design solutions.
- The evaluation and selection of design solutions.

An evaluation is supposed to determine the value of a solution against the specification. This evaluation not only enables the functional and performance development of the product or process to be more manageable, but it also allows a detailed analysis of the problem to be carried out before production.

The following are some of the reasons evaluations are required:

- Requirements have been refined to offer a better understanding of performance, reliability and costs.
- Technical information is gathered or updated.
- General solutions are formed through individual and group work.
- An evaluation directs the processing capability while maintaining an eye on the overall economics of the process or product and the variations in output characteristics, such as safety, serviceability and installation.

Identifying requirements from a design brief

The performance requirements of a product or service, along with supporting technical information, should be made explicit in the design brief as these act as the yardstick by which the product or service is measured, in order to determine whether it is successful or operating beyond its means.

Performance can mean many things, but in general it is a measurement of actual achievement against desired service. The requirements of performance may have operational characteristics, such as speed, durability, quality, repeatability, temperature range, accuracy, etc. Another form of an operational performance requirement is the carrying out of the maintenance schedule applied to a product or service. This gives systematic feedback as to the serviceability of the product, and enables more complete and accurate evaluations to be made.

Fitness for purpose

Once a design specification has been drawn up and the design produced to match that specification, then a further look at the product or process with an eye for fitness for purpose should be used. This casts an overall look at evaluating just how applicable the product or service is in meeting the requirements laid down in the specification.

The following series of questions could be asked to aid the decision-making process:

- What does it do?
- What does it cost?
- What else can do the job?
- Does it meet the requirements?

In an evaluation context, each of the above questions could be asked of a variety of elements to decide its fitness for purpose. The following gives an example of elements considered for evaluation.

Ergonomics requires that a human interface and product combine and conform to certain standard dimensions, therefore the evaluation of what the product or process does and costs, in terms of interaction, needs careful examination to see that it matches the requirements of the design brief.

Aesthetically the product or process must look apt, and its appearance must reflect the design specification not only in shape, size and form but also in colour, texture and overall performance. (For example, a car service that occurs in a noisy workshop smelling of deisel oil may provide an excellent service, but will it attract many customers? A possible solution here would be a separate environmentally clean customer reception hiding the workshop area.)

Quality plays an important part in the role of fitness for purpose, not often in the actual fitness itself but more in the overall control of the product. An example of this is in the choice of material for the conducting of electricity. Gold by nature of its excellent conductivity properties would be the first choice of any designer except for its high costs. Thus a trade-off of costs against performance, giving the highest possible quality product, is arrived at; in this case copper is the usual alternative material (gold is used on products where performance is paramount).

A service, on the other hand, may require that a process places a high emphasis on quality and may be reflected in the price of such, as in the case of having a routine maintenance schedule built into the sales package of a product.

Size and weight of designs may vary due to technological advances in production methods or to the development of new materials.

As developments occur it is wise to re-evaluate existing designs in order to maintain an up-to-date product or process. (An example of this was the development of the microchip which has transformed the communications industry to name but one. Another future development may be the arrival of the superconductor operating at room temperatures, bringing innovation to existing designs.)

Maintainability, serviceability or repairability are terms that are often interchangeable, as they describe the diagnosing and repairing of products.

However, in today's climate of the throwaway philosophy – the one in which it is easier to replace than to repair – designers have to be very aware that products still have to be easy to diagnose, disassemble and repair, and this takes extra effort in the design process. A clear indication of this is when performance is easily checked and maintenance involves the use of the smallest possible variety of tools and equipment.

Standards have to be met and legislation complied with in order to produce effective goods. Therefore stringent examination of all requirements have to be checked, especially if the intended market is international. What may fulfil design safety requirements in this country, may fall short in others, often on insignificant minor points. With this in mind the designer should always consult the relevant standards organisation or have a copy of the standards available.

> **Self-assessment task**
>
> When evaluating a design or a service the following elements may require testing to see that they conform to national standards. List the relevant national standards for:
>
> (a) safety
> (b) quality control
> (c) transportation

Reliability

A definition of reliability is that of a system performing to its intended function. This is usually associated with dependability, with successful operation and with the absence of breakdowns or failures. A manufacturer of home appliances has to be concerned with reliability because frequent failures would result in customer dissatisfaction. Also reliability is important when safety is concerned, especially to eliminate failures; for example, the electronics designer must safeguard against electrocution just as the aviation designer must guard against systems failure on aircraft. Thus reliability is connected to failure through the probability that a system will perform properly for a given specified period under set conditions. Reliability can be measured as a function of time and against a possibility of a failure rate.

The nature of failure varies and may be early failures mainly due to teething troubles, random failures which are very difficult to predict or fatigue failures which are often caused through misuse or old age. The behaviour of these failure rates tend to be formed from similar characteristics. These characteristics are often described as the bath tub curve, as it is shaped like a bath tub with a higher initial failure rate smoothing out to a useful working lifespan, then tending to increase again in failures as they become aged.

Figure 5.5 plots the failure rate against time in graph form with the following typical results:

- Electronic hardware (Fig. 5.5(a)) tends initially to have a few teething problems, then stabilises over a useful working life, until the components break up or degenerate through ageing.
- Computer software (Fig. 5.5(b)) has a high initial failure rate (commonly known as bugs) but as time progresses they are gradually ironed out until they run almost indefinitely without problems (they tend to be superseded by more advanced programming).
- Mechanical equipment (Fig. 5.5(c)) again has a high initial failure rate, but testing and resetting quickly lower the failure rate. However, due to ageing and wear, the number of breakdowns tend to rise steadily until it is uneconomical to maintain.

Reliability plays an important role in the efficiency, profit, market share and life cycle of product or processes.

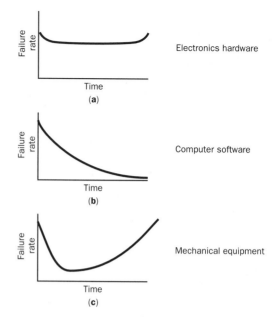

Figure 5.5 Failure rate/time graph

Increasing the reliability will increase manufacturing or servicing costs, but will reduce the costs of reworking or scrap. There comes a point at which the cost of improving reliability cannot be met by the savings made in rework or reducing scrap (Fig. 5.6). Reliability is known as meeting customer requirements over a period of time.

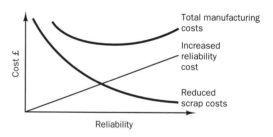

Figure 5.6 Optimum reliability/cost curve

The main method to evaluate designs for reliability is to employ a technique known as Failure Modes and Effect Analysis (FMEA). This lists the possible modes by which the components or process may fail and traces the consequences of each mode of failure, indicating:

- Cause of failure.
- Possible effects.
- Probability of occurrence.
- Criticality of failure (four levels: negligible, marginal, critical and catastrophic).
- Possible actions to reduce failure rate.

Costs

Detailed accounts of the proposed costs of the design or process must be compared to the actual costs incurred. They may be broken down into three main areas:

- specification costs
- production costs
- maintenance costs

Prototypes place a cost on the product or service that often is difficult to estimate, because it is very seldom known exactly what time or resources it will take to match the specification. Generally a working budget is agreed first and the designer tries to stick within its limits. At this stage there is no pay-back of any money spent in research and development of the design; this has to be added to the selling price, often spread over a period of several years.

There are then production costs to consider. This is an area in which a designer may evaluate a design based purely on economic production and not reliability or function. This cost-cutting scenario is often at the expense of quality and should be used sparingly. Cost savings should never be made at the expense of safety. The life of the whole component or service should be considered when deciding upon costs, for example, it may be prudent to sell a product at a slight loss initially in order to gain market share, only to recoup this at a later stage.

Maintenance schedules and servicing costs have also to be considered, as have decommissioning costs (i.e. currently the decommissioning cost of a nuclear reactor is upward of £100 million).

Technical information relevant to design briefs

Information is a commodity that very often needs translating to fit a purpose, because any information received has to be processed and transmitted in a form that is suitable to aid the design process. This information may come from: market analysis, trend studies, patents, technical journals, research, customers, design catalogues, experiments, standards, legislation, company visits or CD ROMs.

The evaluation of information can be categorised into the following areas:

- *Value* – the importance to the receiver of the information.
- *Reliability* – is the information correct?
- *Precision* – that is, clarity of the content.
- *Form* – that is, how is it presented (graphic or numeric)?
- *Quality* – the refinement of the detail within the information.
- *Originality* – an indication of the authenticity of the information.

Obtaining technical information relevant to a design brief involves looking at various criteria, such as material information, serviceability factors, functional information and processing capabilities, each of which are explored further below.

Materials

The vastness of information available about materials may fill a small library; however, it is primarily the strength of materials and their other properties that interest a designer most. The following is a list of the most commonly used materials in the design of mechanical products (defined under the AISI, the American Iron and Steel Institute system):

Steels and irons
  1020
  1040
  4140
  4340
  S30400 (stainless)
  S316 (stainless)
  01 tool steel

Grey cast iron

Plastics
  ABS
  Polycarbonate
  Nylon 6/6
  Polypropylene
  Polystyrene

Aluminium and copper alloys
  2024
  3003 or 5005
  6061
  7075
  C268 (copper)

Other metals
  Titanium 6-4
  Magnesium AZ63A

Ceramics
  Alumina
  Graphite

Information regarding these materials usually refers to their properties, which normally consist of the following:

- tensile strength
- yield strength
- endurance limit
- elongation
- modulus of elasticity
- density
- hardness
- coefficient of thermal expansion
- melting temperature
- thermal conductivity
- cost per unit weight
- cost per unit volume

Serviceability

Defined as useful and durable, incorporating serviceability demands that great care and consideration has to be given to how the products are to be used and maintained at the outset of the design stage.

The forethought that needs to be taken into account includes:

- routine maintenance
- repair and replacement
- access
- time factors

Functionality

When solving or developing design solutions, the term *function* may relate to the performance of the task to which the product has to eventually conform in order to be useful. This involves ensuring that the designed system has a clear and reproducible input/output relationship.

An example of functionality is that of a fuel gauge, where liquid is introduced and removed from a container frequently while the actual quantity of liquid in the container at any one time may be measured. This results in a liquid system having a flow of material with the function being the measurement and indication of the actual quantity of liquid (Fig. 5.7).

Designs are based on building an overall function from a series of other functions. This modular approach may come in one or several sizes, stages or finishes. A typical schematic diagram of a basic modular function system is shown in Fig. 5.8.

The overall function may be derived from a series of varying types of other functions and each may have differing claims as to how essential they are to the overall function. There are five basic forms of function:

1. Basic functions are essential to a system and, in general, form a basic recurring event.
2. Auxiliary functions are applied by joining or locating essential types of functions.
3. Special functions complement the task-specific basic functions; they are not essential and are of a possible type.
4. Adaptive functions are required when a system needs to be adapted to other systems and allow for unpredictable circumstances. These may be of either essential or possible types.
5. Customer-specific functions are unpredictable, as the customer may adapt or utilise other functions outside the scope of the design.

A simpler example of functionality may be extracted by considering a domestic vacuum cleaner. Its prime function is to clean floors, but it also has secondary functions such as being safe to use, store dust, being quiet, looking good, being inexpensive, etc.

If we look further into these secondary functions, combinations of ideas of how to solve the design problem develop, such as:

- clean floor – suck, blow, sweep, wash, beat, mop
- being safe – low weight, insulation, smooth edges
- store dust – fabric bag, paper sack, plastic container, dust pan
- being quiet – insulation, ear plugs, manual operation
- looking good – colour, shape, size, texture
- being inexpensive – income range, competition, value for money

By listing the primary and secondary functions of the vacuum cleaner, features that a customer may require are highlighted. These features are known as *attributes*. There are many attributes required in a finished product. Some of the more

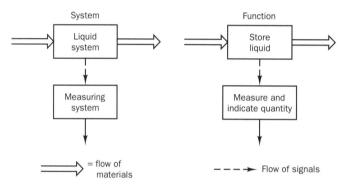

Figure 5.7 Function diagram of a liquid system

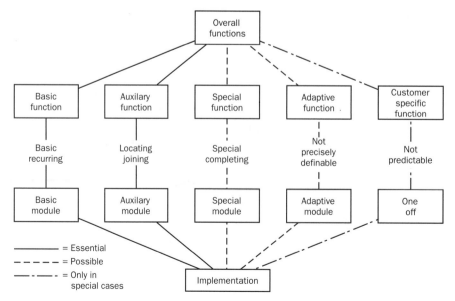

Figure 5.8 Schematic diagram of a modular function system

popular ones are listed in Table 5.5, together with consideration factors used by the designer to develop the attributes further.

Table 5.5 Product attributes and consideration factors

Attribute	Consideration factors
Appearance	shape, size, colour, texture
Capacity	size, force, movement, direction
Cost	initial, running, maintenance, trade-in, development, safety
Durability	abuse, misuse, accident, corrosion, humidity
Interchangeability	rapidity, accuracy, multi-role, modules
Life	first line, backup
Maintainability	continuous, regular, sporadic, none
Performance	force, velocity, pressure, energy
Portability	lift, transport, orientation
Reliability	failure rates, repeatability
Safety	electrical, mechanical, chemical noise
Simplicity	technology, manufacture, maintenance, use
Serviceability	in situ, remote
Weight	amount, distribution

Of all the attributes listed in Table 5.5, there are five which are the most common requirements in general designs:

- cost
- performance
- appearance
- reliability
- safety

Often designers place these five attributes in a ranking order of importance, so that borderline decisions can be made. An example of this is in the design of a new CNC machine tool, one in which performance and reliability are of prime importance, while cost, safety and appearance are less critical.

> **Self-assessment task**
>
> Prioritise the following products and processes using the five prime attributes of cost, performance, appearance, reliability and safety:
>
> (a) A washing machine.
> (b) A pair of hospital crutches.
> (c) A child's toy.
> (d) A car alarm.
> (e) A domestic smoke detector.

Process Capability

A designer must have a basic understanding of the process capability of a manufacturing process in order to be able to justify utilising the most appropriate methods available. Process capability is a statistical examination of the output of a process in a numeric form so that a graphical representation may be compared to that of the standard bell-shaped frequency distribution curve. Any deviation from the frequency curve will offer a description of the capability of the process to perform within its intended capacity or accuracy. It will not tell you what is wrong but it will indicate drifts in working patterns against a preset norm.

Generating design solutions

By varying the techniques used in generating a design solution, more possibilities open up that may have been neglected if only one method was used. Some of the techniques that may be employed are listed below:

- shifting boundaries
- brainstorming
- new combinations
- checklists
- objectives tree
- performance specification

Shifting boundaries

Many designs start with preconceived ideas, especially about boundaries or constraints, such as manufacturing methods, costs, types of material, sizes, etc. Each of these constraints is connected to another, some are loose connections, others very strong, but by changing one you can affect the others.

This may be represented by drawing the conventional solutions linked together and identify any further possibilities, including risky solutions or improbable solutions, as extensions of the conventional solutions (Fig. 5.9).

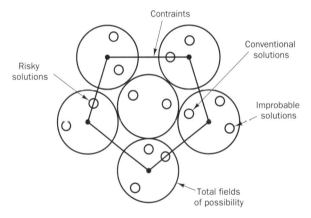

Figure 5.9 The interrelationships of conventional solutions to design problems

> **Self-assessment task**
>
> There is a requirement to develop a method of cleaning the windows in a multistorey office block from the inside. Spend 10-15 minutes designing a solution then follow the steps below:
>
> 1. List the major constraints of your design.
> 2. Highlight any constraints which are well documented, e.g. costs, standard components.
> 3. Try to add to the list constraints which are not well documented.
> 4. Add others that you may know from experience, e.g. market preferences, competitors' successes.
> 5. Choose one from this list (preferably not costs) and modify it by using a hypothetical circumstance.
> 6. Identify any new possibilities that open up.
> 7. Develop proposals for these new possibilities.

Brainstorming

This is often seen as unproductive play activity. It may be a licence to fantasise or have silly ideas, but it opens the mind to possibilities that may be deliberately suppressed. The real work done in brainstorming comes from the rational analysis of the ideas generated.

Brainstorming is a technique best employed in a group format (although it may be done solo) and has some basic rules:

- Criticism of ideas is not allowed.
- Produce and record as many ideas as possible.
- Use group activity whenever possible (4–8 people).

The initial question starting the brainstorm should not be too vague as to include irrelevant material, or too narrow to limit the range of ideas.

Self-assessment task

Consider that you are working for an automobile company and have been requested to look at alternatives to the conventional windscreen wiper. The brief for this task is to use the brainstorming technique. Spend up to 5 minutes producing as many ideas as possible to answer this question: 'How can we develop the conventional windscreen wiper?'

New combinations

When two or more established designs or elements are joined into one new device then it may be said to be a new combination. These may be attributes (performance) or geometric (shape), combining to give one new form.

An example of such a device is found in the connecting of a washing machine to the cold/hot water household pipe. A conventional solution to joining the two pipes together would be as in Fig. 5.10.

An analysis of the attributes is given in Table 5.6.

Table 5.6 Analysis of a pipe-joining problem

Form	Pipe	T-connector	Tap
Attribute	copper: carries water under pressure, must be drained	copper: connected after pipe cut	plastic: on/off valve

The draining of the system causes a problem to the plumber, as does cutting the pipe exactly and ensuring a leakproof joint. In single words this gives:

Pipe – Cut – Connector – Tap – Seal

Combining Cutter/Connector/Seal/Tap offers the solution in Fig. 5.11.

Checklists

This, as the term implies, is a list, but it is a list of words set to trigger new or different ideas.

- *Combine* – join units, purposes or ideas.
- *Reverse* – turn it upside down, backwards, inside out, make positive negative, stop the moving, start the stationary.
- *Rearrange* – interchange components, other layouts, sequences, change pace.

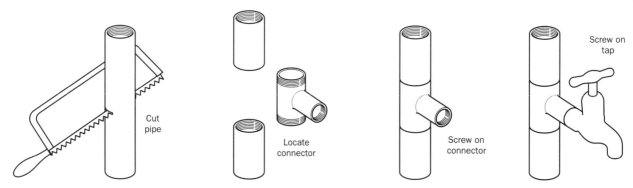

Figure 5.10 Traditional solution to join two pipes

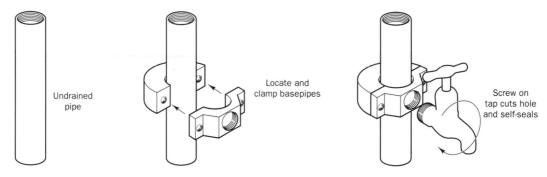

Figure 5.11 Alternative cutter/connector/seal/tap solution

- *Substitute* – other material, processes, ingredients.
- *Magnify* – frequency, strength, height, length, thickness.
- *Minify* – subtract, make smaller, lower, shorter, lighter.
- *Modify* – change motion, colour, sound, form, shape, smell.

Objectives tree

This is used to clarify customer and other requirements and may be useful in identifying alternative designs. This process develops a list of objectives which have differing levels. An example of an objectives tree for a lawnmower that is required to be safe, reliable, convenient to use and simple to make is shown in Fig. 5.12.

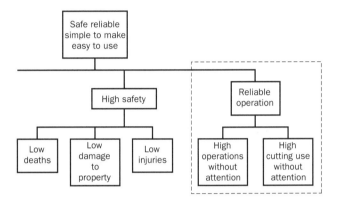

Figure 5.12 Objectives tree for a lawnmower

Performance specification

Good communication with the client or customer is essential for the production of a performance specification. Not only are the boundaries established but a specification requirement list of product alternatives, product types and product features are generated and may represent the ultimate reference source. The list may provide alternative, more economic, items for the most detailed elements, i.e. temperature control, speed, capacity, etc.

Self-assessment task

The performance attributes of a domestic mixing tap are as follows:

1. To have various water flow rates.
2. To have various water pressures.
3. To have various pipe connections.
4. To have various fitting requirements.
5. To be independent of external heating.
6. To have various loads to operate tap.

This general description is then transferred into a specification: 'Control the water temperature from the water drawn from the household hot/cold mains supply by a one handed light operating mixer tap.'
   Your task is to create a detailed performance specification of the above.

Evaluating design solutions

For evaluation to occur, consideration must be given to both the technical and economic characteristics of the design, while maintaining industrial and environmental safety requirements. A simple checklist of headings may assist in the evaluation (Table 5.7).

Table 5.7 Design evaluation check-list

Heading	Example
Function	Has it met the design specification?
Safety	No additional safety measures required, safety guaranteed?
Ergonomics	Good form design, no strain or impairment of health?
Production	No expensive equipment required, small number of simple components, few production methods?
Quality control	Simple and reliable procedures with few tests?
Assembly	Easy, quick, no special requirements?
Operation	Simple, long service life, easy handling?
Maintenance	Little, simple, easy inspection and repair?
Costs	Low, no special running or associated costs?

Processing capability

The term *processing capability* describes the measurement of maintaining the variation of a product, a process or service about a preset value.
   The designer employs mathematics to assist in the measurement of process capability in the form of *normal distribution curves* (Fig. 5.13). This relies on meeting specific values within a tolerance band and is very important in determining the capability of any process. The width of the normal curve can be broken down in to *standard deviations* (the larger the standard deviation the fatter the curve and hence more variation about a mean; the smaller the value the thinner the curve with less variation about the mean). This smaller value gives an indication of a more capable process.

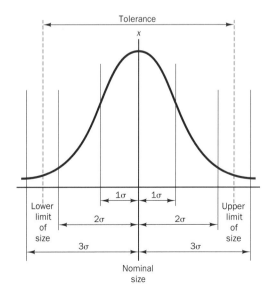

Figure 5.13 A normal distribution curve

Time

Time is very important to the engineering process; the development time of designs, or lead time, may range from hours to years. If too long is spent in the design stage competitors may steal the lead in advancing a product or process. A current example of this occurs in the computer technology industry, so much so that as a process or product is produced ready for the retail market, it has already been surpassed in the design stage by the next generation. An estimated percentage breakdown of man-hours spent on the development and evaluation of designs may be seen in Table 5.8. As can be seen from the table, the greatest amount of time is devoted to the conventional activities – that is, the firming up of design concepts involving calculations and layouts.

Table 5.8 Time spent in evaluation and design

Stage	Proportion of time (%)
1. Clarifying the task	10
2. Identifying essential problems	1
3. Establishing functions	4
4. Searching for solutions (i.e. brainstorming)	19
5. Combining solutions	3
6. Preliminary concept calculations/designs	25
7. Preliminary concept layouts	35
8. Evaluating concept designs	3
Total	100

Economics

Evaluation of cost, ease of assembly, reliability and maintainability are of equal importance to the overall design as the functional and safety characteristics.

Estimating the production costs can be the most difficult task a designer faces. It is important that an estimate is made early in the design process, so that it may be compared to the original requirements. In general, as the design is refined so should the cost estimate; as the design develops, the cost estimate must converge on the final cost. The total cost of a product is based on two broad areas, *direct costs* and *indirect costs*. Direct costs relate to the actual production of products and may be split further into material, labour and expenses. Indirect costs are cost not directly related to the production of a product, and are often termed *overheads*.

Figure 5.14 shows how the designer has an impact on the direct costs, as tooling, materials, labour and the purchase of parts have an influence on the final manufacturing costs.

Output characteristics

Thought has to be given not only to the design of the product or process, but to the whole life cycle, including installation, operation, maintenance and disposal.

Installation requires the designer to compile detailed instructions for unpacking, making the necessary connection for power, support services, environmental issues and any start-up procedures and testing requirements.

Operational requirements generated by the designer include, actual operational instructions (covering a normal range of activity), emergency and shutdown operations and any fault-finding instructions.

Maintenance instructions and procedures, including diagnostics, need to be documented during the product design stage. This enables preventative maintenance, failure analysis and repairs to be a part of a useful serviceable product or process life.

Disposal or the decommissioning of product and processes have to be considered at the design stage and in the overall evaluation of the design. This part of the design cycle is often overlooked and may result in disastrous consequences. An example of this has become evident now that some of the older nuclear plants have served their allotted serviceable time and require decommissioning. The initial plant designs failed to take into account the real dangers of exposure to toxic contamination, thus new sophisticated and expensive disposal procedures have now been adopted to safeguard society.

Safety

Throughout the design process the designer must observe that a considerable concern for safety is uppermost. The evaluation stage of the design must endorse this by attempting to single out attributes of the design and subjecting them to critical analysis just on the ground of safety.

Portability

Once a general design has been conceived there is often an evaluation aimed at an attempt to minimise it – that means trying to make a more compact version of the design. This is mainly done in order to satisfy one of two criteria:

- reduce costs
- to make the design portable

Making a design portable may offer it an extended lifespan or give it an increase in application. The evaluation and redesign of computers and their components led from a computer that once filled a normal family sized room to one that sat on top of a desk (semi-portable as it needs an independant power supply) then to one that is truly portable, i.e. a lap top.

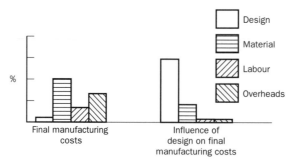

Figure 5.14 The designer's impact on direct costs

5.3 Use technical drawings to communicate designs for engineered products and engineering services

Communication is the cornerstone of engineering for it conveys thoughts, ideas and concepts through to final solutions. The most widely used form of communication, especially as an international language, is that of graphical communication which is used extensively in engineering as drawings.

Using graphical methods to communicate final design solutions

There are many types of drawing methods and conventions, so a common procedure has been adopted in order to standardise the communication process. This standard is laid down in the British Standards publication 'Engineering drawing practice' BS 308: 1972.

Part 1: General principles
Part 2: Dimensioning and tolerancing of size
Part 3: Geometrical tolerancing

It outlines four main types of drawing methods, such as:

- detailed drawings
- working drawings
- subassembly drawings
- general arrangement drawings

This standard of communication is supplemented by other methods, such as:

- sketching
- circuit diagrams
- flow diagrams
- charts and graphs

Supplementary methods of communication through drawings

Let us first look at the supplementary methods of communication employed by the engineer.

Sketching

This is a widely used form of communication and is employed very rapidly utilising very little instruments and costing next to nothing to produce. The engineer needs to be able to sketch to be able to:

- Form more concrete visual ideas.
- Convey ideas, detail or information to others.
- Record items for later use.
- Aid the production of working drawings.

There is no need to be an artist in order to communicate by sketching, but there are some simple aids to sketching that may prove useful:

Straight lines
For long lines (over 150 mm) the placement of two shorter end lines then joined together usually produces the desired effect (Fig. 5.15(a)). Another method is to break the long stroke of the line up into shorter sections (Fig. 5.15(b)).

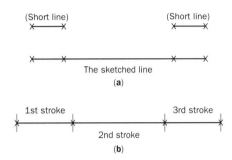

Figure 5.15 'Short section' method of drawing a long straight line

Curved lines
These present a different problem, that of trying to keep a curve line rotating about an axis. One particular method to achieve this is to draw up a box-like shape of similar size to the finished curve or circle and complete the curve or circle in quadrant sections (Fig. 5.16(a)). A similar method uses four equidistant marks at right angles to each other, which are then joined together (Fig. 5.16(b)).

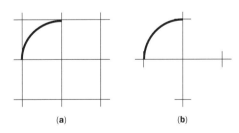

Figure 5.16 Drawing a curve freehand

For a smoother circle try drawing another four similar distant marks placed at 45° in each quadrant and then draw through the eight marks.

An approach to sketching clean lines is to draw box-like shapes faintly first, in good proportion starting with the smallest items then completing a bold outline (Fig. 5.17). This keeps good proportion with strong bold lines indicating the essential parts of the drawing and no fuzzy like lines confusing intricate detail.

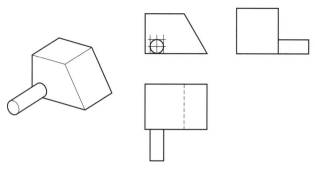

Figure 5.17 Rough sketches in proportion

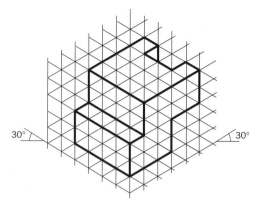

Figure 5.18 Sketching on isometric graph paper

Self-assessment task

Using the method mentioned above, sketch these four components.

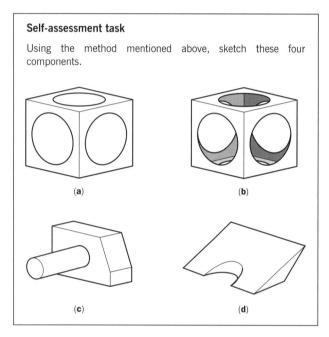

A very useful aid to sketching is the use of lined graph paper, especially isometric graph paper which has lines drawn at 30° to the horizontal. This enables sketching to be reasonably quick and better proportioned (Fig. 5.18).

Circuit diagrams

This type of drawing is used as a working drawing, yet it often looks like a young child's stick-like drawing. There are two key elements to remember when reading circuit diagrams:

1. A symbol is used to represent a component (i.e. electronic, pneumatic or hydraulic). This symbol has no relationship to the component's size, shape, colour or orientation, but only its function.
2. The symbol is positioned within the drawing so that a clear and concise diagram which is easy to follow is produced. This position bears no connection to the layout of actual components used in manufacture.

Typical symbols used in electronic circuitry (Fig. 5.19) enable designers to have a common set of symbols from which to

Component	Symbol
Basic resistor	─▭─
Permitted alaternative	─/\/\─
Variable resistor	─▭─
Variable resistor – preset	─▭─
Voltage divider with moving contact	─▭─
Capacitor	─╢╟─
Polarised electrolytic capacitor	─╢╟─
Variable capacitor	─╫─

Figure 5.19 Symbols for resistors and capacitors

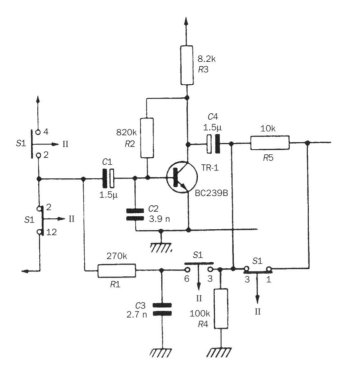

Figure 5.20 Diagram showing the connections between components

draw upon when designing. This then leads to less misinterpretation when designs are passed back and forth between designers.

Note that the construction between the connection of components (box-like shapes) are shown as joined by thin continuous lines (Fig. 5.20). This again is a convention adopted by designers in order to represent the connections made within a circuit without having to draw in full any detail such as soldering points or screwed connectors.

Flow diagrams

These are often box-like or family-tree shaped and are used to explain a process in logical steps. They are used extensively in computer programming and command structuring. A typical flow diagram is shown in Fig. 5.21, which describes the step-by-step approach of answering a telephone.

Charts and graphs

Technical information essential to the production or servicing of products which require referencing need a communication method that is quick, unambiguous, easy to understand and international in language. A standard means of achieving this is by the use of charts and graphs, the most visual form of presentation. There are several common forms of chart, including:

- pie chart
- bar chart
- pictograms

The four main methods of communicating through drawings mentioned earlier (detailed drawings, working drawings, subassembly drawings and general arrangement drawings) require draughting skills and an understanding of the conventions used for an effective transfer of information.

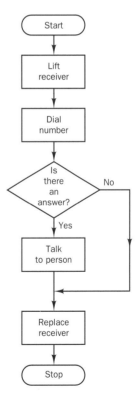

Figure 5.21 A flow diagram

Main methods of communication through drawings

Detailed drawings

These are sometimes known as component drawings and are usually a drawing of a single part, containing any information necessary for the manufacture or service of that part, this includes being fully dimensioned (Fig. 5.22)

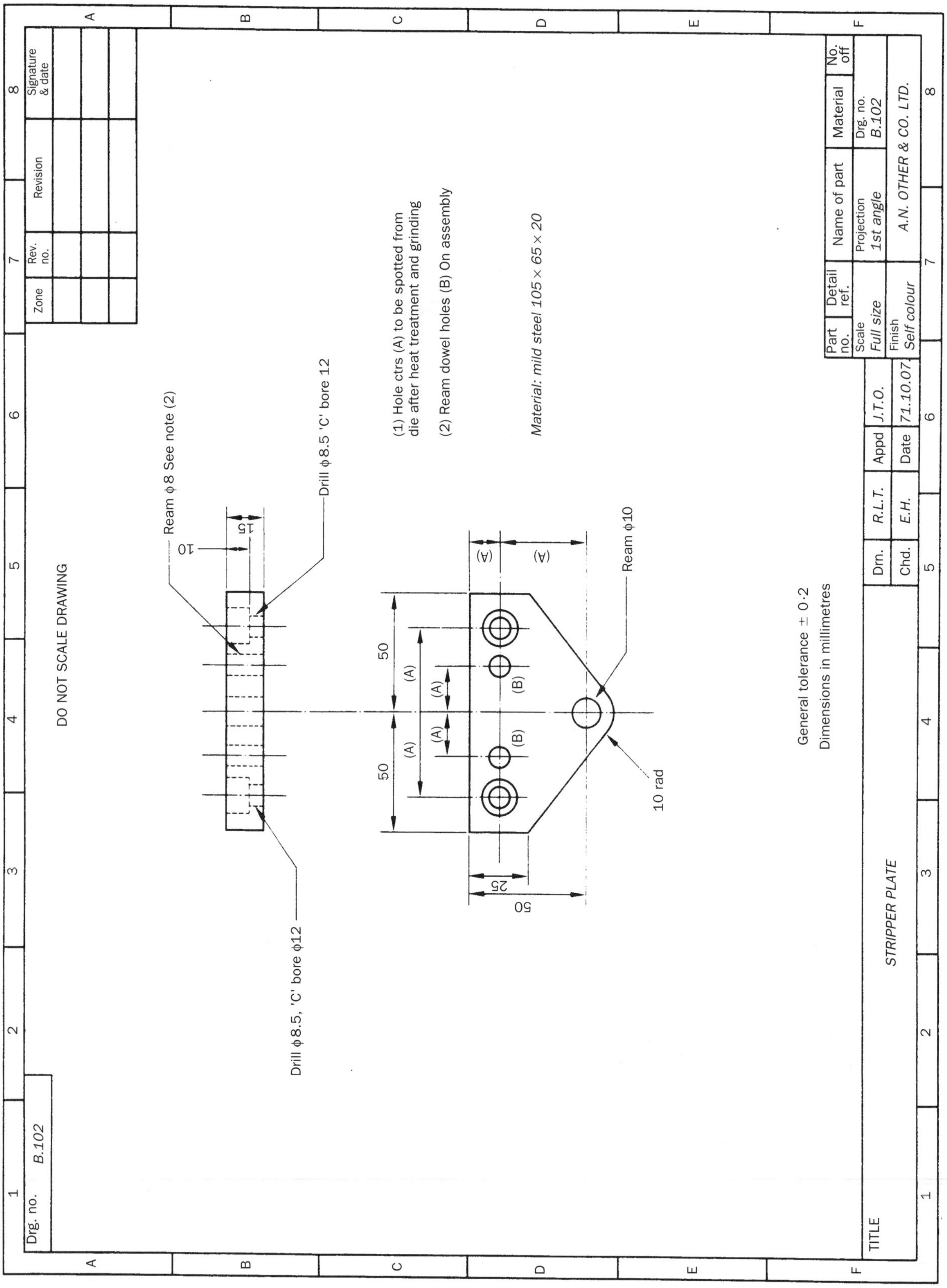

Figure 5.22 A detailed drawing of a component

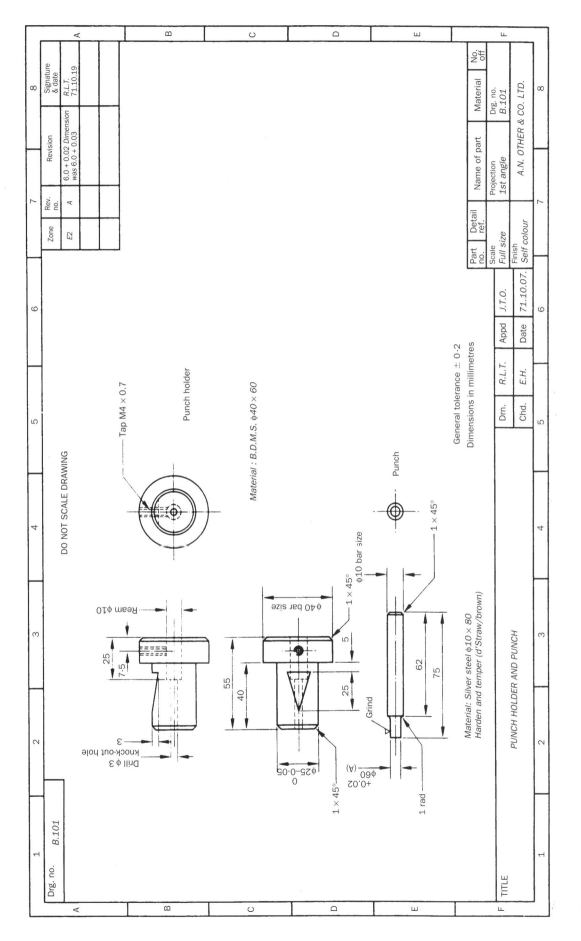

Figure 5.23 A collection of detailed drawings forming a working drawing

Working drawings

These differ from detailed drawings in that they are a collection of detailed drawings collated and placed on one drawing sheet, still requiring to be fully dimensioned (Fig. 5.23).

Subassembly drawings

A subassembly drawing shows a collection of parts fitting together (Fig. 5.24), but this covers only a small part of the overall arrangement. An example of this is the drawings of the fitting together of a plastic model aircraft wing, which would show only the associated wing parts and not the body or the nose cone or cockpit of the plane. These types of drawings often have parts lists and cross-referencing information to enable the assembly to be made easier. They are quite often sectioned to enable internal features to be seen.

General arrangement drawings

The whole layout or arrangement is shown in general arrangement drawings, which are also known as general assembly drawings (Fig. 5.25). They usually only indicate overall dimensions and final assembly instructions (i.e. if shafts are right handed or left handed, vertical or horizontal, etc.).

Very little of the internal detail is indicated on general arrangement drawings as this could offer too much information on one drawing and thus cause confusion.

Interpreting the detailed drawing

The main focus for manufacture comes from the communication of the detailed drawing, which must contain universal standards and conventions if it is to be understood effectively. There are various methods of conveying this information, the following offer some of the more widely used.

Projections

The form of projection extensively used in engineering is that of orthographic projection. This represents three-dimensional objects or components by two-dimensional (flat) views. There are two common conventions, first-angle projection and third-angle projection (Fig. 5.26), which consist of projected views called:

- front view
- side view
- plan view

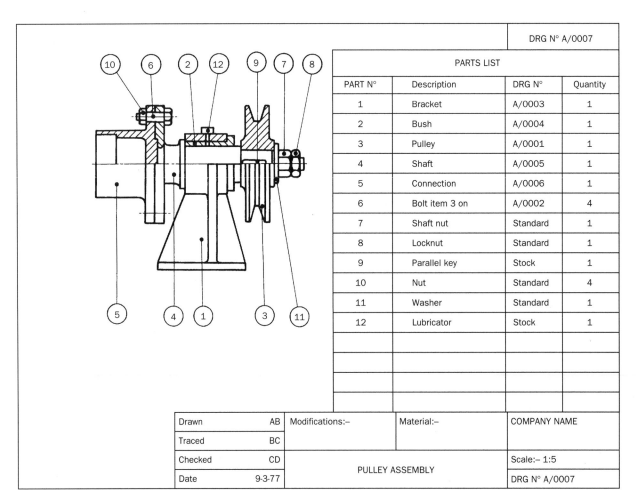

Figure 5.24 Typical subassembly drawing

Unit 5: Design Development

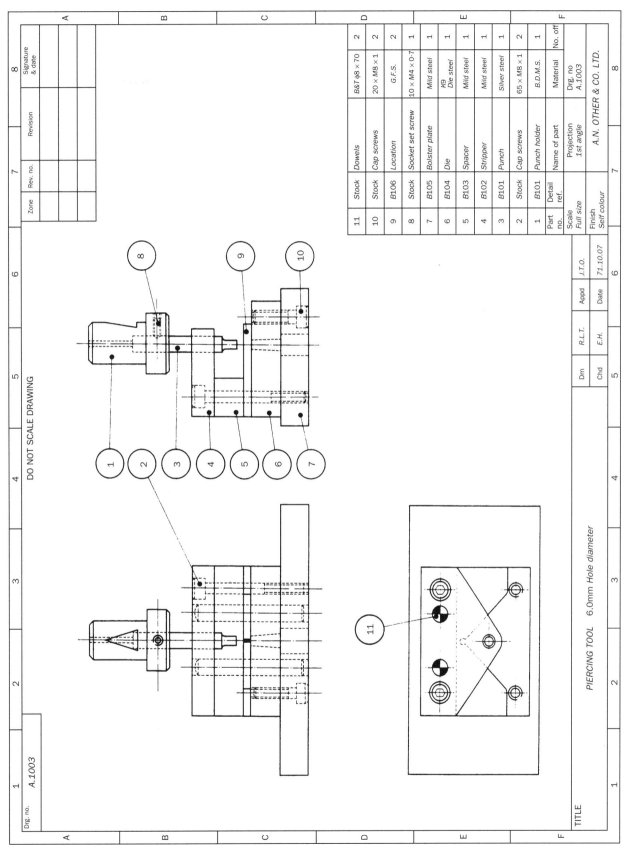

Figure 5.25 General arrangement drawing

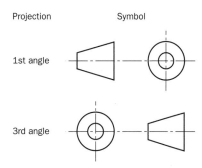

Figure 5.26 Orthographic projection

Typical examples of the layout of first- and third-angle projections are shown in Fig. 5.27.

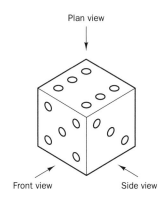

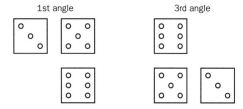

Figure 5.27 Typical examples of first- and third-angle projection

Self-assessment task

Using first-angle projection draw the plan view, side view and front view of the shape below.

Isometric projection

This projection gives a three-dimensional look to a shape as the projection is based on three main axes. Of these three axes (Fig. 5.28), one is a vertical axis and the other two are inclined at 30° to the horizontal at either side of the vertical. This may give a pictorial view which appears a little out of proportion (due to the lack of a vanishing point), but that can be remedied by slightly reducing the inclined axes to approximately 27°.

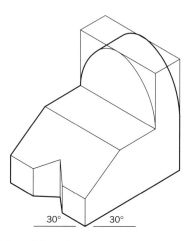

Figure 5.28 Isometric projection

Circles are difficult to produce accurately in isometric form. A simple method of obtaining good results requires that a circle is first drawn as an orthographic view then divided up into a number of equally spaced labelled sections (Fig. 5.29). The 30° axis for the isometric projection is made and equally divided labelled sections are also made; the corresponding lengths are then transferred from the orthographic view to the respected isometric view (e.g. 1–b_1, 2–c_1, etc.). The marks made are then all joined together freehand and result in an isometric circle (Fig. 5.30).

Sectioning

The interior features of a component may be difficult to show even using hidden detail, especially if there is extensive detailing causing confusion among hidden detail lines. This can be rectified by drawing a sectional view, which is achieved by taking a cut with an imaginary saw through part of the component and removing this cut section (Fig. 5.31). The view that remains will show solid material where the cut was made. This part of the component is hatched with thin lines at 45° to the horizontal to indicate it is a cut-away part of a solid object.) There are certain components or parts of components which are never hatched, such as ribs, bolts, washers, studs and screw threads.

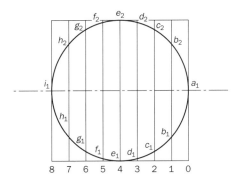

Figure 5.29 Orthographic view

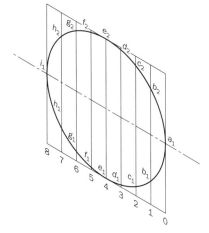

Figure 5.30 Isometric view

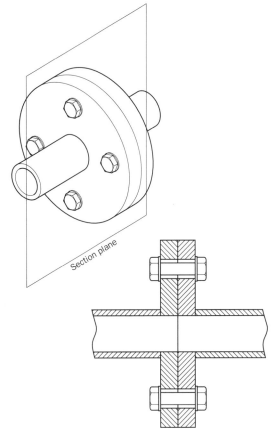

Figure 5.31 A sectional view of a component

Computer-aided design (CAD)

This is just an extension of the design engineers' tools for it is they who remain the designers and makes the decisions. There are many claims that CAD increases productivity; some are true but others fall far short of the early optimism generated by the introduction of CAD. Symbolic or schematic drawings are usually significantly faster to draw, as is any part which is repetitive, but in general mechanical applications take about the same time as manual methods.

There are benefits arising from the use of a CAD system, such as:

- Storage and recall of large numbers of drawings.
- Uniformity of drawings.
- Existing drawings are easier to modify.
- Bills of materials list are generated automatically.

There are also some disadvantages of employing a CAD system, such as:

- High cost of equipment which is soon out of date.
- Training requirements for designers.
- Maintenance of equipment and associated costs.
- A systems manager is often a requirement.

Using graphical methods to communicate final designs for engineering services

An appreciation of how to draw, the skills required and the conventions used, is essential for all service engineers, as they must be able to read and understand drawings and diagrams, often as a localised part of the process or system (i.e. coin rejection system of a vending machine). To enable them to gain a rapid understanding of the the workings of a system there are certain types of drawing convention more suited to their needs, such as:

- exploded views
- schematics

Exploded views

An exploded view may reveal more information than a sectioned view about the assembly of components, especially if there are a number of separate components. They may be represented either in an orthographic projection (Fig. 5.32) or as a pictorial isometric projection (Fig. 5.33). Both these forms are used extensively as part of assembly instructions or in maintenance manuals.

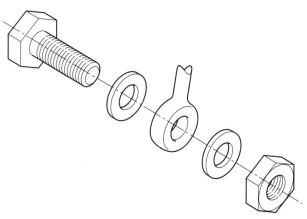

Figure 5.32 Exploded view: orthographic projection

> **Self-assessment task**
>
> Construct logic circuit schematics using AND and OR gates only to give an output of:
>
> (a) $X = A$ OR B & C OR D & E OR F
>
> (b) $X = A$ & B OR C & D OR E & F
>
> (c) $X = A$ OR B OR C OR D & E OR F
>
> (d) $X = A$ & B & C OR D OR E & F

Figure 5.33 Exploded view: isometric projection

> **Self-assessment task**
>
> Draw an exploded isometric view of a common ball point ink pen. Typical components are nib, ink tube, outer case, end stopper and cap.

Schematics

These type of drawings convey information similar to flow diagrams in that they use shapes to represent components (Fig. 5.34). The shapes are usually from a group of symbols which form a convention (i.e. logic gates BS 3939, hydraulic components, etc.). Many of these are now already pre-drawn on computer-aided design (CAD) packages, in symbol libraries ready to be inserted on a drawing by just selecting its name.

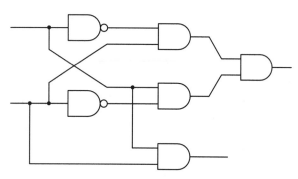

Figure 5.34 A schematic drawing

Drawing standards and conventions

The requirement for engineers to communicate designs from minute precise features such as microchip circuits to elaborate construction projects like bridges, meant that conventions and standards had to be established.

In the case of mechanical engineering, BS 308 fulfils that role, while BS 3939 enables the electronics designers to fulfil theirs.

The standard convention BS 308 includes elements such as:

Part 1: General principles

- Types of line.
- Layouts of drawings.
- Projections.
- Sections.
- Conventional representation of common features.
- Abbreviated terms used.

Part 2: Dimensioning and tolerancing

- General principles of dimensioning.
- Tolerancing of dimensions.
- Dimensioning common features.
- Surface texture symbols.

Part 3: Geometrical tolerancing

- Symbols.
- Interpretation of form and position.
- Maximum metal condition tolerancing.
- Co-ordinate and position tolerances.
- Typical applications.

Examples of the above are shown in Fig. 5.35.

Parts lists for technical drawings

The term 'parts list' is sometimes known as a Bill of Material (BOM); it acts as an index to the product or process. These are usually applied to assembly drawings and consist of six pieces of useful information:

- An item number or letter.
- A part number identifying the component for stock control, purchasing and manufacture.

Example	Type of line	Line width mm	Example of application
A ————	Continuous (thick)	0.7	Visible outlines and edges
B ————	Continuous (thin)	0.3	Fictitious outlines and edges Dimension and leader lines Hatching Outlines of adjacent parts Outlines of revolved sections
C ∼∼∼	Continuous irregular (thin)	0.3	Limits of partial views or sections when the line is not an axis
D – – – –	Short dashes (thin)	0.3	Hidden outlines and edges
E —·—·—	Chain (thin)	0.3	Centre lines Extreme positions of movable parts
F —·—·—	Chain (thick at ends and at changes of direction, thin elsewhere)	0.7 0.3	Cutting planes
G —·—·—	Chain (thick)	0.7	Indication of surfaces which have to meet special requirements

- *Dashed lines.* Dashed lines should comprise dashes of consistent length and spacing, approximately to the proportion shown in the examples in the table.
- *Thin chain lines.* Thin chain lines should comprise long dashes alternating with short dashes. The proportions should be generally as shown in the table but the lengths and spacing may be increased for very long lines.
- *Thick chain lines.* The lengths and spacing of the elements of thick chain lines should be similar to those of thin chain lines.
- *General.* All chain lines should start and finish with a long dash and when thin chain lines are used as centre lines they should cross one another at solid portions of the line. Centre lines should extend only a short distance beyond the feature unless required for dimensioning or other purposes. They should not extend through the spaces between views and should not terminate at another line of the drawing. Where angles are formed in chain lines, long dashes should meet at corners. Arcs should join at tangent points. Dashed lines should also meet at corners and tangent points with dashes.

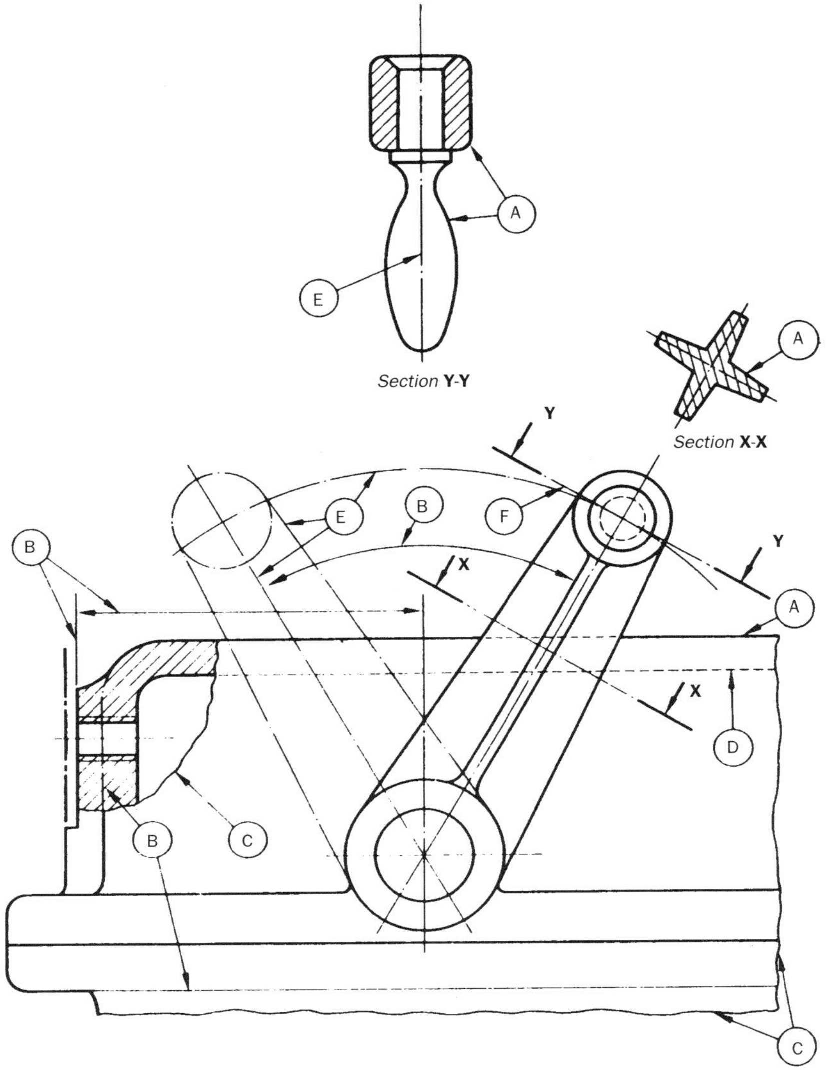

Figure 5.35 Types of line and their applications: thicknesses and proportions

- The quantity required.
- A brief description or title of the component.
- The material the component is made from.
- The source of the component.

A typical bill of material is shown in Table 5.9.

Table 5.9 Example bill of materials

Item	Part	Qty	Description	Material	Source
1	Z-1123	2	Top head	Mild steel	Brown's
2	Z-1124	2	Base	Cast iron	Foundry
3	CT-345	1	Seal	Rubber	Smiths
4	–897	4	M10 bolts	Mild steel	Stock

5.4 Unit test

Test yourself on this unit with these multiple-choice questions.

1. When identifying customer requirements, maintenance of the product has to be considered. Which of the following is a maintenance requirement?

   (a) marketing
   (b) servicing
   (c) ventilation
   (d) display

2. Which of the following design terms relate to the appreciation of beauty or good taste?

   (a) quality
   (b) ergonomics
   (c) aesthetics
   (d) geometry

3. When designing, careful consideration has to be made regarding costs. Which of the following is NOT a major contributor to this design aspect?

   (a) simplification
   (b) standardisation
   (c) elimination of complexity
   (d) detail

4. Which of the following is regarded as a universal standard?

   (a) BSI
   (b) DIN
   (c) CEN
   (d) ISO

5. Which of the following is regarded as a European standard?

   (a) BSI
   (b) DIN
   (c) CEN
   (d) ISO

6. To maintain safety standards within the workplace, the layout of the environment is significant. Under the Factories Act 1961, the general minimum requirement of floor space to be provided to each person is?

   (a) 3.7 square metres
   (b) 4.7 square metres
   (c) 7.3 square metres
   (d) 7.4 square metres

7. Noise is a form of pollutant and designers have to be aware of thresholds. Which of the following would be the maximum recommended exposure time to 90 dBA of noise?

   (a) 20 hours
   (b) 10 hours
   (c) 8 hours
   (d) 4 hours

8. Which of the following terms BEST describes the force that involves a twisting motion?

   (a) bending
   (b) tension
   (c) compression
   (d) torsion

9. Designers have to be aware of the differing types of energy available. Which of the following makes use of vibrating air?

   (a) chemical energy
   (b) sound energy
   (c) nuclear energy
   (d) electrical energy

10. Environmental considerations make the designer aware of power sources. Which of the following is NOT a CLEAN power source?

    (a) wind
    (b) wave
    (c) solar
    (d) oil

11. Which of the following is NOT a major operational requirement of performance?

    (a) durability
    (b) repeatability
    (c) size
    (d) speed

12. Evaluating design briefs requires categorising information. Which of the following areas would determine if the information presented is correct?

    (a) reliability
    (b) value
    (c) precision
    (d) quality

13. Designs are often built based on a series of functions. Which function applies the joining or locating of other functions?

    (a) basic function
    (b) adaptive function
    (c) auxiliary function
    (d) customer specific function

14. Designers considering factors such as: size, force, movement and direction would place these as being within which attribute section shown?

    (a) appearance
    (b) cost
    (c) capacity
    (d) reliability

15. Evaluating design solutions enable a designer to categorise functions. Which function would BEST describe, good form design with no strain or impairment of health?

    (a) ergonomics
    (b) safety
    (c) production
    (d) quality control

16. Output characteristics of designs mean thought has to be given to the whole life cycle of the product. Which stage of this life cycle would the designer give consideration to emergency operations?

    (a) installation
    (b) operational requirements
    (c) maintenance
    (d) decommissioning

17. When drawing which of the following BEST describes a collection of parts fitting together?

    (a) detail drawings
    (b) projections
    (c) sub-assembly drawings
    (d) general arrangement drawings

18. Mechanical designers communicate through the standard convention of BS 308. This is broken down into 3 sections, one of the sections concerns Geometrical Tolerancing. Which of the following elements of convention is from that section?

    (a) maximum metal condition
    (b) abbreviated terms
    (c) tolerancing of dimensions
    (d) surface texture symbols

19. Electrical & Electronic graphical symbols conform to BS 3939, the symbol represented by a circle with a cross in its centre is which component?

    (a) a battery
    (b) an aerial
    (c) a filament lamp
    (d) a signal lamp

20. Within BS 308 a line which is drawn as intermittently dashed, represents the application?

    (a) hidden detail
    (b) cutting planes
    (c) hatching
    (d) projection lines

Engineering in Society and the Environment

This unit consists of three main elements.

- Investigate the effects of engineering on society.
- Investigate the effects of engineering on the working environment.
- Investigate the effects of engineering on the physical environment.

The first element considers the effects of engineering on society and how it effects the economy on a local, regional, national and European scale. The first section also considers the effects of engineering technology on the social environment, on employment patterns and on working conditions.

The second element considers the effects of engineering on the working environment more deeply, including the ergonomics of the working environment. It also considers the precautions that must be taken against risks in the working environment.

Finally, the third element considers the effects of engineering activities on the physical environment, for example, the effects of the materials used and the effects of the waste products of manufacturing activities. It also considers how the functions of environmental legislation affects engineering activities.

6.1 Investigate the effects of engineering on society

After reading this element the student will be able to:

- Identify the contribution of each engineering sector to the economy using quantitative data.
- Describe the effects of engineering technology on the social environment.
- Explain the effects of engineering technology on employment patterns.
- Explain the effects of engineering technology on working conditions.

The contributions by the engineering sectors to the economy

Basic economics

The terms gross national product (GNP) and gross domestic product (GDP) appear frequently in the financial pages of the newspapers. To understand these terms let's look at a small family unit and then see how the economic principles involved can be expanded to a national scale.

For this example our family consists of a father who is unemployed but receives an income from the redundancy money he deposited in a building society together with money inherited from a relative. He also does part-time gardening. A mother who designs and produces knitwear at home to augment their income. A son who is in his last year at school but works part-time at a weekend on the forecourt of a local petrol station to save up for a motorcycle. Their individual and total incomes are as shown in Fig. 6.1.

		£
Father	Investment income from building society	3,600
	Part-time gardening	4,050
Mother	Income from making and selling knitwear	4,500
Son	Part-time job at a petrol station	900
		13,050

Figure 6.1 The total annual income (gross) for the family

Another way of looking at the same figures is to consider the values of the outputs (products) of the individual family members. The mother makes her knitwear garments and sells them. It is easy to see that her income results from the products (outputs) she sells. In the sense of this example, services which are rendered and paid for also count as products. Therefore the part-time work their son does at a weekend also counts as a product. Again, the father is doing the building society a service by depositing his redundancy money and his inheritance in a monthly income account. He has loaned them the money in the return for an interest payment. This service also counts as a product in this sense. He also provides a part-time gardening service for his neighbours and friends for which they pay him. The values of the individual products and the total product value for the family is shown in Fig. 6.2.

		£
Father	Loan of capital to the building society (service)	3,600
	Part-time gardening (service)	4,050
Mother	Sale of knitwear (goods)	4,500
Son	Sales skills (service)	900
		13,050

Figure 6.2 The total or gross family product per annum

Another name for the total value of the family's product could be *gross family product*. In this context, *gross* means total with no deductions. If we did this exercise for every individual and family in the UK who are producing goods and services and then added them all together we would have the *national product*.

As well as looking at the incomes and outputs (products and services) of individuals and families, we can also look at the national product of the goods themselves. Let's take the knitwear produced by the mother in our example. The sheep farmer recovers his costs and makes a profit when he sells the fleece from his sheep to the yarn manufacturer. The yarn manufacturer turns the fleece into knitting wool and recovers the production costs and makes a profit by selling the wool at a higher price than that paid for the fleece. The difference between the price paid for the fleece and the selling price of the wool made from it is the *added value*. The mother sells the knitwear for more than she pays for the wool. This increases the added value still further. This happens at each stage in the chain from the initial exploitation of a natural resource (the fleece) to the retail sale of the knitted garment in a shop. The sum of all the added values for the garment is the *national added value* for that garment. This is shown in Fig. 6.3. If we add together the national added values for all the knitwear garments manufactured in the UK, we then have the national added value for that industrial sector.

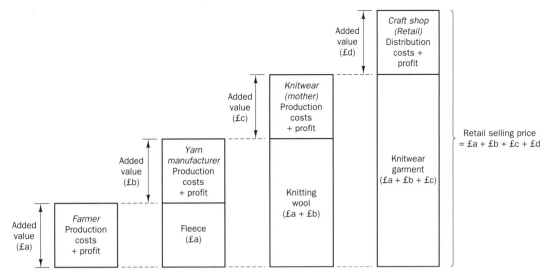

Figure 6.3 National added value for a garment

> **Self-assessment test**
>
> 1. Calculate the retail cost of a knitwear garment, given that:
>   (a) The farmer sells a fleece for £25.
>   (b) The added value of the yarn manufacturer is £15.
>   (c) The production costs and profit for knitting the garment is £80.
>   (d) The craftshop mark-up is £60.
> 2. Calculate the *national added value* for this type of garment if 30,000 persons in the UK are making and selling them.

Note how the added value for the garment is the same as the retail selling price or *product value* in cash for the garment. However, as well as producing knitwear or providing services our family also buys food, clothes, energy and all the other things that they need. Thus goods, services and the money that pays for them are in constant circulation. They form a closed system as shown in Fig. 6.4. The goods and services are produced to satisfy the consumers' demand for such goods and services. The consumers also work in the industries that produce the goods and services. They are paid wages and use this money to buy other goods and services. Thus the money also circulates as shown in Fig. 6.4.

If production does not satisfy consumer demand, there will be too much money chasing too few goods and prices and profits will rise causing *price inflation*. If production exceeds consumer demand, there will be difficulty in selling all the products made. Prices and profits will have to fall and there will be *price deflation*. This will cause some producers to go out of business, so reducing the level of supply and the balance between supply, and demand will be restored.

Gross national product (GNP)

We saw earlier that the gross family product was the sum of the monetary value of all the goods and services produced by the members of our imaginary family. Similarly the *gross national product* for the UK can be can be found by adding

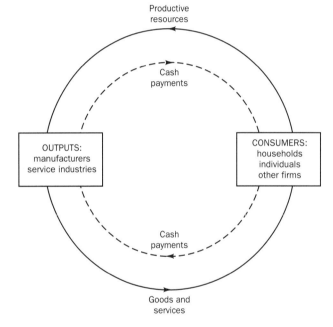

Figure 6.4 Example of a closed system

together the monetary values of all the goods and services produced by the UK. As previously, the word 'gross' means the sum of all the goods and services produced with no deductions for loss in value of the plant, equipment, buildings, etc. This loss in value being the result of wear and tear brought about by normal use.

We have to be a little careful about the word 'national' in this context. It includes:

- Outputs from resources within the UK owned by citizens of the UK.
- Outputs from resources abroad but owned by citizens of the UK.

It does not include:

- Outputs from resources within the UK but owned by people from other countries who are not citizens of the UK.

Therefore we can define GNP as the total monetary value of the outputs of all the resources owned by the citizens of the UK no matter where those resources may be.

Gross domestic product (GDP)

This time we total up the monetary values of all the outputs produced within the boundaries of the UK no matter who owns the resources – UK citizens or foreigners. This is the *gross domestic product*. However, this time, we do **not** include any outputs produced overseas even if the resources are owned by UK citizens. When determining the health of the UK economy, the GDP is usually considered more important than the GNP. The government issues statistical information on the performance of the UK economy via the *Central Statistics Office*. This information is compiled in the 'Blue Book' which is entitled *United Kingdom National Accounts*. An example of the information provided is given in Table 6.1.

Gross national product and gross domestic product per head

Both these figures are produced in a similar way depending upon whether we use the GNP or the GDP as our starting point.

GNP per head
= (GNP) / (number of persons employed in producing the outputs)

GDP per head
= (GDP) / (number of persons employed in producing the output)

Both these calculations are a measure of *productivity*. Since the latter calculation reflects the production within the UK, it is a more useful measure of the efficiency of the UK industry and commerce. Table 6.2 is based upon government statistics and shows the relationship between output, employment and productivity for the UK over a number of years. The figures shown are not actual quantities but are an *index of performance*. That is, a particular year is chosen to represent 100 per cent as a basis for calculation and comparison. The following years' figures are represented as a percentage increase or decrease compared with the base year.

Table 6.2 Output, employment and productivity for the period 1990–95

Year	Output	Employment	Output per head (Productivity)
1990	100.0	100.0	100.0
1991	100.5	100.1	100.4
1992	100.6	100.0	100.6
1993	100.5	99.5	101.0
1994	99.9	99.4	100.5
1995	90.5	90.0	100.6

Self-assessment task

State which of the following conditions represents a rise in productivity and which represents a fall in productivity.

1. Output rises more rapidly than employment.
2. Output rises more slowly than employment.
3. Output falls more rapidly than employment.
4. Output falls more slowly than employment.

Employment

Figure 6.5 shows the structure of a typical manufacturing company. You can see from the figure that some of the persons employed are actually making the products sold. Other persons are providing essential backup services such as selling the products, writing the letters, working out the wages and doing the accounts. The persons actually making the products are called the *direct labour force* since they are directly employed in manufacture. The persons employed in the backup services are called the *indirect labour force*.

Sometimes the persons directly employed in one industry may yet be indirectly employed by another. For instance the lorry driver delivering the goods made by the company in Fig. 6.5 is working directly for the road haulage firm employing him, but he is working indirectly for the manufacturing company whose goods he is transporting.

Self-assessment tasks

1. With reference to Fig. 6.5, list which of the elements (boxes) of the company represent direct labour. Give reasons for your choice.
2. Do all the remaining elements represent indirect labour? If so, why?

Table 6.1 Sample extract from the CSO *Blue Book: United Kingdom National Accounts (1994 edition)*

(Figures are in £million)	1983	1984	1985	1986	1987	1988	1989	1990	1991	1992	1993
All industries											
Income from employment	169,847	181,406	196,858	212,380	229,832	255,634	283,454	312,358	329,609	342,215	352,896
Gross profits and other trading income	86,225	92,843	101,904	105,600	121,785	136,220	151,124	154,251	140,013	145,570	162,193
Rent	18,857	19,816	21,875	23,848	26,155	29,904	33,730	38,569	44,707	49,193	52,872
Imputed charge for capital consumption	2,498	2,619	2,830	3,068	3,307	3,634	4,005	4,391	4,363	4,207	3,942
less Stock appreciation	−4,204	−4,513	−2,738	−1,835	−4,727	−6,375	−7,061	−6,131	−2,010	−1,832	−2,359
less Adjustment for financial services	−11,893	−12,688	−12,827	−14,789	−15,677	−17,589	−23,493	−24,552	−20,782	−23,326	−23,741
Statistical discrepancy (income adjustment)	−105	−1,170	—	—	—	—	—	—	—	—	317
Gross domestic product	261,225	280,653	307,902	328,272	360,675	401,428	441,759	478,886	495,900	516,027	546,120

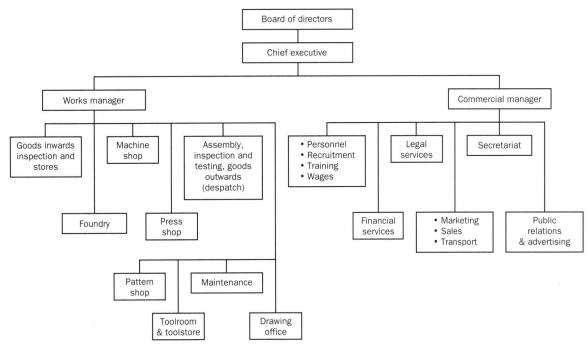

Figure 6.5 Structure of a typical manufacturing company

Exports

Exports are the goods and services we sell to other countries. There are two sorts of exports:

- *Visible exports* are actual goods that you can see and handle, such as motor cars, machinery, clothing and foodstuffs, that are shipped abroad and for which we receive payment.
- *Invisible exports* are services for which we are paid by overseas countries. Examples of invisible exports are: financial services such as banking and insurance. Tourism is also an invisible export since the people from foreign countries visiting the UK bring and spend their money here. This is a very important invisible export and brings much foreign currency into this country. Technical training and consultancy services for our overseas customers are another form of invisible export no matter whether the service is provided in their own country or within the shores of the UK.

Imports

These are the goods and services we buy from overseas countries, such as cars from Europe and the far east, airliners from America and wine and food from France. In fact any activity that takes money out of the UK represents an import. When we go abroad on holidays we take money out of the UK into the country we are visiting. This transfer of money out of the UK makes our holiday as much an import as buying goods from that country. Similarly, buying a holiday-home abroad also takes money out of this country and counts as an import even though the building remains in the country in which it was built.

Balance of payments

Whether, as individuals, we buy goods by credit card, by hire–purchase or for cash, we eventually have to pay for the goods from our earnings. Therefore our personal income has to be equal to, or greater than, our expenditure or we get into financial difficulties. The same applies to nations, the financial value of their exports must equal or exceed the financial value of their imports.

In the UK the value of our imports usually exceeds the value of our visible exports. Fortunately our invisible exports in the form of banking, insurance and other financial services, together with tourism and inward investment makes up the balance and keeps the country solvent. So that businessmen and financiers can see the state of our trading balance, the government publishes monthly *balance of payment* figures. Because of delays and difficulties in collecting the necessary data, these figures are only a rough guide in the short term and have to be revised and corrected from time to time.

Sectors of the engineering industry

The engineering industry can be divided up into a number of sectors. Each sector makes its own contribution to the national economy. We can consider the main sectors to be as follows:

- *Chemical engineering* is responsible for the development, production and marketing of plastics, lubricants, fuels, paints, fertilisers, pharmaceuticals, etc. Leading firms in this sector include ICI, British Petroleum, Glaxo and Fisons.

- *Mechanical engineering* is responsible for the development, production and marketing of machine tools, metal fasteners, gas and steam turbines for power stations, ball and roller bearings, hydraulic and pneumatic equipment, etc. Some leading firms in this sector are: Rolls-Royce, GKN, Cincinnati Milacron, Berox Advancemill Co. Ltd, William Asquith (1981) Ltd, and GEC-Althsom, etc.
- *Electrical and electronic engineering* is responsible for the development, production and marketing of electrical generators and motors for industrial and domestic requirements, wires and cables, substation transformers and switchgear, computers, radio and television sets, video-recorders, hi-fi systems, radar and navigational aids, etc. Leading firms in this sector are: GEC, BICC, ICL.
- *Civil engineering* is responsible for the development, production and marketing of the infrastructure projects that our industrial society depends upon, such as site development (demolition, reclamation and the opening up of 'green field' sites), roads and motorways, bridges, docks, factory buildings, power stations, stormwater drainage and sewers, etc. Leading firms in this sector are: Balfour-Beatty, Bovis and Wimpey.

Some sectors are highly interdependent. For example:

- *Automobile (road vehicle) engineering* is responsible for the development, production and marketing of cars, lorries, buses, vans, motorcycles, etc. Automobile engineering is often considered as belonging to the mechanical engineering sector, since road vehicles are made largely of metal and contain mechanical engineering units such as engines, gearboxes and differentials, as well as the body shell and subframes and suspension. However, a modern car is also dependent upon the chemical engineering sector for such products as rubber for tyres, plastics for the trim and many, increasingly, exterior body panels, lubricants and fuels. It is also dependent upon the electrical and electronic sector for starter motors, generators, batteries, lights, engine management computers and electrical control equipment. Motor vehicles are also dependent upon the civil engineering sector for the roads upon which they run. Leading automobile companies are: Rolls-Royce, Rover, Honda, Ford, Toyota, Nissan, Peugeot, and their suppliers such as Lucas Industries PLC.
- *Aeronautical (aerospace) engineering* is responsible for the development, production and marketing of fixed wing aircraft and helicopters for passenger and military purposes, space-rockets and satellites, missiles, etc. Leading aerospace manufacturers are: British Aerospace (fixed wing) and Westland (helicopters). They are concerned mainly with airframe design and construction, and the overall assembly. Aeronautical engineering is dependent upon the products and activities of other engineering sectors (see 'Self-assessment test' below).

In this section only leading companies with active manufacturing facilities within the UK have been quoted. Many are multinational and there are many other leading companies in these engineering sectors who are active overseas.

> **Self-assessment test**
>
> Find out and explain briefly how the aerospace engineering sector is dependent upon:
>
> (a) The mechanical engineering sector.
> (b) The electrical and electronic sectors.
> (c) The chemical engineering sector.
> (d) The civil engineering sector.

Having examined the products and activities of various main sectors of the engineering industry, let's now see how they contribute to the national economy. Figure 6.6 shows some important facts about the engineering industry in the UK. From Fig. 6.6(a) we can see that for the period 1990 to 1996 engineering exports exceeded home demand (in this figure q = quarterly period). The figures have been adjusted so that we do not have to take into account the changing value of the pound. It looks as though the engineering industry is making a big contribution to the balance of payments. However, now look at Fig. 6.6(b). Here we see that, for the same period of time, the value of the goods imported by the engineering industry cost more than the value of the goods exported. In fact the engineering sector produced a balance of trade deficit.

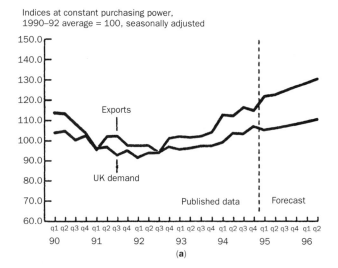

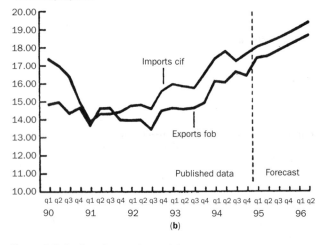

Figure 6.6 Engineering products: (a) exports and UK demand; (b) exports and imports

Since the engineering sector imports more than it exports why bother to have an engineering industry?

- It provides gainful employment for a very large number of people who would be unlikely to find alternative jobs at an equivalent rate of pay.
- All the goods it contributes to the home market would have to be imported if the industry closed down. This would worsen the trade deficit even further.

Table 6.3 compares engineering sales, exports, imports and employment for a period from 1980 to 1995. It breaks these down into various sectors. These are not necessarily the same as our earlier breakdown. There are no hard and fast rules for this. The sectors can be broken down and/or grouped together to suit the need of the statistics being collected for any given purpose. However, the table has been included because it gives a sense of scale to the industry. Column 1 shows that the value of the total sales of engineering goods (at home and abroad) has steadily increased. Column 2 shows that the value of engineering exports has also increased over the same period. However, column 3 shows that the value of engineering imports has, overall, outstripped the value of engineering exports except in 1980. Column 4 shows that the increase in sales has been achieved with a reduced workforce.

Be careful, because over the same period of time the value of the pound has depreciated. More goods have to be sold to achieve the same income from our exports and goods imported cost more in real terms. For example, if the value of the pound halved in comparison with other currencies, we would have to

Table 6.3 Engineering sales, exports, imports and employment

Trade	Year	Sales (£ billion)	Exports (£ billion)	Imports (£ billion)	Employment (000s)	General prices index (1995 = 1.000)	1995 purchasing power (£ billion)
Electrical and instrument engineering	1980	17.07	5.63	5.18	863	0.445	38.3
	1990	41.03	19.54	22.76	682	0.836	49.1
	1991	40.51	20.20	22.75	642	0.888	45.6
	1992	40.42	21.26	25.26	596	0.926	43.6
	1993	44.21	25.54	29.18	576	0.954	46.3
	1994	52.16	30.28	31.80	581	0.976	53.4
	1995*	56.99	33.66	35.18	586	1.000	57.0
Mechanical engineering	1980	21.50	6.90	3.53	1021	0.445	48.3
	1990	37.89	11.84	10.26	655	0.836	45.3
	1991	35.80	11.06	9.31	609	0.888	40.3
	1992	35.14	11.19	9.79	567	0.926	37.9
	1993	35.94	11.87	9.52	537	0.954	37.7
	1994	38.55	13.19	10.81	527	0.976	39.5
	1995*	40.98	14.21	11.76	534	1.000	41.0
Metal goods	1980	9.35	1.01	0.77	509	0.445	21.0
	1990	17.20	1.91	2.41	368	0.836	20.6
	1991	15.95	1.99	2.36	335	0.888	18.0
	1992	15.42	2.12	2.38	313	0.926	16.7
	1993	15.66	2.22	2.40	306	0.954	16.4
	1994	15.87	2.52	2.76	314	0.976	16.3
	1995*	16.58	2.71	2.97	318	1.000	16.6
Motor vehicles	1980	10.54	3.26	3.24	448	0.445	23.7
	1990	23.21	8.05	12.97	267	0.836	27.8
	1991	22.50	9.46	10.65	245	0.888	25.3
	1992	23.82	9.90	12.70	240	0.926	25.7
	1993	24.55	9.27	14.74	218	0.954	25.7
	1994	27.61	10.45	16.81	214	0.976	28.3
	1995*	30.51	13.01	17.39	217	1.000	30.5
Other transport equipment	1980	6.54	2.42	2.17	384	0.445	14.7
	1990	15.31	7.87	6.37	264	0.836	18.3
	1991	14.64	7.82	5.50	250	0.888	16.5
	1992	14.95	7.33	5.24	221	0.926	16.1
	1993	15.52	7.10	5.27	205	0.954	16.3
	1994	15.45	7.24	6.15	191	0.976	15.8
	1995*	15.52	7.28	6.25	182	1.000	15.5
Total engineering	1980	65.01	19.21	14.89	3225	0.445	145.9
	1990	134.63	49.20	54.78	2237	0.836	161.0
	1991	129.40	50.53	50.57	2081	0.888	145.7
	1992	129.75	51.78	55.36	1937	0.926	140.1
	1993	135.88	56.00	61.12	1841	0.954	142.5
	1994	149.64	63.68	68.32	1826	0.976	153.3
	1995*	160.59	70.86	73.56	1837	1.000	160.6

*Forecast figures
(*Source*: Central Statistical Office, Employment Department and EEF estimates; EEF staff forecasts (1995))

export twice as many goods to achieve the same income. At the same time we would have to pay twice as much for the same level of imports. On the other hand, a low value for the pound makes our exports cheaper and encourages other countries to buy our goods. This stimulates our economy and provides more work for the engineering sector. It is the job of the treasury officials and the Chancellor of the Exchequer to maintain a balance between all these factors.

> **Self-assessment task**
>
> From Table 6.3 it can be seen that between 1980 and 1995 the value of the total engineering sales rose substantially, while the number of persons employed in the engineering sector almost halved over the same period.
> Explain briefly how this apparent increase in sales value could be achieved with a smaller salesforce. There are two main factors you should consider.

Economic scale

So far we have only considered some basic economic principles on a small scale with our imaginary family and on the larger national scale. We need to consider the effects of the engineering sector on economies at the local, regional, national and European levels as well.

Local scale

If we look at maps of towns, countries or continents we will find that the spread of industrial activity of any kind is uneven. This is because in the early days of industrial activity, the transportation of people and goods was slow and difficult. Industries grew up where raw materials were available and where there was a market within easy reach. Thus engineering activity grew up near sources of coal to power the steam engines that drove the early factories. Coal could also be turned into coke to smelt iron ore. Iron ore and limestone would also be required for the smelting process. For example, all these were found together in that area of the West Midlands known as the 'Black Country'. Towns prospered wherever industries were set up. They grew around the factories, since the workers had to be within walking distance of their place of employment. Hence many engineering firms are still found within towns and cities.

However, such firms are at a disadvantage for the following reasons:

- Their factory sites are constricted by adjoining properties and expansion is not usually possible except by expensive demolition and redevelopment.
- Transportation of plant and materials to the firm and manufactured goods from the firm is restricted by traffic congestion.
- Access for the workforce may be restricted by inadequate public transport and lack of parking facilities for personal transport (cars).

As a result, progressive and expanding firms have tended to move to industrial estates on 'green field' sites outside towns. Such sites have easy access to the motorway system, and the railway network. They also have the advantages of room to expand and the fact that noise and other forms of pollution are removed from towns together with heavy traffic. Let's now look at the economic effects of such movements.

Many engineering firms that chose to remain in the town and city centres became stifled, slowly declined and closed down. The migration of successful and progressive companies from the original city and town centre sites to the suburbs and beyond has resulted in dereliction, unemployment and social deprivation within such towns and cities.

Although the migration from the town and city centres to their outskirts has often brought economic gain and affluence to the suburbs, it has also resulted in an ever-expanding 'urban sprawl' and a conflict of interests between farming and industry for the use of 'green belt' land.

> **Self-assessment tasks**
>
> The town of Corby, in Northamptonshire, grew up alongside substantial iron-ore deposits and the iron and steel industry that used these deposits.
> 1. Find out what the effect was on the town of Corby when the deposits became uneconomical to work and the associated iron and steel works closed down.
> 2. Find out what Corby has done to attract and develop new industries and how successful it has been in recovering socially and economically.

Regional economy

Engineering and associated industries grew up around towns such as Dudley, Brierley Hill, Stourbridge and Willenhall. All these towns are in the West Midlands. Their local economies are to some extent interdependent and combine together to become a regional economy. Just as there has been a migration of local economies from town centres to the suburbs and beyond into the green belt, so there has been a similar migration from the traditional regional centres of industrial activity to rural and semi-rural areas where there is room for large-scale development. An example is the Toyota car plant built recently at Burnaston in Derbyshire. Just as the migration of industry from town centre sites has resulted in dereliction, unemployment and deprivation, the same has happened on a larger, regional scale. The recent industrial recession hit the Black-Country region of the West Midlands particularly badly. Large companies that were household names some twenty years ago and even whole industries have closed down. The sites have been levelled and are being redeveloped for other purposes. One of the most notable has been the development of the Merry Hill shopping complex on the site of the Round Oaks Steel Company near Dudley.

Local and regional authorities have had to combine together to demolish and redevelop derelict industrial sites with the aid of funds from central government and the European Union. New industries have been invited to set up in these areas to take advantage of an improved infrastructure and a pool of unemployed, skilled labour.

National economy

An industrial map of the UK shows an uneven spread of industries. As has been mentioned earlier, these are largely the result of the coal-dependent industries of the nineteenth century growing up on and around the major national coal fields. This led to a migration of labour from less prosperous areas and additional industries grew up to service their needs. This further developed the uneven regionalisation of the engineering and associated industries. However, many of the heavy engineering, ship building and other 'smoke stack' industries have now closed down due to lack of demand and also due to fierce competition from the Far East (Pacific Rim economies). Fortunately there has been inward investment from the Far East and Europe and traditional industries have been replaced with lighter, high technology industries such as the manufacture of silicon chips and electronic equipment.

These industries are no longer dependent on the coal fields and can be sited anywhere there is a suitable labour force and good communications. Initially many of these new companies have tended to site their operations mainly around the Thames Valley for easy access to Europe. A more recent trend has been the building of high-technology industries in the once prosperous but now more derelict areas of South Wales, the North East of England and in Scotland and Northern Ireland. Table 6.4 shows the trends in employment in the UK. It can be seen that the engineering sector is still the biggest employer of labour in the UK.

> **Self-assessment tasks**
>
> Look at Table 6.5. It shows the number of men and women employed in the main engineering sectors in the various countries of the EU. The numbers represent the percentage of the total working population for each country so employed.
>
> 1. Which country has the highest percentage of its workforce employed in metal manufacturing and engineering?
> 2. Which country has the least percentage of its workforce employed in metal manufacturing and engineering?
> 3. Where does the UK come in this league?
> 4. Of the sectors shown for the UK, which employs the most persons and which employs the least?

Economy of the European Union

The same factors that governed the regional and national growth of UK industry are largely responsible for the uneven distribution of industry across Europe. The larger industrialised areas were established in the eighteenth and nineteenth centuries in the countries of England, France, Germany and northern Italy. Most of this industry was sited and still is sited within a triangle formed by London, Milan and Hamburg. Areas outside this triangle, such as Spain, Portugal, southern France, southern Italy and Greece are much less industrialised.

Much of the industrial activity that used to be concentrated in Europe, North America and Japan, has now moved to the 'tiger economies' of the Pacific Rim countries where wages are much lower and conditions of employment less regulated. Such countries are Hong Kong, Taiwan, Korea, Thailand and Indonesia. China is also becoming increasingly industrialised and a major player in the world economy. For example shipbuilding, which used to be a major industry in the UK, has now virtually finished. Only ships for the Royal Navy and oil platforms are built here because they are of strategic importance. The building of merchant shipping is now mainly centred on the Far East. The knock-on effect has been a corresponding downturn in the demand for steel and also in heavy engineering for ships' engines and ancillary machinery.

However the movement is not wholly one-sided. The rapid development of industrial activity in the Far East has resulted in wage inflation and labour shortages. The one feeds off the other. This has resulted in many Far Eastern multinational companies opening up plants in Europe including the UK. Many Far Eastern companies prefer to come to the UK as its labour costs tend to be lower than for mainland Europe. They can then use the UK as a spring-board for their exports into mainland Europe. Sansung have recently agreed to build factories in South Wales to manufacture silicon chip integrated circuits and also complete electronic devices. It is estimated that this will provide employment for up to 6,000 persons. So the coal-mining and heavy industries of this region will be replaced with modern high technology, light industry. European employees by economic activity can be seen in Table 6.5.

Effects of engineering technology on the social environment

Domestic and leisure

It is difficult to separate these two uses for modern technology since many of the devices are equally important to both applications. The car is essential for going to work and visiting customers, yet it is equally important for shopping, family outings and holidays. The telephone is equally important for household management as it is for chatting to friends and relations. The computer can also be used for games, education, household management and working from home. The television can also be used for entertainment and for business by obtaining financial and business news via Ceefax and Teletext services.

Developments in electronic technology have greatly influenced the sophistication of domestic appliances. Silicon-based temperature sensors that are more sensitive, give closer control and are more reliable than the bimetal strip thermostats that they have replaced in many applications. Control by dedicated computers using microchip technology enables a wider range of programmes to be available to the user. Although such control systems are extremely complex, their use in domestic appliances has made such appliances easier to control.

Refrigerator

One of the earliest examples of modern technology to influence domestic practice was the introduction of the refrigerator to the home. Although widely used abroad and particularly the USA before the Second World War, domestic refrigerators were not widely used in the UK until after 1945. By maintaining foods at a constant temperature not exceeding $+4\,°C$, they may be stored without deterioration for several days. Prior to the use of refrigerators, perishable foods such as milk, butter and meat had to be bought on a daily basis or kept in a cool cellar. The sealed door of a refrigerator prevents insects such as flies coming into contact with the food and spoiling it.

Table 6.4 Employees in employment: by Standard Industrial Classification and gender (1981 and 1994)

Regional trends	Proportion of total workforce within each Standard Industrial Classification by gender (%)																							
	Agriculture, forestry and fishing				Energy and water supply				Metals, minerals and chemicals				Metal goods, engineering and vehicles industries				Other manufacturing				Total manufacturing			
	1981		1994		1981		1994		1981		1994		1981		1994		1981		1994		1981		1994	
	M	F	M	F	M	F	M	F	M	F	M	F	M	F	M	F	M	F	M	F	M	F	M	F
United Kingdom	2.2	1.0	1.8	0.6	4.9	1.0	2.2	0.7	5.9	2.1	4.0	1.4	18.4	6.5	13.4	3.6	10.9	10.6	10.3	7.2	35.2	19.2	27.7	12.2
North	1.8	0.4	1.6	0.3	9.6	1.3	3.3	0.9	10.8	2.4	7.1	1.5	19.0	6.0	14.2	3.4	8.7	10.1	10.8	8.5	38.6	18.6	32.1	13.4
Yorkshire & Humberside	2.0	0.9	1.8	0.5	9.9	1.2	2.3	0.6	9.9	2.7	6.7	1.7	15.4	4.8	12.7	2.8	12.6	14.8	13.4	9.3	38.0	22.3	32.8	13.7
East Midlands	2.8	1.5	2.2	0.8	10.0	1.1	2.1	0.5	5.6	2.6	4.6	1.8	20.6	5.9	17.0	3.7	14.8	21.1	16.4	14.7	41.0	29.6	38.1	20.2
East Anglia	6.7	3.9	4.1	1.9	2.5	0.6	2.5	0.7	3.7	1.2	2.7	0.7	15.7	5.6	12.7	3.6	13.3	12.7	12.9	8.0	32.7	19.5	28.2	12.3
South East	1.2	0.8	0.9	0.5	2.4	0.9	1.8	0.7	3.2	1.8	2.1	1.2	16.2	6.6	10.0	3.1	9.5	7.2	7.6	4.8	28.9	15.6	19.7	9.2
Greater London	—	—	—	—	—	—	1.5	0.7	—	—	1.0	0.6	—	—	5.2	1.7	—	—	7.3	5.0	—	—	13.6	7.3
Rest of South East	—	—	1.7	0.9	—	—	2.0	0.6	—	—	3.0	1.6	—	—	14.1	4.2	—	—	7.8	4.7	—	—	24.9	10.6
South West	4.1	1.5	3.2	1.0	2.7	0.9	1.8	0.7	3.8	1.3	3.0	0.7	18.6	5.5	13.9	3.3	11.2	8.5	9.8	5.1	33.6	15.3	26.7	9.0
West Midlands	1.7	1.0	1.6	0.7	3.9	1.0	1.6	0.6	8.3	4.0	6.0	2.2	32.4	13.3	23.6	7.1	9.1	8.5	9.7	6.7	49.8	25.7	39.2	16.0
North West	0.9	0.4	0.9	0.4	3.9	1.0	2.0	0.7	7.5	2.6	4.8	1.8	19.8	6.2	15.2	3.4	14.1	13.0	12.4	7.8	41.4	21.8	32.4	13.0
England	1.9	1.0	1.6	0.6	4.6	1.0	2.0	0.7	5.8	2.3	4.0	1.4	19.2	6.9	13.7	3.7	11.0	10.5	10.4	7.1	36.0	19.6	28.1	12.2
Wales	3.3	1.1	2.9	0.7	10.1	1.4	2.7	0.8	11.3	2.2	8.2	1.4	13.6	6.1	13.0	4.9	7.2	8.0	10.9	7.4	32.1	16.3	32.0	13.7
Scotland	3.4	0.7	2.2	0.4	5.7	1.0	4.1	0.8	5.0	1.2	2.4	1.1	15.5	4.6	12.0	3.4	11.1	11.9	9.2	7.5	31.6	17.7	23.7	12.0
Northern Ireland	5.6	1.5	5.9	0.8	3.1	0.6	1.9	0.3	3.9	0.6	3.1	0.6	11.3	3.4	8.5	2.1	13.6	14.4	12.6	9.8	28.9	18.4	24.2	12.5

Table 6.5 European employees by economic activity (Eurostat 1991)

Economic activity	Proportion of population employed with each economic activity (%)												European average
	Belgium	Denmark	Germany	Greece	Spain	France	Ireland	Italy	Luxembourg	Netherlands	Portugal	United Kingdom	
Energy and water	1.6	1.0	1.8	2.0	1.4	1.4	1.7	1.5	1.3	1.2	1.5	2.5	1.7
Mineral extraction, chemicals	5.4	2.1	5.7	3.9	4.4	3.5	4.2	4.2	8.8	2.9	5.1	3.5	4.3
Metal manufacture, engineering	10.2	8.4	18.2	4.3	8.8	10.5	8.0	8.8	3.7	7.4	5.6	10.7	11.5
Other manufacturing industries	10.2	11.1	10.2	15.0	12.6	9.4	12.67	13.4	7.2	9.2	20.5	9.1	10.9
Building and civil engineering	5.9	5.7	6.8	8.4	10.6	7.1	7.4	9.3	9.9	6.8	9.1	5.2	7.3
Total employment	33.3	28.3	42.7	34.1	37.8	31.9	34.0	37.1	30.8	27.5	41.8	31.0	35.6

Freezer

This is an extension of the refrigerator. Food is stored at −18 °C or lower and is frozen solid. Perishable foods such as raw and cooked meats, etc., may be stored for three to six months. The main benefit is the cost savings of buying in bulk and the availability of a wide range of food stuffs on hand in the house when visitors call. Space and cost savings can be achieved by the use of combined 'fridge-freezers' which share a common set of machinery.

Both refrigerators and freezers now have 'sealed for life' integrated motor-compressor units that require no maintenance. If they fail in service, they are simply replaced as a unit. The cabinet temperatures are more likely to be controlled electronically and some units give a readout of the actual temperature in the cabinet, and audible warning if a fault causes the temperature to rise above a preset, safe temperature.

Washing machines

Washing machines that contain dedicated computers to ensure a correct wash program for each load have eased the chore of household laundry. They adjust themselves automatically to changes in the pressure and temperature of the water supply and, in some machines, to the load as well. This ensures consistent results. Some machines also incorporate a tumble drying cycle so that clothes can be dried on wet days. Tumble drying is essential for people living in flats who do not have the facility to hang out their washing to dry, and for people living in or near industrial towns where air pollution can soil their freshly washed clothes.

Dish-washing machines

These are increasingly used by the busy family. By washing at higher temperatures than can be tolerated when washing by hand, the dishes are more effectively cleaned and sterilised. Drying is by evaporation of the moisture inside the cabinet of the machine in sterile conditions.

Microwave ovens

High frequency electromagnetic (radio) waves are generated by a 'magnetron' and induce electric eddy currents in the food. The passage of such currents through the food causes it to heat up and cook. A peculiarity of high-frequency electric currents is that they tend to travel on the surface of a conductor. Therefore the eddy currents induced in the food also tend to heat the surface of the food first. This produces the same effect as cooking in a traditional radiant heat oven. The fact that the heat is generated in the food reduces the amount of energy required to cook a given mass of food. This reduces both the cooking time and the energy cost. Foods such as stews need to be stirred from time to time to ensure uniform heating.

Food processor

This can liquidise, blend, mix, chop and mince foods by use of various accessories. The motor speed is electronically controlled to suit the process being used and it automatically maintains the speed selected despite changes in the texture of the food. The food processor has removed many of the physical chores of cooking and produces more consistent results.

Vacuum cleaner

Modern materials technology has enabled microporous filters to be used in place of paper and cloth bags. This has resulted in dust and insects, such as house mites, being more effectively removed with less chance of them being recirculated. This is very important to people with chest complaints and allergies. Improved design and materials for the fan motor has resulted in more powerful suction with no increase in physical size.

Communications

Telephones, television, video recorders and radios have all benefited from modern technology. The telephone network offers an increased range of services. Digital exchanges are more reliable than the electromechanical exchanges and are more compact. Breakdowns can be corrected more quickly by plug-in replacement printed circuit boards. Long-distance calls are clearer and less subject to interference by the use of fibre optic cables. Intercontinental calls by satellite are as easy to make as dialling your neighbour. Mobile phones which link into the telephone network by radio are essential to people on the move, such as sales persons and maintenance technicians.

Television has also improved with the introduction, some years ago, of better definition pictures in colour. Text services are available and satellite services can bring programmes direct from abroad, as well as providing a wider range of programmes. Cable television provides even more services while, at the same time, freeing up sections of the radio spectrum for other uses. The video recorder allows programmes to be recorded for future viewing. They can be preset to record various programmes at different times on different channels while the owner is away from home.

Transistor technology has enabled radios to be reduced in size and battery energy consumption has also been very much reduced. This has resulted in the development and increased use of personal radios that will clip onto the waistband of the user's trousers or skirt and can be played via a lightweight headset without causing noise pollution and annoying other people.

Music

Stereo hi-fi units have been around for many years but were revolutionised by the introduction of 'compact discs' The music is stored digitally on the disc and is read by a laser light unit which produces a digital signal. A dedicated computer converts the digital signal into an analogue signal that can be amplified and played through loudspeakers. A miracle of micro-technology has enabled all this equipment to be compressed into the case of a personal CD player no bigger than a personal radio. In this application the sound is reproduced through a stereo head set.

Self-assessment tasks

Using the headings for the domestic and consumer appliances just discussed, carry out a survey of not more than 50 homes in your street to find out how many of these appliances, etc., each family owns.

1. Which is the most popular item? Why do you think this is?
2. Which is the least popular item? Why do you think this is?
3. How many houses had computers? What did they use them for?
4. What positive impact have computers had on family life?
5. What negative impact have computers had on family life?

Health

Anyone who has watched the many hospital and medical practice dramas on television will be aware of the many aspects of engineering technology used in health and medicine, particularly in the field of diagnostic equipment.

Endoscope

This uses optical fibre technology for carrying out internal examinations visually without the need for intrusive surgery. For example, an endoscope consisting of a fine, flexible tube carrying glass fibres with a light at the end can be inserted down a patient's throat for visually examining the lining of a patient's stomach for ulcers. The outer end of the instrument may have an eyepiece or a camera. The end inside the patient is steerable so that the consultant can control the zone being examined.

X-ray equipment

Over exposure to X-ray radiation is extremely dangerous to both patient and the radiographer. Considerable development has been made to obtain better pictures with lower radiation levels. To speed up the process, the X-ray film processing equipment has also been automated. Electronic image intensifier equipment also enables the levels of radiation to be reduced still further.

Ultrasonic scanning

Also known as 'ultrasound' since the sound wave used are above the highest frequency that the human ear can hear. A hand-held transducer, which looks like an inverted mushroom, is pressed onto the patient's body at the point under examination. Bursts of high-frequency sound waves are emitted from the transducer and bounce back from the internal organs of the patient. These reflected sounds are received by the transducer and are fed back to a computer that analyses the reflected signal and compares it with the transmitted signal. This enables a moving picture of the organs to be built up. The results of the scan can be saved as digital information for future reference. This technique is very safe as there are no harmful radiations. It is widely used for examining unborn babies.

Computerised axial tomography (CAT or CT)

This equipment creates a three-dimensional image by taking low, and safe, intensity X-rays from a camera that rotates around the body while travelling along the body. The images are effectively 'slices' of the body and they are computer enhanced and combined together to form a 3D image of the internal organs of the body and any tumours that may be present. It is possible for the organs to be seen functioning. The images can be saved digitally for future reference.

Magnetic resonance imaging (MRI)

This is a similar scanning process but, instead of using X-rays, the body's tissues are examined magnetically. MRI can distinguish between the different tissues of the body by comparing the different magnetic properties of their atoms. This technique can detect tumours as small as a pea. Again the images can be saved digitally for future reference. All the techniques discussed so far combine sophisticated engineering hardware with computer technology. However, it is not only in diagnostics that engineering and computer technology is used in medicine. Here are some further examples.

Kidney dialysis equipment

These machines act in place of the natural kidneys and cleanse the patient's blood of the toxic substances constantly building up in the human body. The patient is coupled to the machine and his or her blood is pumped through the machine where it is cleansed and fed back into the patient's body. This may need to be done at regular intervals or, in extreme cases, continuously. The only permanent cure is a kidney transplant.

Pacemakers

These aid patients with heart conditions. The pacemaker is inserted into the chest cavity and produces electrical impulses that stimulate the heart and makes it beat regularly and strongly. The device contains its own batteries and microelectronic circuitry in a self-contained package. It has to be replaced every few years when the batteries become weak. Some pacemakers contain miniature computers that can sense the load being placed on the heart by physical work being done. The computer adjusts the strength and frequency of the stimulating pulses accordingly.

Life support equipment

Life support equipment such as heart–lung machines and respirators are available to maintain the essential bodily functions during surgery and subsequently during intensive care nursing. Such equipment combines highly sophisticated precision engineering with computer technology for the control systems.

Monitoring equipment

Microelectronic equipment is used to monitor a patient's condition during surgery and subsequently during intensive care. The vital signs monitored can be: blood pressure, heart beat, respiration and temperature. As well as presenting a visual readout on screen, audible warning tones can alert the medical staff when things start to go wrong. When the patient is sufficiently recovered to move about, *ambulatory biomonitors* can transmit data on the patient's body functions to a nurse's workstation by low-power radio transmissions.

In addition to the clinical equipment described above, precision engineering is required to make the artificial replacement joints for arthritis sufferers. Artificial limbs that are masterpieces of ingenuity and technology offering high degrees of mobility have also been developed. Artificial hearts are also being developed to overcome the shortage of suitable transplant organs.

Physiotherapy equipment can range from the trainer's cold sponge on the sports field to the highly sophisticated equipment in the physiotherapy department of large hospitals. This equipment is required for the rehabilitation of seriously injured accident victims. Exercise machines to maintain and improve the muscle-tone of healthy people are also widely used under supervision in health clubs.

Self-assessment tasks

The effects of technology and engineering on the provisions for healthcare just discussed refer mainly to the facilities found in large hospitals.

1. Consult your doctor (GP) to find out how developments in technology have influenced the equipment he uses in his practice.
2. Consult your dentist to find out how developments in technology have influenced the equipment and techniques he uses in his practice.

Effects of engineering technology on employment patterns

Every advancement in technology has had profound effects upon employment patterns. At the commencement of the industrial revolution, the mechanisation of spinning yarn and weaving cloth destroyed the many cottage industries engaged in these activities and threw many self-employed people out of work. They had no alternative but to move from their country cottages into the towns where the new factories were being built. The accommodation being provided for them was often of low quality, the hours of work were long and the wages paid were barely at subsistence levels. Many resisted this change and during the 'Luddite' riots that ensued attempts were made to smash the new machines. However, the march of progress continued despite their efforts.

In our own lifetime it has been the computer and information technology in all its guises that has resulted in many companies being able to 'downsize' their workforces while increasing their productivity. Many highly paid skilled jobs have been lost and, unfortunately, many of the alternative jobs that have been created are low skilled and low paid. This has adversely affected the national economy for, at a time when more and more people require social benefit payments for unemployment or to subsidise low wages, the tax income to the Treasury, that pays for these benefits, has fallen as fewer people pay income tax or pay at a lower level because of reduced earnings. Figure 6.7 shows how the overall level of employment in the engineering industry has declined.

Because of the volume of production, the car industry and the suppliers associated with that industry have seen some of the most dramatic changes. The tedious and repetitive operations carried out by production line workers has given way to automated production using robots. In the electronics industry, the manual wiring up of circuits has given way to the printed circuit board loaded with components by 'pick and place' robots. Automatic flow soldering fastens the components in place and makes the connections.

Where such manufacturing techniques are not applicable a team approach is preferred. The team is given goals to achieve in respect of quality, quantity and target dates. The old concept of single skill workers with prescriptive job-specifications that led to restrictive practices and demarcation disputes is not applicable to team work. Team work requires

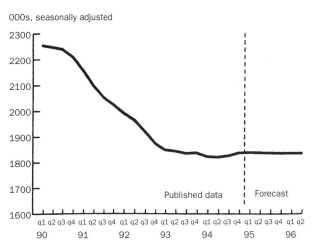

Figure 6.7 Engineering employment

flexible, multi-skilled workers. The team is free to share the work out among the members as they see fit in order to achieve or improve upon the goals set. Members of the team who have difficulties are helped by their workmates and become more productive than they would be working on their own. Slackers who are not pulling their weight are quickly straightened out by the other team members.

Maintenance is another field where multi-skilling is essential. Many machines are now computer controlled. Therefore the maintenance engineer must be capable of not only repairing a mechanical breakdown but must also be capable of diagnosing electrical and software faults and carrying out routine repairs to the machine's control systems. In all these examples of multi-skilling, firms have had to provide retraining programmes and support for their workers in the initial stages. The most successful firms are those that have analysed their needs and taken retraining at all levels of their activities most seriously.

At a higher level, the design and development of a new product will involve a project team of highly qualified professional engineers working under the leadership of a project director. Regular report-back meetings and presentations by individual members quickly identify areas of difficulties and enable the full weight of the expertise available to be brought to bear on the problem. This ensures a quicker and more satisfactory solution to the problem than leaving the individual to struggle in isolation. Figure 6.8 shows how the pattern of graduate recruitment has changed as the mechanical engineering industry has declined and the electronic industry has developed.

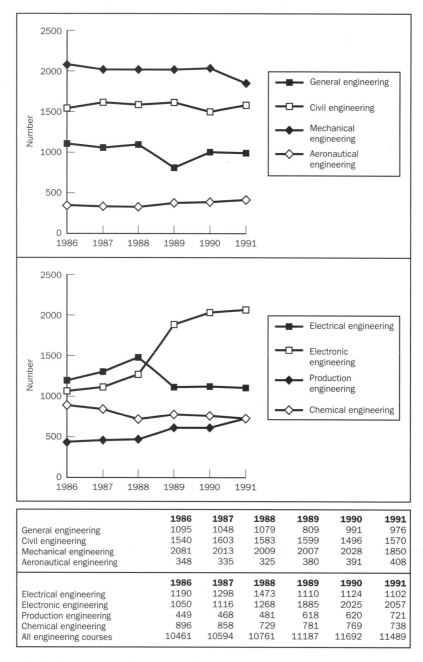

	1986	1987	1988	1989	1990	1991
General engineering	1095	1048	1079	809	991	976
Civil engineering	1540	1603	1583	1599	1496	1570
Mechanical engineering	2081	2013	2009	2007	2028	1850
Aeronautical engineering	348	335	325	380	391	408
	1986	**1987**	**1988**	**1989**	**1990**	**1991**
Electrical engineering	1190	1298	1473	1110	1124	1102
Electronic engineering	1050	1116	1268	1885	2025	2057
Production engineering	449	468	481	618	620	721
Chemical engineering	896	858	729	781	769	738
All engineering courses	10461	10594	10761	11187	11692	11489

Figure 6.8 First degree graduates, home, men and women, engineering and technology courses at universities and polytechnics in UK (*source*: Central Services Unit (Polytechnics and Universities Output 1986–91))

> **Self-assessment tasks**
>
> Approach a small to medium sized manufacturing company that has been making the same or similar product or range of products over a period of, say, 10 years. Try to find out:
>
> 1. How changes in technology have changed their methods of manufacture
> 2. What effect this has had on:
>   (a) productivity
>   (b) the size of their labour force
> 3. What further changes they expect to make over the next few years.

The same flexibility is also needed among the office staff who need to be able to handle word-processing software, spread sheet software, accountancy software and management software with equal facility.

The effects of engineering technology on working conditions

At the start of the industrial revolution, little thought was given to the conditions under which people worked. The hours were long and conditions dangerous. Accidents leading to severe injury and death were frequent. Children were pressed into work as young as eight years old and education for the working masses was minimal or non-existent. Wages were paid on an hourly basis for the time they attended work, or for the work they produced (piece-work). There was no holiday pay, no sick pay and no compensation for injury or death. Fines (stoppages) from meagre pay packets for trivial offences were frequent.

Such conditions could not be tolerated and the trade unions, employment legislation and a more enlightened society gave rise to the improved working conditions that apply today. One of the most important, recent pieces of legislation is the Health and Safety at Work, etc., Act. This will be considered in Element 6.2. Such improvements did not come into being without considerable opposition from the employers and their parliamentary representatives.

The trade union movement grew up to try to counterbalance the power of the employers and the movement fought hard for better working conditions, shorter working hours, a decent wage and better social conditions for its members. It must be remembered that at this time social unrest and revolution were rife on the Continent of Europe. Therefore, the ruling classes, who dominated Parliament, were fearful of the masses banding together in Britain. They feared that the unrest brought about by the industrial revolution might lead to full scale revolution and civil war, as had happened in France. This fear led to legislation called the *Combination Acts* of 1799 and 1800. These Acts made it illegal for workers to combine together and form unions.

These Acts were repealed in 1824. In 1825 Parliament gave trade unions the bare rights to exist. In 1852 an amending measure gave workers the right of peaceful picketing. These rights remained in force until 1871, when the existing arrangements were swept away by the Trade Union Act and the Criminal Law Amendment Act. This legislation was retrogressive. It put limits on a worker's right to strike and removed the right to peaceful picketing. The Criminal Law Amendment Act was replaced in 1875 by the Conspiracy and Protection of Property Act. This reinstated the right to strike and the right to peaceful picketing. Various legislation has been enacted at intervals up to the present day. Such legislation has extended or curtailed the rights of workers depending upon the political complexion of the government of the day. Individual unions still fight for their members' working conditions and wages both locally and nationally. They have links with unions in other countries through international labour organisations. For many years workers sought representation in Parliament through their own members rather than having to depend upon the good offices of sympathetic members of the established political parties. However, before we can consider the birth of the Labour Party we have to consider the formation and role of the Trades Union Congress (TUC).

The Trades Union Congress came into being in Manchester in 1868. Its membership was based on the north country unions. Other areas had separate conferences of locally amalgamated unions. This continued until 1871, when the repressive measures of the legislation of that date (discussed above) made it necessary for these local amalgamations to be replaced by a national body. The Trades Union Congress took over the responsibility for co-ordinating the unions' fight for better working conditions on a national basis. The structure of the TUC remained unchanged until its general council was called upon to organise the general strike of 1926.

This strike was defeated and the general council of the TUC tried to become more conciliatory in its approach to industrial relations (Mond–Turner conversations). Arrangements were made for regular consultations with the employers' organisations, but this was not particularly successful. During the Second World War the government set up various consultative bodies with the employers and the unions (as represented by the TUC). This was done in an endeavour to improve industrial relations and increase productivity. Again the exercise was not particularly successful. The TUC is now generally recognised as representing wide-ranging working-class interests. However, it does not have any enforceable authority over the unions who comprise its membership and it can only exert moral pressure. In fact some unions are not members of the TUC.

The British Labour Party came into existence in 1900 and it was, at first, known as the Labour Representative Committee. This committee was set up as a result of a resolution passed at the Trades Union Congress of 1899. It fought its first general election in 1900 in 13 constituencies and returned 2 members to Parliament. The aim of giving working people their own voice in Parliament was thus achieved. After the general election of 1906 the Labour Representative Committee changed its name to the Labour Party and formed itself into a separate party within the House of Commons. The unions continued to sponsor members at every general election and by-election and still do to this day. They are the major source of funding for the party.

Just as the power and influence of the employers resulted in their employees banding together into unions, so the growing power of the unions resulted in the employers banding together as a counter-force. The National Federation of Associated Employers was founded in 1870 and from 1890 there was rapid growth in powerful combinations among both employers and employees. In fact, by 1936, some 1,820 employers' associations had been registered and were dealing with wage bargaining and labour matters in general. This may seem a very large number compared with the number of trade unions registered at that time. However, it must be remembered that the employers' organisations represented many different trades or sectors of the industry. On the other hand, the unions were usually common to, and spread across, many different sectors of the industry.

The employers' organisations not only concerned themselves with negotiations with the trade unions, they also offered a useful forum for the exchange of information and ideas concerned with technical developments. Nowadays, they are widely used as a means of interpreting the constant flow of EU regulations and directives, and in assisting their members in understanding and implementing such information. Like the unions, they also form pressure groups to try to encourage favourable legislation. Employers' organisations can be divided into three main groups.

- *Trade-based* organisations with common interests in the organisation of their sector of the industry and in the advance of technology. Such an organisation is the Engineering Employers' Federation. This organisation has its headquarters in London and co-ordinates – on a national level – the Engineering Employers' Associations that operate locally. Among their many activities on behalf of their members, the local associations are concerned with local wage bargaining, with training and retraining, and with employers' interests within the local community, as well as promoting technological advance and organising trade exhibitions.

- *General groupings* of employers such as the Confederation of British Industries (CBI), the Chambers of Commerce, and The Institute of Directors. These general groupings, which represent many industries, balance the role of the TUC. They also act as pressure groups to encourage favourable legislation which promotes the requirements and advancement of British Industry at home and abroad.

- *Technical Development Associations* such as the Copper Development Association (CDA), the Cast Iron Research Association (CIRA), and the Motor Industry Association (MIRA). Bodies such as these provide a forum for the exchange of technical information as well as carrying out fundamental research on behalf of their members.

In addition, organisations such as the Royal Society for the Prevention of Accidents (RoSPA), the British Safety Council and the Institution of Occupational Health are constantly working to improve working conditions, reduce accidents and reduce the causes and effects of industrial diseases. Employers are wise to co-operate with these organisations since the courts are empowered to exact heavy penalties in the event of accidents or chronic ill-health. In addition a healthy workforce is more productive and time lost through accidents reduces productivity and profits.

Self-assessment tasks

1. Describe briefly the legislation affecting the trade unions that has been enacted since 1975, and the effect this has had on the power and influence of the trade unions.
2. Describe briefly the work of:
   (a) the CBI
   (b) the Chambers of Commerce.

6.2 Investigate the effects of engineering on the working environment

After reading this element the student will be able to:

- Identify features of the working environment, giving examples.
- Describe the effects of engineering applications on features of two working environments.
- Identify ergonomic considerations in the working environment.
- Explain precautions against risks in the working environment.

Features of the working environment

A safe and comfortable environment is an efficient and productive one. This section will, in turn, consider comfort, health, productivity and efficiency. Reference will be made to the appropriate legislation as it affects the working environment.

Comfort

Comfort in the workplace can be both mental and physical. A worker free from concerns and distractions will be more relaxed and will be able to concentrate on the job in hand. This, in turn, will result in better productivity and less chance of errors resulting in faulty work and accidents. One of the ways of reducing workers' concerns is adequate job training so that they know:

- What to do.
- How to do it.
- The targets to achieve.
- Potential hazards and how to avoid them.
- The correct procedures in the event of an emergency.
- Where to get guidance if things start to go wrong.
- How they fit into the overall structure of the business.
- To whom they are immediately responsible.

As well as being mentally comfortable an employee must also be physically comfortable. A tired worker is a careless worker and that is when faulty goods are made and accidents happen. The basic essentials for an employee's physical comfort are:

- Warm and dry working conditions that are neither too hot nor too cold.
- Protection from adverse weather conditions when working on site.
- Clean air to breathe. It must be free from fumes, dust, unpleasant odours and hazardous contaminants.
- Freedom from noise pollution or the provision of comfortable ear protectors.
- Provision for sitting at the workstation when the nature of the work permits.
- Adequate and clean washing and toilet facilities sited reasonably near the working area.

Health

Health and safety in the working environment is covered by extensive legislation both national and international. The hazards of the workplace affect not only the employees but can also affect visitors to a firm and the area surrounding a firm. For example, where a firm is located in a town, neighbouring householders may be affected by noise, dust pollution and unpleasant odours. A firm located on a 'greenfield' site might contaminate adjacent pasture with toxic effluents. These may then get into the food chain by animals grazing on such pastures. This section is concerned with health problems. These are the long-term effects of manufacturing on employees and the environment. Accidents will be dealt with later.

While most health problems are related to manufacturing processes, even office workers can be affected from a badly planned and badly organised working environment. Typical problems are:

- Back ache from poor seating and inadequate exercise.
- Eyestrain from using computers for too long at a stretch and from poor lighting.
- Repetitive strain injury from making the same movements continuously (e.g. keyboarding).

Most of these problems can be avoided by frequent changes of activity.

The legislation concerning health and safety in working environments is published in a range of booklets and guides. You do not need a detailed knowledge of all the rules and regulations but you should be aware of the:

- Range of such legislation and associated publications.
- Main purposes of the legislation.
- Agencies responsible for the enforcement of the legislation.
- Availability of the booklets and guides.

Typical of the more important booklets and guides are:

- A guide to the Health and Safety at Work etc., Act 1974.
- The Reporting of Injuries, Diseases and Dangerous Occurrences Regulations 1985.
- The Workplace (Health, Safety and Welfare) Regulations 1992.
- The Health and Safety (Display Screen Equipment) Regulations 1992.
- The Management of Health and Safety at Work Regulations 1992.
- The Manual Handling Operations Regulations 1992.
- The Control of Substances Hazardous to Health Regulations 1988.
- The Noise at Work Regulations 1989.
- The Electricity at Work Regulations 1989.
- The Ionising Radiations Regulations 1985.
- The Pressure Systems and Transportable Gas Containers Regulations 1989.

These can be purchased from Her Majesty's Stationery Office (HMSO) which has branches in all major towns and cities, from Chambers of Commerce and Industry, and from leading booksellers. They may be referred to in the reference sections of all major public libraries, also college and university libraries. Such publications are starting to be held on CD ROM and on the Internet. These regulations, and others, are enforced by the inspectors of the Health and Safety Executive and Local Authorities. Such persons have wide powers and firms not complying with the legislation can expect to find themselves being penalised through the Courts of Law.

> **Self-assessment task**
>
> Briefly describe the main items of information you can obtain from the booklets and guides listed above.

Productivity

The purpose of any business is to make a profit for the benefit of the proprietors, in a private company, or the shareholders in a public company. To do this, the company must provide a service or product for which there are customers and which is better and/or cheaper than any similar services or products offered by competitor companies. To achieve these aims and to still make a profit, it is necessary to reduce the *unit cost* of production. This can only be achieved by increasing *productivity*. Productivity can be increased in three ways:

- Maintaining the level of output but reducing the unit cost of manufacture.
- Increasing the level of output for the same unit cost of manufacture.
- Increasing the level of output and reducing the unit cost at the same time.

Remember that increasing the level of output is only possible if the market demand is sufficiently buoyant to absorb the extra output.

There are various ways in which the unit cost of manufacture can be reduced. The simplest is to reduce the level of wages paid. However, this is hardly conductive to good labour relations and would most likely be counter-productive since the output of a dissatisfied labour force would fall. Alternatively the working environment can be made more convenient so that maximum output can be achieved with minimum effort.

This applies equally to a workshop production area, an office worker's desk or the flightdeck of an aircraft. For instance, the introduction of desktop computers has enabled an office worker to word process documents, call up data from sales files, purchase files, a customers' data base, and manufacturing schedules at a key stroke. Previously this would have meant making 'phone calls and having to move about the office to find the files in various cabinets.

In the production workshop manually operated machines may have been replaced with computer-controlled machines. Such machines work not only work faster and more consistently but, if they are loaded and unloaded by industrial robots, they can work round the clock on a 'lights-out' basis (they don't have to see). Remember that such a solution involves heavy capital expenditure which must be recovered. This will have to be offset against increased profits made as a result of the increased productivity. There must be sufficient product demand to warrant such a high level of capital expenditure.

Unfortunately the demand for increased productivity and the solutions outlined above have resulted in a labour force that is smaller but better qualified and more flexible. This has been brought about by 'downsizing' the labour force (redundancy) and retraining schemes. It has also resulted in severe problems of unemployment, a problem made worse by the recent downturn in world trade and the availability of cheap goods from the Far East. In some instances, companies with factories in various parts of the UK have been able to concentrate an adequate level of manufacture on a single site by a combination of automation and 'downsizing' their labour force to suit a smaller and more competitive market. This has been disastrous for some communities where the town has grown up around the company and no alternative employment is available. It affects everyone in the community since there is less money to be spent in the shops and on recreation facilities. There is also the loss of orders for small firms servicing the major company.

Efficiency

Productivity and efficiency tend to go hand in hand. Maximum output for minimum effort is one way of defining efficiency as far as productivity is concerned, providing there is no loss of quality as a result. This can be achieved through automation or by making manual operations as convenient as possible. Techniques such as 'work-study' are used to measure the movements made by a worker, the time taken and the effort used. By laying out the 'workstation' as conveniently as possible and making the working environment as pleasant and as comfortable as possible, substantial savings can be made for little cost. Figure 6.9 shows a typical office workplace as

1. Adequate lighting
2. Adequate contrast, no glare or distracting reflections
3. Distracting noise minimised
4. Leg room and clearances to allow postural changes
5. Window covering
6. Software: appropriate to task, adapted to user, provides feedback on system status, no undisclosed monitoring
7. Screen: stable image, adjustable, readable, glare/reflection free
8. Keyboard: usable, adjustable, detachable, legible
9. Work surface: allow flexible arrangements, spacious, glare free
10. Work chair: adjustable
11. Footrest

Figure 6.9 Typical office workplace (*source*: HSE booklet *The Health and Safety (Display Screen Equipment) Regulations* 1992)

recommended in the HSE booklet: *The Health and Safety (Display Screen Equipment) Regulations 1992*. That is, the workstation must be *ergonomically* designed. Ergonomic considerations will be dealt with later; first, let's consider two case studies to see how two working environments have changed.

> **Self-assessment task**
>
> Briefly describe how you think your place of work or study could be better arranged to be more comfortable, convenient and safe so that you could work more efficiently.

The effects of engineering applications on features of working environments

Engineering personnel are required to work in a variety of locations throughout the world:

- Inside or outside buildings.
- In factories, workshops, offices, building sites, farms and quarries.
- On land, sea or in the air.
- Above and under ground.
- In very hot and very cold conditions.
- In extreme weather conditions.

Some jobs may require work in one specific place (workstation), at a desk, control console or at a machine. Other jobs may involve movement between sites (e.g. maintenance engineers, installation engineers, commissioning engineers, etc.). It is sometimes difficult and even impossible to influence the working environment as, for example, electrical engineers restoring overhead transmission lines in blizzard conditions.

Nevertheless, if engineering personnel are to work efficiently and achieve levels of productivity that will be profitable, then attention must be paid to those features of their workplace that affect their health, safety and comfort and convenience of working (ergonomics). We must now consider how engineering activities can affect these features.

Workplace safety, comfort and health

A safe, comfortable and healthy working environment is usually an efficient and productive one

Safety

We considered the prevention and protection against accidents in an earlier section, let's now consider the cost of accidents to industry. Studies have shown that the financial losses resulting from industrial injuries may amount to 37 per cent of profits in any one year.

Humanitarian reasons alone should be sufficient reason for an employer to try to eliminate injuries resulting from accidents. However, the economic incentives are equally compelling. Although health, safety and accident prevention measures cost money, these costs are relatively small compared with the costs involved when accidents resulting in injury and ill-health do happen. The costs associated with accidents can be:

- Legal costs and compensation payments to injured parties.
- Time taken in preparing defence documents.
- Medical and treatment costs.
- Lost working time of the injured person(s).
- Lost working time of other employees if HSE closes the process down.
- Cost of temporary labour.
- Repair and replacement of damaged plant and equipment.
- Replacement of lost or damaged materials.
- Compensation associated with environmental damage in the event of chemical spillages and the release of toxic vapours into the atmosphere.
- Loss of production, business interruption, delivery rescheduling and customer dissatisfaction.

Lighting

Natural and artificial lighting should be of sufficient intensity to enable the work to be carried out without eye-strain and for persons to move about without risk. The lighting should be uniform without glare or shadows. This applies particularly to stairways. Natural lighting should, preferably, be from the north to avoid glare and heat. Individual local lighting should also be provided at workplaces where fine work is being carried out.

Ventilation

The indoor working environment should be well ventilated so that stale air is replaced at a sufficient rate to avoid drowsiness. The recommended rate is 5 to 8 litres of fresh air per second per person. Special precautions must be taken where the process produces heat and dust, as in foundries, or unpleasant odours and toxic fumes as occur when painting, electroplating and using adhesives. An increasingly common problem is 'legionnaire's disease' caused by the growth of bacteria in the warm and moist conditions of ventilation cooling towers. Guidance on the avoidance of this problem can be obtained from the HSE.

Temperature

The temperature of indoor working environments should normally range between 16 and 20 °C. The lower temperature is more suitable for light physical work such as operating a machine and the higher for office work and instrument assembly. Lower temperatures may be more comfortable for heavy physical work, but should not fall below 13 °C. Thermometers must be provided at various places in the workplace to ensure that these conditions are being achieved. Space heating in the winter and air conditioning in the summer should be provided to maintain comfortable working conditions at all times.

Noise

Excessive noise must be avoided. The pitch of the noise as well as its intensity are factors in causing fatigue and poor concentration leading to errors and accidents. Noise is also wasteful of the energy that causes it and energy costs money, which is another reason why noise should be minimised. If the process is such that the generation of noise is unavoidable, then ear protectors (ear-muffs) must be issued by the management and worn by the workers.

Space

Offices and workshops should be of sufficient size to provide adequate free space between the workstations so as to allow easy access and safe movement of goods and people. The number of persons allowed in a given space and the proportions of that space is laid down in the workplace regulations. As a guide, the volume of the empty room in cubic metres should be divided by eleven. This will give the maximum number of persons permitted. This does not apply to sales kiosks, vehicle driving cabs and large auditoria. In practice, fewer than the maximum permitted number of persons may be able to use the room or workshop due to the space taken up by furniture, machines and equipment. When calculating the volume of the room or workshop use the actual ceiling height up to and including three metres. Ceiling heights above three metres are taken as three metres for the sake of the calculation.

Workstation comfort

Workstations should be arranged so that the tasks can be undertaken safely and comfortably. The worker should be able to reach his or her machine, tools and materials easily without bending and stretching. This applies not only to a machine operative and an assembly worker in a workshop, it also applies equally to an office worker at his or her desk or computer. Where possible the worker should be able to perform his or her duties sitting down. In this case the seating should be adjustable to provide the correct posture for the user. Mobile seating should have five castors for safety. The HSE publishes information dealing with seating and the ergonomics of the workplace (see Fig. 6.10 and also refer back to Fig. 6.9).

Lifting and handling

Manual and power-driven mechanical handling devices should be used for moving loads exceeding 20 kg. Such equipment and its associated slings must be regularly inspected and certificated. Each item of equipment must be labelled to show its safe working load (SWL). Only properly trained personnel should use such equipment.

First aid

In all places of work there should be a fully stocked first aid kit. One or more persons in the workplace should be trained in first aid procedures. The location of the first aid kit and the names of the qualified person should be clearly displayed in all the work areas. Larger firms will have an ambulance room with a full-time professional nurse. An accident report book must also be available in which notifiable accidents and injuries must be logged.

Emergency procedures

All employees must be regularly drilled in emergency procedures such as giving the alarm in the event of accidents and fire, and the correct procedure for the evacuation of the premises in the event of fire. Notices must be prominently displayed giving the necessary instructions. The time taken to evacuate the premises and any incidents (such as emergency exits being locked or obstructed) must be logged. Responsible persons should be trained in the use of extinguishers but only if they can be used without personal risk. The professional brigade must always be called out.

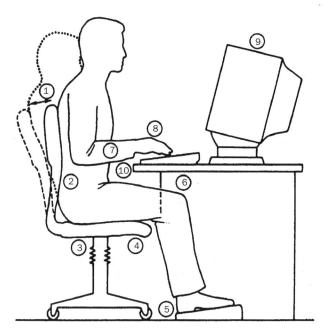

1. Seat back adjustability
2. Good lumbar support
3. Seat height adjustability
4. No excess pressure on underside of thighs and backs of knees
5. Foot support if needed
6. Space for postural change, no obstacles under desk
7. Forearms approximately horizontal
8. Minimal extension, flexion or deviation of wrists
9. Screen height and angle should allow comfortable head position
10. Space in front of keyboard to support hands/wrists during pauses in keying

Figure 6.10 Seating and posture for typical office tasks (*source*: HSE booklet *The Health and Safety (Display Screen Equipment) Regulations* 1992)

Sanitary facilities

Washing and toilet facilities must be available within easy reach of the workplace. The facilities provided will depend upon whether they are to be used by male or female employees, the number of employees and the type of work being performed. For example, pit-head showers would be inappropriate in an office block. Guidance as to the number and type of utensils provided is given in the workplace regulations and local authority building bye-laws. Such facilities must be kept clean, in good condition and in good working order.

Refreshment and rest areas

The facilities provided will depend upon the number of employees and the type of work involved. Rest facilities with provision for sitting down are of particular importance to persons who have to stand to perform their work and where the work is of a heavy physical nature. If the numbers are sufficiently large, a canteen should be provided, otherwise seating in a clean area with a hot and cold drinks vending machine and a chilled sandwich cabinet will suffice. Drinking water from the mains supply should also be provided adjacent to the work area. Disposable cups or a drinking fountain should be used for reasons of hygiene.

Miscellaneous provisions

To prevent falls and injuries from falling objects, overhead catwalks and scaffolding should be properly fenced and 'kickboards' provided. Ladders should be 'footed' by a second worker if they are only being used temporarily, or they should be securely lashed if they are to be in position for any length of time. Tanks containing dangerous substances must be fenced and covered. Inspection pits must also be fenced and covered when not in use. Wherever possible doors should contain transparent panels so it can be seen if anyone is on the other side before throwing them open. Windows should be designed so that people cannot fall through them when they are open.

Workplace efficiency and productivity

As well as a safe comfortable and healthy environment for the personnel of a company, efficiency and productivity also depend upon such factors as:

- workshop layout
- machinery and equipment
- energy usage
- energy waste disposal
- disposal of by-products

Workshop layout

Much can be done to reduce accidents, and increase efficiency and productivity, by the layout of the machines and equipment in a workshop. Figure 6.11 shows a well laid out workshop. You can see that there are ample gangways clearly marked and free of obstructions. The machines are arranged so that bar stock does not protrude into gangways. This arrangement also prevents the operators being distracted by other workers passing close behind them. Grinding machines are arranged so that any grit is not thrown towards other machines where it would damage the slideways and bearings. The positioning of the grinders against an outside wall also facilitates the provision for dust extraction equipment. There is easy access to the emergency stop switches for the whole shop and the fire extinguishers are strategically placed. Scale models of standard machines are available and these can be used in conjunction with scale plans of the workshop floor for experimenting with different layouts in order to achieve the most efficient workshop layout.

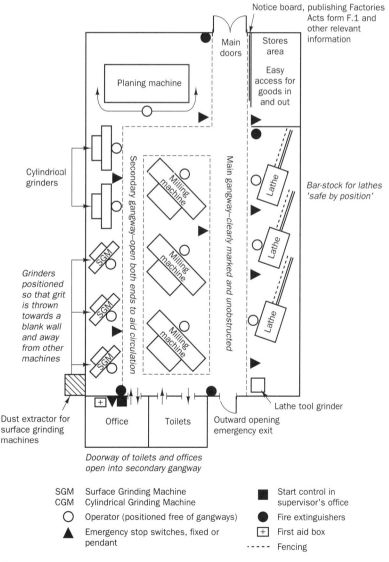

Figure 6.11 Machine shop layout

The layout shown in Fig. 6.11 is suitable for a 'jobbing shop' working on a one-off production basis and also for a small to medium batch size production basis. For large batch sizes and continuous production the layout can be designed specifically for a single product or a small range of products of a similar type. Such a workshop layout is shown in Fig. 6.12. The raw materials flow in at one end of the workshop and the finished components flow out at the other end. The fastest machine or process sets the standard and other machines and processes have to be duplicated or triplicated to keep it fully and continuously fed.

Self-assessment tasks

1. Draw a floor plan of a workshop with which you are familiar.
2. If you are of the opinion that it is already a satisfactory layout, state why you think so.
3. If you do not think that the present layout is NOT satisfactory, design a better layout, explaining the reasons for your changes.

Answer 1 and *either* 2 *or* 3 but not both

Machinery and equipment

Many UK companies and, in fact, whole industries have been lost because there was no policy to keep the machinery, equipment and manufacturing techniques up-to-date. The companies that have survived have invested heavily in computer-controlled machines, industrial robots for handling the materials and work on and off the machines and in automated assembly stations. Such equipment produces work to greater accuracy and standards of quality than the older manually operated machines. They do not get tired, do not make mistakes, are completely consistent and can work 24 hours a day without rest, holidays and without going sick.

The drawing offices have been equipped with computer-aided design (CAD) and drawing facilities, and the commercial offices use word processors in place of typewriters and computerised accounting in place of handwritten ledgers. Production management and other management services will have been computerised and staff will have undergone considerable retraining and will continue to have their skills updated.

Energy usage

Energy is used directly in production to drive the machines in the workshops and the photocopiers and computers in the offices. It is also used indirectly for such purposes as lighting, heating and air conditioning.

Modern premises are thermally insulated to prevent overheating in the summer and the loss of heat in the winter. Most large buildings are subject to periodical 'energy audits' to ensure that energy is not being wasted. Energy is expensive and likely to become more so in the future, so it is in the interests of the occupiers of any premises to use it as sparingly as possible. Also, wasted energy results in the unnecessary release of 'greenhouse' gases into the atmosphere. This contributes to global warming.

Modern buildings are fitted with double glazing to prevent heat loss and eliminate noise. The glass used is often tinted or reflective to keep the sun from overheating the building in the summer and to reduce glare. This reduces the need to provide full air conditioning. To conserve heat even further many large buildings use a recirculating ventilation system. The air

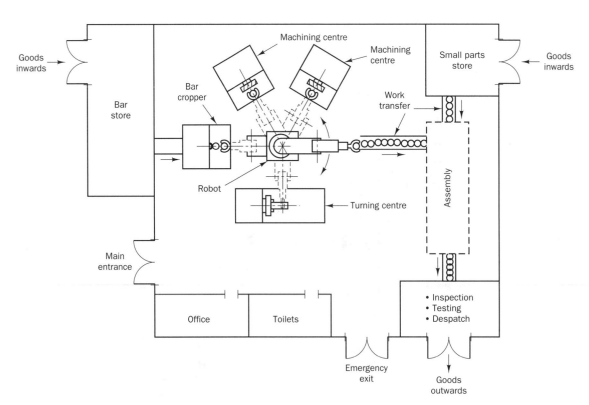

Figure 6.12 Layout for large batch sizes and continuous production

is exhausted from the building, filtered, humidity controlled and revitalised by the addition of a controlled amount of fresh air before being warmed (or cooled) and fed back into the building. Care must be taken in the conditioning of the air to ensure that germs and bacteria are not allowed to accumulate and breed in the system.

Energy waste disposal

Some engineering processes produce heat energy that can easily be lost to the atmosphere. In the early days of the iron and steel industry, the blast furnaces were open topped. Vast amounts of heat escaped to the atmosphere and the soot and grime from the furnaces quickly covered the surrounding fields, turning them black and killing the vegetation. Because of this, such areas of the West Midlands became known as the 'black-country' – a term that is still used to this day. The life expectancy for the people living in these conditions was only about 30 years.

Nowadays, blast furnaces are capped and the waste heat is used to heat the incoming combustion air, so increasing the efficiency of the furnaces. The waste gases from the blast furnaces can also be burnt to heat the furnaces (called 'soaking pits') where, in an integrated iron and steel works, the cast ingots of steel are kept hot, ready for rolling into different shapes.

In some areas the waste heat from power stations is used for district heating schemes. The waste gases from the decaying vegetable matter in landfill waste disposal sites is increasingly used to generate electricity which is fed into the national grid system.

The space heating of modern buildings is by oil-fired or, preferably, gas-fired boilers. The emissions from gas-fired boilers are cleaner than those from oil-fired boilers. Both gas- and oil-fired boilers are easily arranged for automatic control and very efficient combustion. Public buildings and hotels often have sensors to detect when people are in a room. When the room is unoccupied, the level of heating is reduced to save energy. Solid fuel is no longer used in the heating systems of large modern buildings.

Disposal of by-products

The waste of one industry can become the raw materials of another industry. Recycling waste is now big business and a means of preserving the world's diminishing stocks of mineral and other resources. For example, Fig. 6.13(a) shows some die-cast castor components as they leave the pressure die-casting machine. Figure 6.13(b) shows the flash and sprue after clipping. Figure 6.13(c) shows the castors after they have been clipped from the sprue and flash. The sprue and flash represents waste metal that can be melted down and reused in the process.

Sometimes scrap metal from an engineering machine-shop can be sold to a firm specialising in its collection. This scrap metal is then melted down, and turned back into castings and other useful forms of raw material for the engineering industry. The sale of the scrap metal forms a secondary source of income for the producer. The coking of coal and the refinement of crude oils produces a whole range of chemicals that can be used to make plastics, paints, cosmetics and pharmaceuticals. Such chemical by-products are called 'feed-stocks' and are an essential source of income to the petrochemical industry. Without the sale of these by-products you would not be able to afford very much petrol for your motorbike!

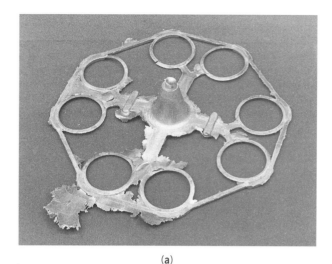

(a)

(b)

(c)

Figure 6.13 Multiple impression casting

Coal-fired electricity power stations produce vast quantities of ash. This can be sold for the manufacture of building materials such as insulation blocks. Unfortunately much of the coal deep-mined in the UK has a high sulphur content and burning it produces the compounds of sulphur responsible for 'acid rain'. To remove the oxides of sulphur from the emissions of the power station furnaces, the flue gases are allowed to react with finely ground limestone to form *gypsum* which is used for making cements, plaster and plasterboards for the building industry. The sale of the gypsum helps to offset the cost of disposing of less useful waste.

Some waste cannot be economically recycled. Some of it is toxic and has to be rendered chemically inert before it can be dumped. This costs money and the cost of waste disposal is becoming an increasing burden on industry as we become increasingly environmentally conscious. There is a limit to how much of this cost can be absorbed by industry and an increasing proportion of the cost of waste disposal will be passed onto the consumer (you and me) in increased prices.

Self-assessment tasks

1. Suggest reasons why solid fuel heating is no longer used in large factories and public buildings.
2. Electricity generating stations in the UK are largely coal, gas or oil fired. What alternative and more economically acceptable methods of generation are available and what are their advantages and limitations?

Ergonomic considerations in the working environment

The Oxford Dictionary defines *ergonomics* as 'the study of efficiency of persons in their working environment'. Compare a veteran car with a modern car. The former paid little attention to the needs of the driver and often placed controls and instruments in inconvenient places. In a modern car we expect to relax in a comfortable seat with all the controls conveniently to hand or to foot so that they can be operated with a minimum of movement and effort. The instruments will be clustered together in front of the driver so that they are visible at all times. With the use of seat-belts, this has become even more important as the driver is physically restrained and cannot reach controls that require major bodily movement.

The same applies to the workplace and each individual's 'workstation'. Let's refer back to Figs 6.9 and 6.10 which show the principles of ergonomics applied to a computer operator. In Fig. 6.10, the seat has been ergonomically designed to enable the operator to sit comfortably and in a medically approved posture that will minimise bodily strain. At the same time the keyboard is conveniently positioned and the foot rest enables the operator to rest his or her feet at a natural angle (at right angles to his or her leg). The seat swivels so that the operator can turn aside to select documents without having to lean or reach. Figure 6.9 shows the whole of the workstation layout.

The same applies to machines in the manufacturing areas of the factory. Not so long ago the controls for a machine had to be placed to suit the requirements of the mechanical linkages rather than the convenience of the operator. Nowadays, electronic, pneumatic and hydraulic controls enable the controls and instruments to be positioned ergonomically to suit the convenience of the operator. Very large machines, such as continuous metal rolling mills, also use closed circuit television so that the operator can monitor the process without having to move from the operating position.

Precautions against risks in the working environment

Health and Safety at Work, etc., Act 1974

It is essential to observe safe working practices not only to safeguard yourself, but also to safeguard the people with whom you work. The Health and Safety at Work etc., Act provides a comprehensive and integrate system of law for dealing with the health, safety and welfare of work people and the general public as affected by industrial, safety, commercial and associated activities. The act places the responsibility for safe working equally upon:

- the employer
- the employee
- the manufacturers and suppliers of materials, goods, equipment and machinery

Health and Safety Commission

The Act provides for a full-time, independent chairperson and between six and nine part-time commissioners. The Commission is made up of three trade union members appointed by the TUC, three management members appointed by the CBI, two local authority members and one independent member. The Commission has taken over the responsibilities previously held by various government departments for the control of most occupational health and safety matters. The Commission is also responsible for the organisation and functioning of the *Health and Safety Executive* (HSE) which has been referred to from time to time earlier in this unit.

Health and Safety Executive

This is a unified inspectorate which combines together the formerly independent government inspectorates such as the Factory Inspectorate, the Mines and Quarries Inspectorate and similar bodies. The inspectors of the HSE have wider powers under the Health and Safety at Work, etc., Act than under any previous legislation. Their duty is to implement the policies of the Commission. Should an inspector find a contravention of one of the provisions of earlier Acts still in force, or a contravention of the Health and Safety at Work, etc., Act, the inspector has three possible lines of action available.

- *Prohibition notice* – If there is a risk of serious personal injury, the inspector can issue a prohibition notice. This immediately stops the activity giving rise to the risk. The activity cannot restart until the inspector is satisfied that the remedial action specified in the notice has been correctly taken.
- *Improvement notice* – If there is a legal contravention of any of the relevant statutory provisions of the Act, the inspector can issue an improvement notice. This notice requires the infringement to be remedied within a specified time.
- *Prosecution* – In addition to serving a prohibition notice or an improvement notice, the inspector can prosecute any person (including an employee) contravening a relevant statutory provision. Finally the inspector can *seize, render harmless, or destroy* any article or substance which the inspector considers to be the cause of imminent danger or personal injury.

> **Self-assessment tasks**
>
> 1. Describe a typical accident, incident, or breach of regulations under which you would expect an HSE inspector to:
>   (a) issue a prohibition order
>   (b) issue an improvement notice
> 2. What action would you expect a factory owner to take in the event of (a) or (b) above?

Thus every employee must be a fit and trained person capable of carrying out his or her assigned task properly and safely. Trainees must work under the supervision of a suitably trained experienced worker or instructor. By law, every employee (and trainee) must:

- Obey all the safety rules and regulations of his or her place of employment.
- Understand and use, as instructed, the safety practices incorporated in particular activities or tasks.
- Not proceed with his or her task if any safety requirement is not thoroughly understood, *guidance must be sought*.
- Keep his or her working area tidy and maintain his or her tools in good condition.
- Draw the attention of his or her immediate supervisor or the safety officer to any potential hazard.
- Report all accidents or incidents (even if injury does not result from the incident) to the responsible person.
- Understand emergency procedures in the event of an accident or an alarm.
- Understand how to give the alarm in the event of an accident or an incident such as fire.
- Co-operate promptly with the senior person in charge in the event of an accident or an incident such as fire.

Workplace risks

Accidents don't just happen, they are caused. There is not a single industrial accident which could not have been prevented by care or forethought on somebody's part. Accidents can and must be prevented. They cost millions of lost man-hours of production every year, but this is of little importance compared with the immeasurable cost in human suffering.

In every eight-hour shift nearly 100 workers are the victims of major industrial accidents. Many of these casualties will be blinded, maimed for life, or confined to a hospital bed for months, never to work again. At least two of them will die. Figure 6.14 shows the main causes of industrial accidents.

Different workshops present different risks (hazards). The more common ones are:

Air pollution
For instance: active or passive smoking in offices, canteens, and workshops, the use of process chemicals such as soldering, brazing and welding fluxes, paint and adhesive solvents and cutting lubricants, dusts created by processes such as grinding.

Hygiene
Personal hygiene is most important. Use a barrier cream to rub into your hands before work or use disposable plastic or rubber gloves. Wash thoroughly with soap and water after work. Avoid using solvents as these can cause skin irritations. Change your overalls frequently so that they can be cleaned.

Temperature
Human beings cannot work effectively if their bodies are too hot or too cold. A correct working environment (temperature and humidity) is essential for efficiency. Extreme temperatures are very dangerous. Persons working in welding bays and near heat-treatment equipment must wear protective clothing to prevent burns. Some processes require very low temperatures involving the use of solid carbon dioxide or liquid nitrogen. Protective clothing must be worn to prevent frost-bite.

Chemical burns
These can be caused by chemical splashes from such sources as alkaline (caustic) degreasing solutions, accumulator acid, and ferric chloride etching agents used for making printed circuit boards. Always wear the protective aprons, gloves and goggles or face-visors provided and handle the containers and equipment with care.

Figure 6.14 Average national causes of industrial accidents

Electric shock

The most common causes of electric shock are shown in Fig. 6.15. The installation and maintenance of electrical equipment must be carried out by a fully trained and registered electrician. An electric shock from a 240 volt single-phase supply (lighting or office equipment) or a 415 volt three-phase supply (most factory machines) can easily kill you. Even if the shock is not sufficiently severe to cause death it can cause serious injury. The sudden convulsion caused by the shock can throw you from a ladder or against moving machinery. To reduce the risk of shock all exposed metalwork must be earthed, portable site equipment should be operated from a low-voltage transformer and/or be 'double-insulated', the supply leads for portable equipment must be regularly checked for damage and all such equipment must be regularly tested and certified safe by a registered electrician.

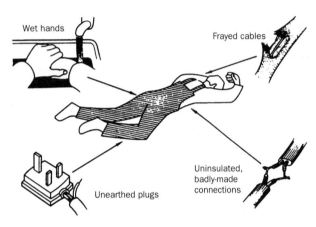

Figure 6.15 Causes of electric shock

Self-assessment tasks

Find out why:

1. An electric circuit should contain a fuse or circuit breaker.
2. Metal-cased electrical equipment must be earthed.
3. Portable power tools should be fed from a low-voltage transformer.
4. A residual current detector (earth leakage isolator) should be incorporated in the supply to all portable power tools.

Crushing

Accidents involving crushing can result from the whole person being trapped between a lorry, or a fork-lift truck, and some immovable object such as a building. It can also occur when heavy loads, being lifted by cranes or other lifting tackle, slip from their slings. The crushing of limbs such as fingers, hands and arms is usually the result of entrapment in unfenced machinery (gears, belts, chains and cutting tools). This can also result in limbs being severed. Transmission guards and cutter guards will be considered in more detail later on p. 193.

Head injuries

These occur when workers are struck on the head by falling objects. Site workers, shipyard workers, steel fabrication workers and others working where overhead cranes are in use should always wear hard safety hats. The temporary and unauthorised storage of heavy objects on high shelves and tops of cupboards is another source of head injuries.

Eye injuries

There are two causes of eye injury. Blinding caused by the penetration of the eye by pieces of metal or other hard and sharp objects during machining operations. Radiation damage by the brightness and heat of oxyacetylene welding or the ultraviolet light of electric-arc welding. Special goggles and visors with filter glasses to remove these harmful radiations should be used.

Foot injuries

These can be caused by treading on sharp objects, by heavy objects falling on, or running over, the foot, or by injury to the Achilles tendon behind the heel of the foot. Industrial safety boots and shoes are reinforced to protect against such accidents and injuries.

Deafness

Continual exposure to noisy machines and equipment can cause permanent deafness. If the source of the noise cannot be removed or the intensity of the noise reduced, then ear-muffs (protectors) must be worn. As has been mentioned previously, noise is wasted energy that has to be paid for. Therefore every effort should be made to avoid noise being generated in the first place, rather than have to deaden it with sound-insulating materials or wearing personal ear protectors.

Cuts

These can range from a small cut that simply requires cleaning and a sticking plaster, to serious lacerations requiring expert medical and surgical attention. All cuts should be properly dressed as soon as they occur to prevent infection. Cuts usually occur through the incorrect and careless use of hand and machine tools in metal-working and wood-working shops.

Self-assessment tasks

1. List what you think should be included in the contents of a first aid box suitable for a small workshop.
2. Examine the contents of a first aid box in a workshop with which you are familiar. In what way does the contents differ from your choice in the above task?
3. What do you think is the reason for this difference?

Strains

These usually result from lifting objects that are too heavy or by using a poor lifting technique. Office workers can also suffer strains from a poor posture when sitting at a workstation. Repetitive strain injury can occur by continually repeating the same action over and over again without a change of activity. Even the comparatively small forces involved in keyboarding can cause such injury

Falls

This is one of the simplest but also the most common cause of industrial accidents. It is caused by poorly maintained floors, floor coverings and stairs. Obstructions to gangways that should be kept clear and oil spillages that have not been cleaned up are other causes of slips, trips and falls. Unsecured ladders should not be used. All falls are made worse if the person concerned is carrying a load at the time.

Prevention and protection

We have just looked at some common causes of industrial accidents. Let's now look at how they can be prevented or how we can be protected against them.

Clothing

For general workshop purposes a *boiler suit* is the most practical and safest form of clothing. However, to be completely effective, some precautions must be taken, as shown in Fig. 6.16.

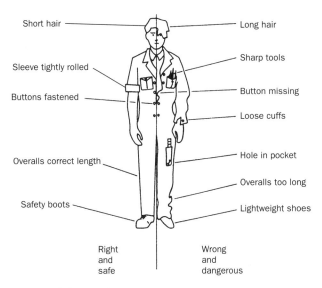

Figure 6.16 Correct dress

Sharp tools

Sharp tools protruding from the breast pockets of the overalls can cause severe cuts to the wrist. In extreme cases this can lead to loss of feel and use of the fingers.

Buttons missing

Since the overall cannot be fastened properly, it becomes as dangerous as any other loose clothing and is liable to become trapped in moving machinery.

Loose cuffs

Not only can loose cuffs be caught up like any other loose clothing, they may also prevent you from snatching your hand away from a dangerous situation if the cuff snags on a piece of the machine.

Hole in pocket

Tools placed in a torn pocket can fall through onto your feet. Although this may not seem to be potentially dangerous, nevertheless it could cause an accident by distracting your attention at a crucial moment when operating a machine.

Overalls too long

These can cause you to trip and fall, particularly when negotiating stairways.

Lightweight shoes

The possible injuries associated with lightweight and unsuitable shoes are:

- Puncture wounds caused by treading on sharp objects.
- Crushed toes caused by falling objects.
- Damage to your Achilles tendon due to insufficient protection around the heel and ankle. Figure 6.17 shows how these injuries occur when wearing lightweight shoes and how they can be prevented by wearing industrial safety shoes complying with BS 1870.

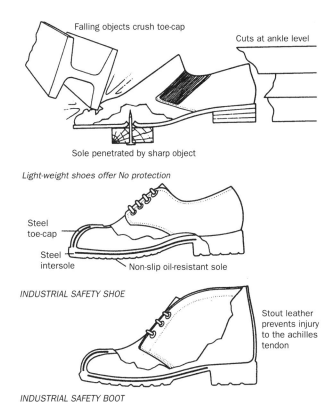

Figure 6.17 Safety footwear

Long hair

Long hair is liable to get caught in machines in which the cutter or work rotates, such as drilling machines and lathes. This can result in the hair and scalp being torn away which is extremely dangerous and painful. Permanent disfigurement will result and brain damage can also occur. Long hair is also a health hazard as it cannot be kept properly clean and free from infection under workshop conditions. A cap should be worn as shown in Fig. 6.18.

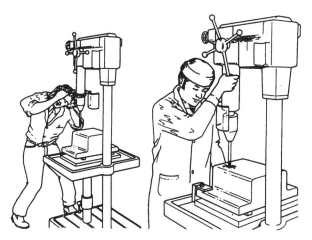

Figure 6.18 The hazard of long hair

Head protection
When working on site or in a heavy engineering erection shop involving the use of overhead cranes, your workshop cap, designed to keep your hair from becoming entangled with rotating cutters, does not provide sufficient protection. Under such conditions, all persons must wear a safety helmet complying with BS 2826. Figure 6.19(a) shows such a helmet. Safety helmets are made from high-impact-resistant plastics or from fibre glass reinforced polyester mouldings. Such helmets can be colour coded for personal identification purposes. They are light in weight and comfortable to wear, yet have a high resistance to impact and penetration. To eliminate the possibility of electric shock, safety helmets do not have any metal parts. The materials used in the manufacture of safety helmets must be non-flammable and their electrical insulation resistance must be able to withstand 35,000 volts. Figure 6.19(b) shows the harness inside a safety helmet. This must provide ventilation and a *fixed safety clearance* of 32 millimetres. The entire harness must be easily removable for regular cleaning and sterilising. It is fully adjustable for size, fit and angle to suit the individual wearer's head.

Eye protection
While it is possible to walk with the aid of an artificial leg, nobody has ever seen out of a glass eye. Therefore, eye protection is possibly the most important precaution you can take in a workshop. Eye protection is provided by wearing goggles or visors. These must be suitable for the activity taking place. For example, when welding, special goggles (oxyacetylene welding) and visors (arc welding) must be worn. These have special lenses to filter out the harmful rays. The correct filter lens for the process must be used. Figure 6.20(a) shows general-purpose goggles and visors for use when machining or using chemicals. Figure 6.20(b) shows typical gas-welding goggles and Fig. 6.20(c) shows arc-welding visors.

Eye injuries fall into three categories:

- Pain and inflammation due to dust and abrasive grit getting between the eye and the lid.
- Damage due to exposure to ultraviolet radiation (arc welding) and high-intensity visible light. Particular care must be taken when working with laser equipment.
- Loss of sight due to the eyeball being pierced or the optic nerve severed by flying splinters of metal (swarf) when machining, or by the blast of a compressed air jet.

Hand protection
Your hands are in constant use and, because of this, they are constantly at risk when handling dirty, oily, greasy, rough, sharp, hot and possibly corrosive and toxic materials. When you merely want to protect your hands from oil, dirt or chemicals yet retain the accuracy of feel to perform precision tasks, you can either wear close-fitting disposable rubber or plastic gloves (like the ones dentists use) or use a barrier cream. The gloves give the best protection and leave your hands clean when you discard them. The barrier cream is rubbed into your hands before work and fills up the pores of your skin with a cream that is water soluble and mildly antiseptic. When you wash, the cream is dissolved out of the pores of your skin and the dirt is carried away with it. The antiseptic prevents infections.

For heavy industrial use, more substantial gloves are required and a typical range is shown in Fig. 6.21. In general terms, plastic gloves are impervious to liquids and should be worn when handling oils, greases and chemicals. However, they are unsuitable and even dangerous for handling hot materials which may melt them. Leather gloves should be used when handling sharp, rough and hot materials. When handling sheet metal leather gloves are available with 'chain mail' reinforcement for the palms.

So far we have only considered protection against accidents but, 'prevention is better than cure', so we must now consider accident prevention.

Behaviour
In an industrial environment foolish behaviour such as pushing, shouting, throwing things and practical joking by a person or group of persons cannot be tolerated. Such actions can distract a fellow worker's attention and break his or her concentration. This invariably leads to scrapped work, serious accidents and even fatalities. Such foolish behaviour observes no safety rules, takes no regard for safety equipment, can defeat the safe working procedures, and defeat the painstaking work of the safety officer.

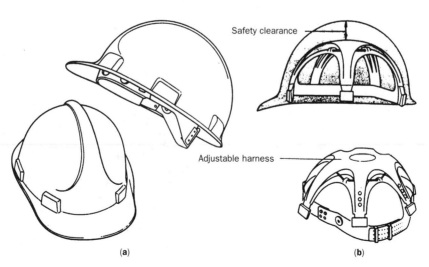

Figure 6.19 Safety helmets: (a) a typical fibreglass safety helmet made to BS 2826; (b) safety helmet harness

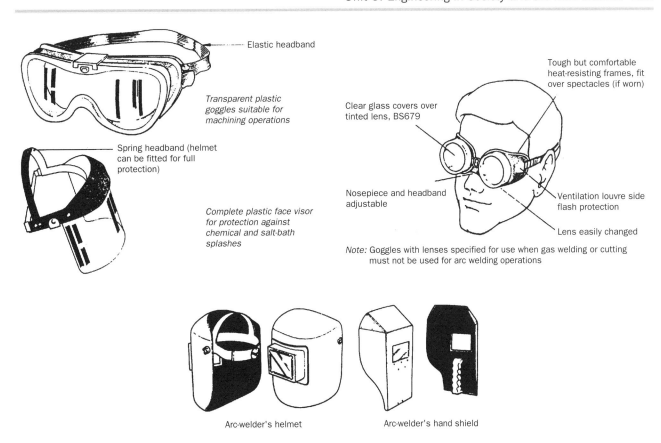

Figure 6.20 Protection equipment: (a) general-purpose goggles and visor; (b) the essential features of gas-welding goggles (c) arc-welding eye and head protection

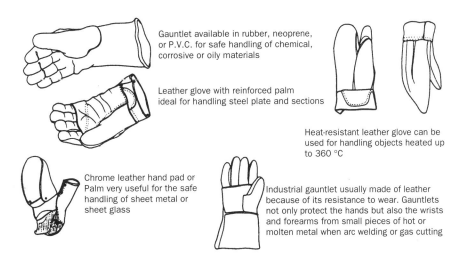

Figure 6.21 Hand protection

Lifting and carrying

As was shown in Fig. 6.14, lifting, carrying and the movement of materials generally is the biggest single cause of factory accidents. Manual handling accidents can be traced to one or more of the following causes:

- Incorrect lifting technique.
- Carrying too heavy a load.
- Incorrect gripping.
- Failure to wear the correct protective clothing.

Figure 6.22(a) shows the wrong technique for lifting which can lead to ruptures, strained backs, sprains, slipped discs and other painful and permanent injuries. Figure 6.22(b) shows the correct way to lift. The back is kept straight and the lift comes from the powerful leg and thigh muscles. Figure 6.22(c) is a reminder that the load being carried must not obstruct forward vision. This will help you to avoid injuries resulting from falling. To carry the load, keep your body upright and hold the load close to your body.

192 Advanced GNVQ Engineering

(a) The incorrect way to lift

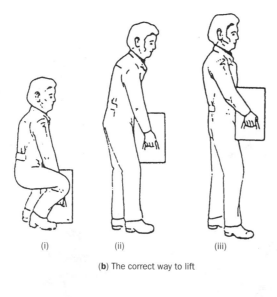

(i)　　(ii)　　(iii)

(b) The correct way to lift

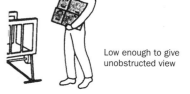

Low enough to give unobstructed view

(c) The correct way to carry

Figure 6.22 Lifting and carrying technique

Protective clothing should be worn when lifting and carrying. Crushed toes caused by dropped loads can be avoided by wearing safety shoes or boots. Cuts and splinters can be avoided by wearing suitable gloves. Burns can be avoided when handling caustic and corrosive liquids by wearing rubber or plastic gloves and aprons as shown in Fig. 6.23.

Transmission guards
By law, no machine can be sold or hired out unless all gears, belts, shafts and couplings making up the power transmission system are guarded so that they cannot be touched while they are in motion. Sometimes guards have to be removed in order to replace, adjust or service the components they are covering. Only authorised persons should remove guards or covers. If servicing or repairing a machine you must:

1. Stop the machine.
2. Isolate the machine from its energy supply.
3. Lock the isolating switch and keep the key in your pocket so that the supply cannot be turned on again until the work has been finished.
4. If this is not possible, remove the fuses and keep these in your pocket until the work has been finished.
5. Replace the guards and covers when the work is complete.
6. Reinstate the supply to the machine and test it.

If an 'interlocked' guard or cover is removed an electrical or mechanical trip will prevent the machine from operating until the guard or cover has been closed again. Figure 6.24 shows some typical transmission guards.

Rubber or plastic gloves
Rubber or plastic apron

Rubber or plastic boots

Figure 6.23 Precautions when carrying corrosive liquids

(a)

(b)

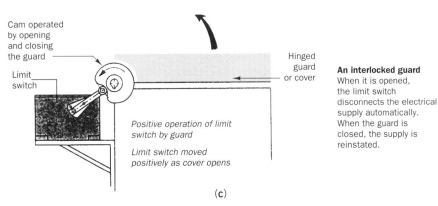
(c)

Figure 6.24 Transmission guards

Cutter and chuck guards

The machine manufacturer does not always supply these guards because of the wide range of work the machine may be called upon to do, and the correspondingly wide range of guards required.

- It is the responsibility of the owner or the hirer of the machine to supply their own cutter guards.
- It is the responsibility of the setter and/or the operator to make sure that the guards are fitted and operating correctly before operating the machine, and to use the guards as instructed. It is an offence in law for the operator to remove or tamper with the guards provided.
- If you are ever doubtful about the adequacy of a guard or the safety of a process, consult your instructor or your safety officer without delay. A range of cutter and chuck guards are shown in Fig. 6.25.

Self-assessment task

Chuck and cutter guards are provided to prevent a machine operator coming into contact with moving chucks, work and cutters which could cause serious injury. Therefore, why do you think that automatic and CNC production machine tools, which do not require an operator, have their cutting zones totally enclosed while the machine is operating?

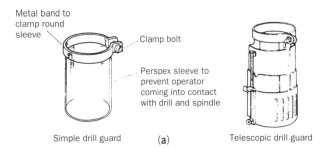
(a)

(b)

Figure 6.25 Cutter and chuck guards: (a) drill guards; (b) milling cutter guard (*continued*)

(c)

(d)

Figure 6.25 Cutter and chuck guards (*continued*): (c) milling cutter guard; (d) lathe chuck guard

6.3 Investigate the effects of engineering activities on the physical environment

After reading this element the student will be able to:

- Describe the effects on the physical environment of the use of materials for engineering activities.
- Describe the short-term and long-term effects on the physical environment of the waste products of engineering activities.
- Describe, with examples, how the functions of environmental legislation affect engineering activities.

Definitions

Before commencing a detailed examination of the effects of engineering activities on the physical environment, it would be as well to define what we mean by some the terms that will be used within the context of this section.

Physical environment
This is the environment in which we live. It consists of the natural environment, the land we live on, the air we breathe, the seas and oceans, plant life and animals including ourselves and the global climate and weather systems. As well as the natural environment, it also includes all the man-made buildings and structures that we have accumulated over the ages.

Effects
These refer to the effects of engineering activities on economic activity, on social organisation, on the pollution and degradation of our physical environment, and on waste disposal.

Materials
These are the materials used in engineering. These may be organic, renewable materials such as timber, or inorganic but naturally occurring materials such as ceramics, and metals that are extracted from mineral ores. They may also be synthetic materials such as plastics which do not occur in nature but are synthesised from the by-products of oil refining and the coking of coal.

Engineering activities
These refer to the production of goods and services, the servicing and maintenance of such goods and services and to materials handling.

Waste products
These include toxic and non-toxic waste, also degradable and non-degradable waste. They do not refer to by-products which can be 'sold-on' as the raw materials of other industries.

Functions
These include the prevention of pollution, regulations, compensation in the event of an offence, sanctions and the definition of responsibilities.

The effects on the physical environment of the use of materials for engineering activities

The range of materials used in engineering are shown in Fig. 6.26. The use of these materials has many effects on the physical environment. Generally these effects degrade the environment. The use of timber has seen the cutting down of many of the world's forests and the destruction of important wildlife habitats. Only recently has the danger of destroying this resource been recognised and efforts are being made to use only woods from quick growing trees that can be replanted. That is, forest management as a renewable resource, rather than a natural resource to be exploited and destroyed.

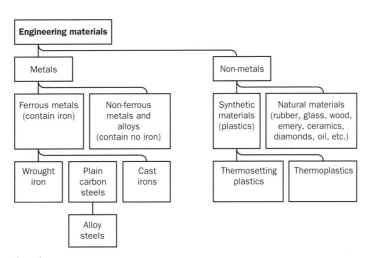

Figure 6.26 Materials used in engineering

Open-cast mining for minerals and quarrying for sand, gravel and rock must, by its very nature, have a marked impact on the environment. Such activities can only take place where concentrations of useful minerals occur, even when this is near built-up areas. Extraction is bound to leave the landscape marked and create some degree of nuisance to those living near such workings. However, winning and harnessing the world's resources is essential since they form the building blocks for life as we know it today. There are three stages to every mining or quarrying operation:

- exploration and development
- operation and production
- decommissioning and rehabilitation

The three main problems associated with such operations are water pollution, atmospheric pollution from dust, and the destruction of local flora and fauna to make room for the huge pits and the access roads for the sites. For example, the pit at Bingham Canyon in the USA where copper ore is quarried is some 4 kilometres in diameter. Deep mining for coal and other substances can lead to subsidence and create eyesores such as the winding gear and spoil heaps at the pit head.

The problem does not go away when the mineral ores have been dug up; the extraction of metals from their ores is another major cause of pollution. Sometimes the extraction plants are built close to the pit, sometimes such a plant is built on a convenient site where it can be fed from several mines or pits and where there is easy access and a plentiful energy supply for the process. Such plants are large, ugly and noisy. They tend to create dust and noxious fumes.

Usually process and drainage water is stored on the site in lagoons and is constantly recycled so that it does not pollute any adjacent natural water sources. The lagoons are lined so that the water stored in them cannot seep into the ground and cause contamination. Any surplus water that is released from the site has to undergo analysis and chemical treatment to render it safe.

For several years now, companies involved in this type of work have invested heavily in time, money and management energy to reduce the environmental impact and pollution to a minimum. They have invested in new plant and equipment which is more energy efficient, creates less pollution and less nuisance from dust and noise. Care is also taken to landscape sites so that they have less visual impact.

In the past, mines were opened, operated and abandoned with little thought for the future. Today, planning authorities insist that, as part of the terms of a mining lease, the land is returned to its former condition or to a condition suitable for a planned alternative use. There are two broad aims:

- In the short term, the aim is to create a stable and self-sustaining land surface which can resist erosion from wind and rain.
- In the long term, the aim must be to return the land to a condition suitable for other forms of land use, such as native vegetation for wildlife habitat, agricultural and forest crops, or alternative industrial use. For example, when the Argyle diamond mine was opened in Australia, the top soil was removed and stored, also seeds were taken from all the vegetation and stored, so that when the site is eventually closed it can be returned to its original condition.

However, the exploitation and extraction of raw materials for the engineering industry has advantages as well as disadvantages. Regional wealth is created by the processes involved. Many towns grew up around the coalfields of the UK. Not only the mine owners and their employees benefited from the extraction of coal, but the money earned by the mine workers, and spent locally, created a trading community. This provided shops, schools, entertainment facilities, houses, etc. The general infrastructure of roads, transport facilities, gas, water and electricity supplies, sewage and storm water disposal also had to be developed. All these services provided employment and increased prosperity. Unfortunately, when the mine closed as a result of economic pressures or for geological reasons, the regional economy collapsed. Much effort is now required to broaden the trading base of such areas and attract new industries so that they can recover and will never again be wholly dependent upon one industry.

Self-assessment tasks

Find out from the public relations department of a major mining company what steps they take to:

1. Reduce the environmental impact while a mining site is in operation.
2. Restore the site after the mine has been worked out and closed.

The effects of the waste products from engineering activities

These include toxic and non-toxic waste, also degradable and non-degradable waste products which are often costly to dispose of safely. They do not refer to by-products which can be 'sold-on' as the raw materials of other industries and are a source of income to the industries creating them.

- Toxic wastes contain chemicals that are harmful to plant and animal life. They must be made safe before disposal or they must be stored so that they cannot escape into the environment. They must never be poured down the sewers or dumped on landfill sites where they may dissolve in rainwater and find their way into the food chain.
- Non-toxic wastes such as wood, builders' rubble and most household rubbish can be safely dumped in landfill sites. It contains little or no amounts of toxic matter. Industrial waste such as power station ash is also safe.
- Degradable waste is material which will rot down, such as garden waste, wood, and some plastics that are deliberately made so that they can be broken down into safe substances by the bacteria that live naturally in the soil (bio-degradable plastics).
- Non-degradable waste is the rubble from building sites, power station ash and scrap metal that is unsuitable for recycling. This lies dormant in the landfill site and does not rot away.

Engineering activities, by their very nature, tend to cause pollution. This can be dust and fumes released into the air we breathe, liquid effluents released into the rivers, estuaries, seas and oceans, and solid wastes buried in landfill sites. Pollution can also be visual (large factory buildings, storage tanks, chimney stacks, etc.). Engineering activities can also cause noise pollution and low-frequency vibration. Since the safe disposal of such waste costs money, industrialists have tended to disregard the damage that has been caused to the environment at large. Fortunately, most responsible companies are now becoming increasingly environmentally conscious as a result of Government and International legislation, shareholder pressure and the perceived advantages of good public relations for sales, both in the way their businesses are run and in the products they sell.

Energy sourcing

Electrical power generation

The engineering industry is a major user of electricity. Because the UK, for geographical reasons, has only limited hydroelectric generating facilities and because of a national distrust of nuclear generation, most of the electricity we use has to be generated in power stations burning fossil fuels. Although gas- and oil-fired power stations are the most efficient and least polluting, these fuels are relatively expensive and scarce compared with coal. Therefore coal is still the most widely used fuel. When it is burnt, waste products are produced. The waste products from coal are given below.

Carbon dioxide
This is the so-called 'greenhouse effect' gas since it traps the heat of the sun's rays and the heat generated by natural and human activities in the earth's atmosphere causing global warming. By increasing the efficiency of the power stations, the amount of this gas produced by Powergen PLC alone has been reduced from 71 million tonnes per year in 1988 to 54 million tonnes per year in 1995.

Particulates (dusts)
These are no longer released from the power station chimneys but are removed by electrostatic precipitators (ESPs). A diagram of such a precipitator is shown in Fig. 6.27(a). The flue gases from the furnace pass through metal screens electrically charged to a very high voltage. This gives the dust particles a positive charge. The collecting plates have a negative charge. Since unlike charges produce an attractive force, the dust particles are attracted to the collecting plates and are removed from the flue gasses passing up the chimneys. Figure 6.27(b) shows how efficient such devices are in cleaning up the atmosphere.

NOx emissions
These are compounds of nitrogen (nitrous oxide and nitrogen dioxide) and tend to create atmospheric ozone and photochemical smog in the presence of strong sunlight and must be reduced as far as possible. They are also a respiratory irritant. In conventional power stations, the fuel burns with a hot, oxygen-rich flame. This maximises the efficiency of the boiler but also promotes the formation of the harmful NOx emissions. New means of burning fuel are being devised to reduce this problem.

Sulphur compounds
These combine with rainwater to form 'acid rain' that causes widespread destruction of the environment. This can range from the destruction of forests, to the destruction of the stone facing of buildings and the death of freshwater fish in streams and rivers. The problem can be controlled by various measures ranging from the use of low sulphur-content fuels, to full-scale flue gas desulphurisation (FGD). The sulphur compounds are made to combine with limestone to form *gypsum*. Much of this can be sold as a by-product for the manufacture of cements, plasters and plasterboard for the building industry. However, the remainder has to be treated as waste and dumped in landfill sites.

Ash
This is produced by burning coal. Powergen PLC, alone, burnt 20 million tonnes of coal and produced 2.8 million tonnes of ash in 1995. Not all of this ash can be sold as a by-product for making building blocks and much has to be treated as waste and dumped in landfill sites. Like gypsum, it is not toxic, it produces no methane gas and it helps to stabilise the site so it is always welcome by the site managers. The ash can also be used in the manufacture of paints, chemicals, filtration and fire-resistant products. The metals vanadium and nickel can also be recovered from emulsified fuel residues.

Water
This is used in the generation of steam and the cooling of the condensers that convert the waste steam back into water to be recycled back through the boilers. Some water is lost to the atmosphere as water vapour from the cooling towers without causing pollution. This loss is made good from the mains supply or from adjacent rivers after treatment. Any waste water (effluent) returned to rivers, sea or sewers must be at a controlled level of purity. It must also contain sufficient oxygen to maintain aquatic life, have a suitable pH and be at the correct temperature. All this is necessary to avoid upsetting the balance of any ecological system into which it is discharged.

Natural gas exploitation and distribution

Anyone who is old enough to have lived near to an old-fashioned gas works, where gas was extracted from coal, will well remember the dust, dirt and unpleasant smells associated with the process. One hundred and fifty years of gas production has left the sites of old gas works heavily polluted with dangerous and toxic wastes. To clean up these sites for reuse will be a slow, difficult and costly process. Unfortunately, at the time that the pollution took place, we knew no better.

By comparison, the extraction and use of 'natural gas' is much more environmentally friendly. It is the cleanest of the fossil fuels, little or no pollution is caused by its extraction, and it can be piped ashore so that there are none of the risks associated with the operation of oil tankers. However, like all fossil fuels, carbon dioxide is produced when it burns. As has already been stated, carbon dioxide is the principal 'greenhouse' gas. Also the heat generated by the burning of gas contributes to global warming. This applies to all fuels. Therefore the British Gas Council is constantly helping its consumers to become more energy efficient in the use of gas. Methane (natural gas) is also a 'greenhouse' gas, but the amount escaping to the environment from leaks is small compared with other sources and is being constantly reduced. Methane is also produced by rotting organic matter in landfill sites (and garden compost heaps). The sources of methane in the atmosphere are shown in Fig. 6.28.

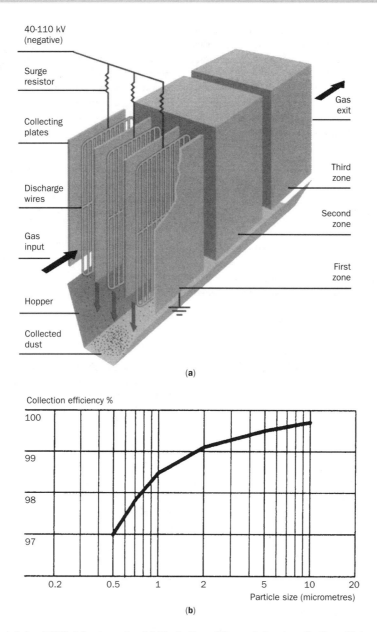

Figure 6.27 Electrostatic precipitator (ESP): (a) schematic; (b) illustration of the variation in collection efficiency with particle size (*source*: PowerGen plc)

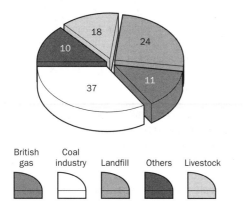

Figure 6.28 Chart showing the percentage of methane emission in the UK in 1992

Oil

Most of the oil reserves of the UK are situated offshore. This has resulted in the operation of oil drilling and operational platforms in the hazardous waters of the North Sea. Some oil is piped ashore but most is brought ashore by tanker. Huge volumes of crude oil are shipped around the world. Small spillages are inevitable and occur frequently but major spillages, although infrequent, are devastating environmentally when they occur. Natural gas is often associated with oil fields and, because many oil fields are sited in out of the way places, the gas cannot be used and is wasted by being flared off. This gives rise to the production of vast amounts of 'greenhouse' gases and heat energy, both of which contribute to global warming.

Nuclear engineering

Nuclear power stations operate without causing any pollution, except in the rare event of a leakage of radioactive materials. In France 75 per cent of all electricity is produced in this way. The remainder is produced by hydroelectric schemes. They no longer burn fossil fuels in their generating plant. The main problem of operating nuclear power stations lies in the technical difficulties and high cost of safely transporting and storing the radioactive waste (spent fuel rods) from the reactors until its radioactivity decays and becomes safe in tens of thousands of years' time. There is also the high cost of safely decommissioning worn-out plant.

Although not a waste problem, the biggest pollution problem associated with the use of nuclear energy is the ever-present danger of the process getting out of control. This is unlikely in the well-managed nuclear plants in Western Europe and the USA but it is an ever-present threat from the badly designed, badly constructed and badly managed nuclear power stations in some other parts of the world, for example the disaster when melt-down occurred in Chernobyl in Russia. This caused radioactive contamination on an international scale.

> **Self-assessment tasks**
>
> 1. Compare the advantages and disadvantages of the following fossil fuels for the generation of electricity: coal, gas and oil.
> 2. State the advantages and disadvantages of using nuclear energy for the generation of electricity compared with the burning of fossil fuels.

Manufacturing industry

Let's now look at some of the causes of pollution and environmental damage associated with the manufacturing industry.

Airborne

This can be in the form or particulates (dust) gases and vapours.

- Smoke is caused by the incomplete burning of fossil fuels for space heating and process heating. It is also produced by steel-making plants, cupola furnaces in iron foundries, and industrial incinerators. It consists of fine particles of carbon and ash combined with various gaseous compounds. Although this form of pollution has been much reduced by such legislation as the Clean Air Act, it is still a prime cause of dirt and respiratory disease in large conurbations and industrial centres. The generation of electricity and the combustion of oil fuels and fuel gases on an industrial scale for process heating is only part of the problem. The exhaust gases from motor vehicles and flue gases from domestic heating also add substantially to the problem.
- Dust and grit are fine particles carried in the air from such processes as grinding, polishing and spray painting. For example, polishing and grinding equipment must be fitted with dust extractors and filters. This is not only to protect the operators but also the people living and working near to the plant. Such plant can never be 100 per cent efficient and despite the best endeavours of industry some dust escapes. Spray-painting operatives have to wear protective clothing and respirators so that they do not inhale particles of paint mist. They work in booths from which the air is constantly exhausted by powerful fans. The air is filtered before being released to the atmosphere. However, the filtering only removes the solids, the volatile solvents still escape, as do the unpleasant odours from the drying paint. This is particularly noticeable when in the vicinity of a metal-finishing plant when stove drying is being used.

Effluent

This refers to the liquid waste from process plants. For example:

- Metal-machining process use cutting lubricants that often contain chemical additives. From time to time the tanks on the machines have to be cleaned out and replenished. The waste-cutting lubricants must not be poured down the drain but must be disposed of safely by a licensed disposal firm.
- Metal-finishing (electroplating) plants use some highly toxic metals and chemical solutions. Great care must be taken to protect the operatives in such plants and, on no account, must the effluent from such plants be poured down the public sewers. Metals are often 'pickled' in acids to remove any oxide film before finishing. The spent acid must be removed by licensed waste disposal firms after being neutralised. The sites of old factories have become severely polluted over the years by chemical spillage and badly stored toxic solids. They cannot be reused for new factories until they have been decontaminated. This is a slow and costly process and it is cheaper and easier to let the old sites remain derelict and build on 'greenfield' sites on the fringes of towns. This destroys arable land and erodes the greenbelt around towns resulting in urban sprawl.
- Process water such as that used in wet (water-washed) spray paint booths must not be released down the public sewers or into rivers without first being treated and subjected to analysis. The purified water must meet stringent requirements so not to disturb or destroy the aquatic ecology of the rivers or the bacterial balance of the sewage treatment plants into which it is eventually discharged.

Solids

This is mainly ash, waste-packing materials, scrap metal and metal swarf, and plastic waste. Wherever possible this should be recycled to preserve the diminishing reserves of the basic raw materials. Non-toxic waste that cannot be recycled can be sent directly to landfill sites. Toxic waste such as chemicals and toxic metals such as lead, cadmium and mercury must only be removed by licensed waste disposal companies who have the facilities to render such substances safe before disposal.

Heat

Heating is necessary in many engineering processes such as the hot rolling of metals, the production of castings in foundries and the sintering of powder metal compacts.

Heating equipment is also used to keep factories and offices at a comfortable temperature in winter. Although heating equipment is more efficient than it used to be and buildings are better insulated to reduce energy costs, much heat escapes to the atmosphere. This adds to the problems of global warming and climatic change.

Chemical industry

The chemical industry faces a number of problems, not only of a practical and technical nature, but also arising from media attention following accidents resulting in loss of life and severe pollution, for example the accident at Bhopal in India. The waste disposal problems of the chemical industry are the same as those already discussed but, in addition:

- The plants tend to be large and unsightly; they consume large quantities of energy and tend to produce large amounts of heat.
- The processes themselves are often potentially hazardous and employ highly toxic, flammable and explosive substances.
- Some of these potentially dangerous substances have to be transported to customers of the industry, and some of the substances are waste products. Only specially licensed companies operating under close government supervision are allowed to transport and/or dispose of these substances.

Recycling

An increasing number of chemical products are now recovered, recycled and reused. This, to some extent, overcomes the need for waste disposal. Recycling can now include plastics such as PTFE and acrylics. New plastics are being developed that are easily recycled or are biodegradable.

Ozone

It is no use manufacturing a product safely if the product itself is unsafe. For example, thousands of aerosol spray cans were made perfectly safely but they contained a CFC propellant. CFC products were also used in refrigerators. When this gas is released into the atmosphere it destroys the *ozone layer* in the upper atmosphere that protects us by absorbing the dangerous ultraviolet rays given off by the sun. Exposure to these rays can result in the development of skin cancers. Although the use of CFCs is now banned, it will take a very long time for the holes created in the ozone layer to repair themselves. At ground level, bright sunlight can react with motor vehicle exhaust fumes to produce ozone. This low level or 'tropospheric' ozone reacts further with the sunlight to produce photochemical or 'summer' smog. This causes breathing difficulties. In the UK this smog has increased by 60 per cent in the last 40 years.

Toxic products

Product stewardship is essential for companies dealing in potentially dangerous substances. It is no good safely manufacturing pesticides and other agricultural crop sprays and weed killers if the users, such as a farmer, his employees and the general public, are made ill by their use.

At one time, highly toxic white lead oxide was used as a pigment in white paints – even in paints used to coat children's cots and toys. Nowadays titanium oxide is used. This is not only a more brilliant white but is chemically neutral and does not affect the human body.

All companies, as well as chemical companies, must be responsible for the effects of the products it sells for the safety, health and welfare of the environment.

Waste disposal

No company, large or small, can just throw its waste away in a haphazard manner. Local authorities only collect and dispose of domestic and light commercial waste. They do not deal with industrial waste. Therefore industry must dispose of its waste through properly licensed waste disposal companies. These will grade the waste collected and dispose of it safely. Sometimes they will have to treat the waste chemically to render it safe. They may also have to incinerate it. Scrap metal is usually sold direct to a scrap metal merchant for recycling.

Let's now examine how our biggest chemical engineering company has improved its waste disposal management over recent years.

ICI is the biggest chemical engineering company in the UK. Figure 6.29 shows how the levels of waste disposal by this company has been reduced since 1990. These wastes can be divided into two categories: hazardous wastes and non-hazardous wastes. Non-hazardous wastes are emitted to the air, some to water (rivers and sea) and some to landfill sites. The figure quoted show the position at the end of 1995 and improvements are being made all the time. Hazardous wastes are treated on site wherever possible to make them safe. They can then be disposed of as non-hazardous wastes. If they cannot be rendered safe on site, they are removed by licensed waste disposal experts.

- Overall hazardous wastes have been reduced by 69 per cent (19 per cent better than the target).
- Hazardous wastes now only represent 3 per cent of the company's total waste.
- Non-hazardous wastes are down by 21 per cent against a target of 50 per cent. This shortfall is due to the fact that hazardous acid waste from the titanium oxide plant is now converted to non-hazardous gypsum before disposal. Gypsum is environmentally benign and can be used in the construction industries for the manufacture of cements, plasters and plaster board. The surplus can be safely dumped in landfill sites. Since it tends to stabilise such sites, it is always welcome by landfill site managers. The spent brine from chlorine manufacture is simply salt water and can be disposed of safely at sea. There is no economically viable way of reducing the volumes of these waste products.

Energy efficiency

The same company has reduced its energy usage by 18 per cent over the same period of time by operating its plants more efficiently, as shown in Fig. 6.30. This has resulted in a corresponding reduction of 'greenhouse' gas emission into the atmosphere and a reduction of waste heat. Both reductions contribute to a corresponding reduction in global warming.

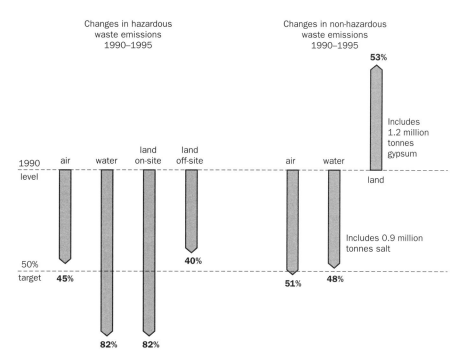

Figure 6.29 Reduction of waste disposal at ICI (*source*: ICI)

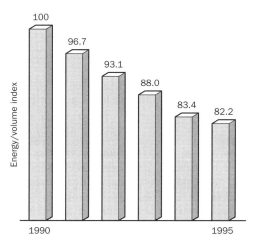

Figure 6.30 Energy efficiency at ICI, 1990–95 (*source*: ICI)

The effects of environmental legislation on engineering activities

Effects on engineering activities

We have already looked at some of the ways in which engineering activities can adversely affect the environment in which we live. Some firms take a responsible attitude towards minimising the damage their activities cause, unfortunately others do not. To protect the environment and, therefore, human society national and international legislation has had to be enacted to control the activities of all firms.

Typical UK Acts of Parliament are:

- Deposit of Poisonous Wastes Act
- Pollution of Rivers Act
- Clean Air Act
- Environmental Protection Act
- Health and Safety at Work, etc., Act.

In addition, there are local bye-laws to be observed and there are the European Union Directives which are activated and implemented through UK legislation in the form of Acts of Parliament or through Statutory Instruments (SIs). This legislation is enforced by:

- Judgements in the UK Courts of Justice in the usual way.
- Judgements in the UK Courts of Justice but based upon European legislation.
- Judgements in the European Court of Justice.

The functions of this legislation are:

- To *prevent* the activities of the industry and its waste disposal from damaging the environment in the first place.
- To *regulate* the amount of damage done by industry, by setting limits on the nature and quantity of pollutants that a factory may release into the environment.
- To *compensate* persons whose quality of life is affected by a company's non-compliance with the appropriate rules and regulation, either deliberately or by accident.
- To *prosecute* companies who do not comply with the legislation in order to bring them into line with best practice and as a warning to other companies who think that they can get away with cutting corners to increase their profitability at the expense of the rest of the community.

New directives and legislation are being introduced regularly in an attempt to reduce environmental damage. Ignorance of this legislation is no defence in law, and companies large and small must keep abreast of the changes. For small companies this can be extremely difficult. They generally have to rely on their trade association publications for summaries of the legislation, its correct interpretation and its correct implementation. Large companies can afford to send representatives from their legal and technical departments to symposia and conferences designed to update their knowledge and clarify the content and implementation of new rules and regulations.

Although this legislation has been enacted with the best of intentions, it has had the effect of making it more difficult and more costly to carry on engineering activities. This has resulted in making the industry less competitive in the world markets. Some industries have closed down in the developed countries and have moved to the developing countries where environmental legislation is less stringent and less strictly enforced.

Some of the effects of this legislation has already been considered in this unit. Coal-fired power stations have had to introduce expensive equipment to remove dust and dangerous chemical substances from the flue gases that are emitted from their chimneys. They can no longer allow hot and de-oxygenated water to flow directly into rivers and the sea. Chemical factories can no longer dump toxic chemicals out at sea or in landfill sites without first making it safe. Water companies are having to improve their sewage treatment facilities. It is no longer acceptable to pour untreated sewage directly into the sea. Motor car exhaust emissions have had to be cleaned up by the use of unleaded petrol, fuel injection in place of carburettors and catalytic converters in the exhaust system. All electrical equipment has to comply with the EU safety regulations. It also has to comply since January 1996 with European Electromagnetic Compatibility legislation. This is designed to prevent the electromagnetic radiation from one piece of equipment from interfering with another piece of electrical equipment. Such interference may cause damage and malfunction. This applies to equipment such as television sets, computers, security alarms, fluorescent lights, electromedical equipment. All new equipment must be tested and certified. The delays and cost of such testing procedures, plus the cost of test equipment and the cost of training personnel in its use and the bureaucracy of obtaining type approval, has already put a number of small electronics firms out of business.

Planning applications

The establishment of new industrial and commercial plant and activities or changes of activities and purpose is subject to planning controls and the local authority planning officers will request a detailed assessment of the processes to be undertaken and their effect upon the environment.

Waste management (within local authority control)

Current regulations are based upon the Control of Pollution Act 1974: section 17 and the Special Waste Regulations derived in 1980 from the powers given in section 17. However, new directives from the European Community are imminent. Local authorities have a three-fold interest in waste management:

- *Waste collection* – Mainly domestic and small commercial concerns. Industrial waste is not included.
- *Waste disposal* – The disposal of all the waste that the collection team hands over to them. This is usually non-toxic and is disposed of in incinerators and landfill sites. Note that waste disposal is due for privatisation in the near future.

- *Emissions to the atmosphere* – This comes within the Environmental Pollution Act of 1990: Part 1.
  - Section A is concerned with large-scale plant such as power stations. It is also concerned with all chemical plant, large and small. Such plant falls outside local authority control and will be considered later.
  - Section B falls within local authority control, and is administered by the environmental health officers through a licensing process – Authorisation for Emission to Atmosphere.

The main method of control is by input to the process and the scale of the process; for example, incinerators must not have a throughput exceeding 1 tonne per hour. There is also legislation on the type of plant to be used and the type of particulate and gaseous substances discharged by any means.

> **Self-assessment tasks**
>
> Find out from your local council:
>
> 1. Where is their local waste disposal site located?
> 2. What sort of waste is sent there?
> 3. How do they deal with any landfill gas (methane) given off by the rotting waste?
> 4. What was the site originally used for?
> 5. How do they intend to restore the site and what do they intend to use the site for when they have finished with it?

Waste management (outside local authority control)

From Monday, 1 April 1996 – vesting day for the new Environment Agency – the officers of Her Majesty's Inspectorate of Pollution have been absorbed into the new agency and are responsible for *Emissions to Atmosphere* and the authorisation for Section A processes under the Environmental Pollution Act mentioned above. That is, all incinerators over 1 tonne per hour throughput, all chemical plant and all other 'specified' processes. The authorisations issued by the agency cover:

- Inputs to the process and the nature of the process itself.
- Polluting substances that may be *emitted direct to the atmosphere* and the volume limits per hour of such pollutants.
- Emissions to sewers and any effects on surface water drainage.
- The nature of solid residues, the need for further treatment before disposal, or the suitability of such residues for disposal direct to landfill sites.

Other industrial processes not covered above, may yet fall within the jurisdiction of the Industrial Pollution Control (IPC) officers of the National Rivers Authority (NRA) – now incorporated into the Environment Agency to provide 'control through consent to discharge'. This lists: temperature, rate of discharge (at a set volume per hour), pH value, sediment, toxicity, and heavy metal content in particular, also the bio-oxygen demand.

The Waste Regulatory Authority (WRA) is now incorporated in the Waste Agency and control is exercised:

- By the issue of waste management licences covering the keeping (storing), treating, transport and disposal of controlled waste.
- Through waste regulation (special) waste management licences and the associated paperwork. This paperwork giving notice of transport must precede every movement and copies must accompany every movement of the waste. This not only applies to persons actively handling the waste but also brokers and intermediaries who may not physically handle the waste.
- By the registration of carriers of waste. Only the Head Office needs to be registered and this registration applies to all branch offices throughout the UK. Section 54 places a *general duty of care* on the carrier of controlled wastes. This includes a requirement of the carrier to take such wastes only to authorised sites. The main controlling aspect of the duty of care is the adequate description of the waste and any special hazards associated with it. This description must accompany any transfer of the waste both physically and in respect of its ownership.

Waste disposal

The scale and problems of waste disposal are enormous. Figure 6.31 puts some facts and figures to this problem.

Landfill presently accounts for 90 per cent of controlled waste in the UK. It is not just a question of finding a big hole and filling it! Before a site can be used it must be properly engineered and prepared, as shown in Fig. 6.32. The following notes refer to a professionally managed site incorporating the requirements of the Health and Safety Executive (HSE) and the Waste Regulatory Authority (WRA).

The officers of the HSE are responsible for ensuring that owners of the site do not put at risk the health and safety of:

- Their employees working at the site.
- Visitors to the site (delivery drivers).
- Persons living and working adjacent to the site.

The officers of the WRA are responsible for ensuring that the well-being of the environment as a whole is not put at risk by operations on the site. For example, no toxic waste must be dumped; no noxious, flammable or explosive gases must be allowed to accumulate; the ground surrounding the site and the water table beneath the site must not be contaminated.

To comply with these requirements the site must be engineered as follows:

- An engineered pit lining is constructed to seal the waste from the surrounding rock, soil strata and water table. Water entering the site must be contained within the site. Capping systems and small working faces restricts the ingress of water.
- Rubbish is deposited in consistent, even layers to strict engineering procedures. This ensures safe decomposition and a stable body of refuse that will become load bearing.
- Decomposing waste can generate landfill gas (LFG) and noxious liquids (leachate). The site must be regularly checked for gas migration and water quality.
- Currently 70 per cent of LFG, which consists mainly of methane, escapes to the atmosphere. Since methane is a 'greenhouse' gas this is undesirable. The rest (30 per cent) is either flared off or is used to generate electricity. Of the 66,000 MW total of electricity generated in the UK, only 32 MW is produced from landfill waste. However, this figure is rising.

Landfill operators must not only provide reassurance of minimal impact on local communities during the site's productive life, but also for many years after it has been filled.

Filled sites offer opportunities for landscaping and redevelopment of public open spaces in areas of former industrial, mining and quarrying dereliction. Restoration is now a key part of landfill management since it returns derelict sites to recreational and agricultural use. Trees are often planted around the perimeter of large modern sites to reduce the visual impact and to act as a noise barrier. They also provide a wildlife habitat.

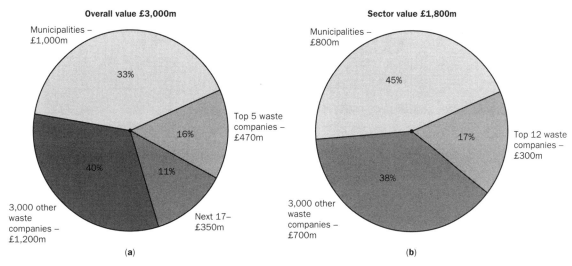

Figure 6.31 UK waste industry sector: (a) market size and share; (b) non-hazardous collection (*source*: BIFFA)

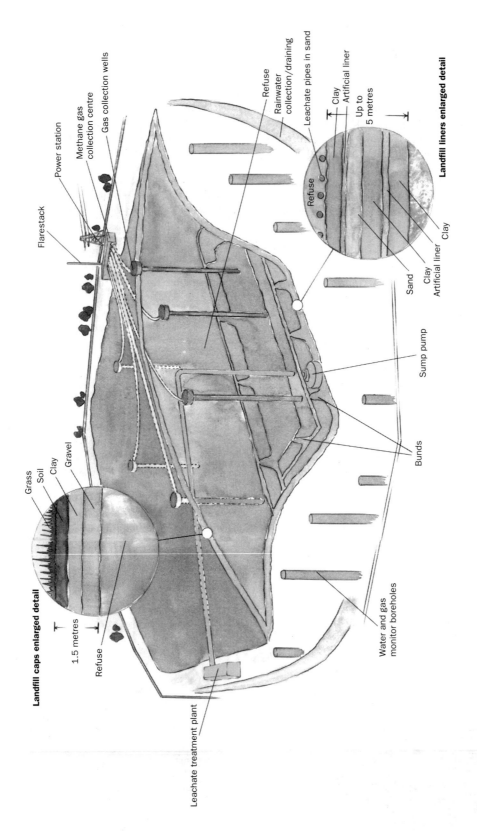

Figure 6.32 Modern, fully engineered, landfill site

> **Self-assessment task**
>
> Find out from the public relations department of a large manufacturing company, producing metal or chemical products, what waste they create, how they dispose of their waste and whether or not they have to take any special precautions.

Briefs

Brief 1: Emergency Procedures

Find out the correct procedures for dealing with the following emergencies.

1. Electric shock

You find a workmate unconscious but clutching a metal-cased portable power tool.

- Describe, in the correct sequence, the immediate actions you must take before attempting to commence resuscitation.
- Describe in detail the correct procedure for mouth-to-mouth artificial respiration and, if necessary, external heart compression.
- Why must there be no delay in commencing the resuscitation procedures and why must it be continued without pause until the patient regains consciousness or the emergency services arrive?

2. Fire

You find that a fire has broken out in a store room.

- Describe, in the correct sequence, the emergency procedures you would take.
- List the more common types of fire extinguisher and the types of fire for which they are used.
- Describe the precautions that must be taken when using fire extinguishers.
- What special precautions should be taken if the fire is adjacent to gas cylinders and solvents (e.g. paint stores or adhesives stores).

Brief 2: Effects of engineering on society

Write a report, using quantitative data, on an industrial region of your choice. Your report should include:

- An explanation of the contribution of each engineering sector (chemical, mechanical, electrical, civil and aeronautical) to the regional economy.
- A brief description of the effects of the application of engineering technology to the social environment (domestic, leisure, health and employment).
- The effects of engineering technology on employment patterns and working conditions (size of the workforce, skill and education requirements, working patterns such as time and place, hazards, safety equipment and health and safety legislation).

The report should be supported by brief general notes on the contribution of one engineering sector chosen from the range of local, national and European Union economies.

Brief 3: Engineering applications

Research and write up case studies for TWO dissimilar working environments (e.g., a production workshop and a commercial office) showing the effects on them of engineering applications. Your case studies should:

- Identify features of each working environment, such as employee comfort, health, productivity, efficiency, heating, ventilation, lighting and noise.
- Describe the effects of engineering on the features identified above.
- Identify the ergonomic considerations of the workplace design, information displays, controls and space relationships.
- Describe the precautions to be taken against risks in each of the environments being reported upon. These should include prevention and protection, air contamination, hygiene and physical danger.

Brief 4: Effects of engineering on the physical environment

Write a report investigating the economic, social, pollution, degradation and waste disposal effects on the physical environment (natural and man-made). The report should cover one engineering activity from each of the following categories:

- Materials (natural and synthetic).
- Engineering activities such as production, servicing, materials handling.
- Waste products such as toxic, non-toxic, degradable and non-degradable substances.
- Functions such as prevention, regulation, compensation, sanctions and definitions of responsibilities.

The report should also include:

- A brief description of the engineering environmental effects the whole range of the use of materials for each engineering activity.
- A brief description of the short-term and long-term environmental effects of the waste products of each engineering activity.
- Brief descriptions supported by one example of each, of how the functions of environmental legislation affect engineering activities.

6.3 Unit test

Test yourself on this unit with these multiple-choice questions.

1. The monetary value of all the goods and services produced in the UK is called the:

    (a) Gross Domestic Product (GDP)
    (b) Gross National Product (GNP)
    (c) Nett Domestic Product (NDP)
    (d) Nett National Product (NNP)

2. Banking services provided by the UK overseas are an example of:
    (a) overseas aid
    (b) visible exports
    (c) cash outflow
    (d) invisible exports

3. Goods and services bought by the UK from sources overseas are called:

    (a) visible exports
    (b) invisible exports
    (c) imports
    (d) inward investment

4. The numerical difference between the monetary values of all forms of imports and all forms of exports for the UK is called:

    (a) inward investment
    (b) the national debt
    (c) the monetary reserve
    (d) the balance of payments

5. The sector of the engineering industry responsible for the design of bridges across roads and rivers is the:

    (a) automobile engineering sector
    (b) civil engineering sector
    (c) mechanical engineering sector
    (d) production engineering sector

6. The quality of long-distance telephone communications have been improved by the use of:

    (a) aluminium conductors in telephone cables
    (b) fibre-optic cables
    (c) plastic insulation
    (d) mobile phones

7. In general, retraining of the engineering workforce has been aimed at:

    (a) flexibility and multi-skilling
    (b) providing single-skill craftspersons
    (c) providing limited skill operatives
    (d) satisfying narrow, prescriptive, job-specifications

8. Frequent changes of activity are desirable to prevent:
    (a) back ache from prolonged sitting
    (b) eyestrain from using computers for too long at a stretch
    (c) repetitive strain injury
    (d) all the above

9. The productivity of the engineering industry has been improved by:

    (a) maintaining or increasing the level of output while reducing unit costs
    (b) reducing the level of output while maintaining the level of unit costs
    (c) maintaining or increasing the level of output while increasing unit costs
    (d) reducing the level of output while increasing unit costs

10. For efficient working, the temperature of the workplace should:

    (a) not fall below 20 °C
    (b) not fall below 16 °F
    (c) not exceed 20 °F
    (d) lie between 16 °C and 20 °C depending upon the type of activity

11. When lifting manually, the load must not exceed:

    (a) 20 ozs
    (b) 20 g
    (c) 20 lb
    (d) 20 kg

12. Energy usage should be:
    (a) kept to a minimum even if this restricts production
    (b) dictated by the needs of maximum production no matter how wasteful
    (c) a sensible compromise between the requirements of production and the environment
    (d) dictated only by the needs of the environment

13. Process waste should be:
    (a) dumped in a landfill site even if toxic
    (b) dumped as cheaply as possible
    (c) treated to make it safe and then dumped
    (d) recycled for reuse wherever possible

14. The requirements of the Health and Safety at Work, etc., Act apply:

    (a) only to employers
    (b) only to employees
    (c) equally to both employers and employees
    (d) only to the largest companies

15. When a mine or quarry used for the extraction of mineral ores has been worked out, the licensed operators must:

    (a) abandon it
    (b) hand the site back to the local authority for reclamation work
    (c) take responsibility for decommissioning and rehabilitating the site
    (d) fill it up by any means including toxic waste

16. By-products are:

    (a) waste products that can be sold on for use as a raw material by another industry
    (b) the products of a subsidiary factory
    (c) non-toxic wastes for tipping in a landfill site
    (d) bio-degradable wastes

17. The 'greenhouse effect' resulting from burning fossil fuels in power stations and road vehicles is caused mainly by:

    (a) water vapour
    (b) nitrogen gas
    (c) carbon dioxide gas
    (d) unburned carbon particles

18. Airborne pollutants from manufacturing industry are still a prime cause of:

    (a) dirt only
    (b) respiratory disease only
    (c) dirt and respiratory disease
    (d) radioactivity

19. Industrial waste must be disposed of through:

    (a) the local authority waste collection service
    (b) licensed waste disposal companies
    (c) dumping at landfill sites
    (d) 'fly' tipping

20. Hazardous waste can be transported:

    (a) by any haulage facility with police escort
    (b) only if every movement is preceded by the appropriate paperwork giving notice of movement and if copies of this paperwork accompanies every movement
    (c) only by companies holding waste regulation (special) waste licences
    (d) only by companies complying with both B and C

UNIT 7

Science for Engineering

The scientific principles underlying engineering are fundamental to an understanding of the subject. The scientific laws and principles explain how engineering systems work, how systems are affected by different variables and changes in those variables.

In this unit the student will learn how to:

- Describe physical systems in engineering in terms of scientific laws and principles.
- Calculate the response of physical systems in engineering to changes in variables.
- Determine the response of physical systems in engineering to changes in parameters.

After reading this unit the student will be able to:

- Identify the scientific laws and principles and the variables and constants that govern engineering systems.
- Describe the relationship between the variables and constants by applying these laws and principles.
- Identify a suitable mathematical model and its key variables for a system.
- Calculate the response of the system to changes in the variables.
- Identify the parameters to be measured.
- Select instruments to measure the parameters indentified.
- Identify possible sources of error when measuring the parameters.

7.1 Static systems

Static systems are the simplest type of scientific or engineering system. As the name implies, they are systems in which there is no motion. Examples include:

- Simple balances, such as an old-fashioned set of kitchen scales.
- A beam supporting a load.
- A steel wire attached to a pendulum weight in a grandfather clock.

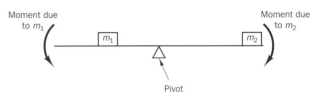

Figure 7.1 A simple balance

Forces

In order to understand these systems, we need to be familiar with the concept of a *force*. A force is anything that causes a stationary object to move or change shape, or a moving object to change its motion (by slowing down, speeding up, or changing direction). Examples are:

- *Gravity* – which causes unsupported objects to fall.
- *Weight* – which is the force exerted by supported objects on their supports.
- *Friction* – which causes a moving car to slow down if the engine is turned off.
- *Impact* – in which collision between two objects has an effect on each. For two snooker balls, the impact force will change their motion, while for two cars in collision the effect is more likely to be a change in shape!

In all static systems, there are forces acting on the various components, but these forces are said to be in *equilibrium*. In other words, the forces act in such a way that they cancel each other out.

In all engineering calculations, it is important to be aware of the units you are using. The SI unit of force is the *newton* (symbol N), named after the great physicist Sir Isaac Newton.

In other words, the further the point of application of a force is from the pivot, the greater the turning force. This is the principle underlying many pieces of engineering equipment, including the simple spanner.

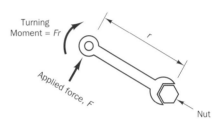

Figure 7.2 Spanner applying force to a nut

Because the force on the spanner is applied at a considerable distance from the nut (the pivot), the turning moment is much greater than if the same force were applied directly to the nut.

For a simple balance, the system is in equilibrium (i.e. balanced), when the clockwise and anticlockwise moments are equal. This statement is known as the *principle of moments*, and it enables us to calculate the mass of an object by using a simple balance, as in the following example.

Simple balance

A simple balance is a system in which a pivoted beam is held in a stationary position by masses placed either side of the pivot.

In Fig. 7.1, the mass m_1 pushing down on the left-hand side of the beam is counterbalanced by the mass m_2 pushing down on the right-hand side. Both masses exert a *force* downwards on the beam, and because the beam is supported at a pivot, the effect of these forces is to *turn* the beam anticlockwise (m_1) or clockwise (m_2). This turning effect of a force is known as a *moment* and is given by

$$M = F \times d$$

where M is the moment in newton metres (Nm), F is the force in newtons and d is the distance from the pivot in metres.

Example 7.1

A balance is set up as shown in Fig. 7.3. It consists of a simple 1 metre long beam with a pivot at its midpoint. A steel block of known mass (500 g) is suspended from the left end of the beam, and the unknown mass is suspended from the other side. The unknown object can be moved along the beam until the system is balanced, and it is found that this occurs when the mass is 30 cm from the pivot.

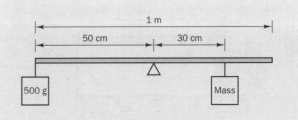

Figure 7.3 Determining the point of balance

The downward force (in newtons) exerted by a stationary object of mass m (in kilograms) is given by

$$F = m \times g$$

where F is the force in newtons, m is the mass in kilograms and g is the acceleration due to gravity.

The acceleration due to gravity is a constant, and is equal to 9.81 metres per second per second (m/s^2). Therefore,

Anticlockwise turning moment = Force × Distance
$$= (0.5 \times 9.81) \times 0.5$$
$$= 4.905 \times 0.5$$
$$= 2.4525 \, \text{N m}$$

By the principle of moments, the clockwise turning moment at equilibrium will have the same value. Therefore,

$$2.4525 = \text{Force} \times \text{Distance}$$
$$= (\text{Mass} \times 9.81) \times 0.3$$

We can rearrange this equation as follows:

$$\text{Mass} \times 9.81 = 2.4525/0.3$$
$$= 8.185$$
$$\text{Mass} = 8.175/9.81$$
$$= 0.833 \, \text{kg or } 833 \, \text{g}$$

Self-assessment task 7.1

1. Using the apparatus described in Fig. 7.3, calculate the mass of an object that balances the system when it is 20 cm away from the pivot.
2. If the known mass is changed to a 1 kg block, how far from the pivot would a mass of 1.25 kg need to be in order to balance the system?

Simply supported beam

The second type of static system we will look at is known as a simply supported beam. This is a beam supported at two points, usually, but not necessarily, the two ends. The beam may carry one or more than one load. This type of arrangement may be used in building, or to provide support to heavy equipment in an engineering workshop (Fig. 7.4).

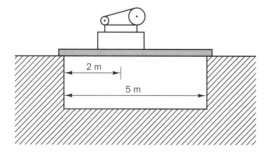

Figure 7.4 A beam in a workshop

Example 7.2

In this example, the beam is supported at each end, and carries the loads from one piece of machinery. In order to clarify the situation, we can represent it by using a simplified schematic diagram, as in Fig. 7.5.

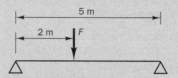

Figure 7.5 A simply supported beam

As the system is static, the downward forces exerted by the load must be cancelled out by an opposing force or forces. The two supports each exert an upward force, known as *reactions*, on the beam, and these counteract the downward force.

The following examples show how we can calculate values of these reactions for various systems.

Example 7.3

The system shown in Fig. 7.6 consists of a 1 m beam supported at each end, with a mass of 50 kg suspended from its midpoint.

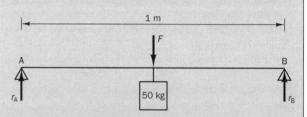

Figure 7.6 Supported beam for Example 7.3

The downward force exerted by the mass is equal to $50 \times 9.81 = 490.5 \, \text{N}$. Therefore the sum of the two reactions, r_A and r_B, must also equal 490.5 N, and as the system is symmetrical we can say:

$$r_A = r_B = 490.5/2 = 245.25 \, \text{N}$$

If the mass is moved along the beam so that it is 40 cm from support A, the calculation is more complicated. The total downward force is still the same, and therefore the total reaction $(r_A + r_B)$ is the same, i.e. 490.5 N. However, because the system is not symmetrical, the reaction at A is greater than that at B (support A bears more of the load).

To calculate the exact values of r_A and r_B we have to consider *moments*, as in the case of the simple balance. Imagine that the beam is fixed at support A, but free to rotate about this point. Because the system is in equilibrium, we can say that the clockwise moment about A is equal and opposite to the anticlockwise moment (Fig. 7.7).

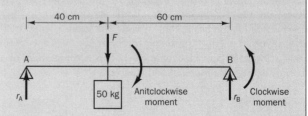

Figure 7.7 System in equilibrium

In this case the clockwise moment is the turning force exerted by the mass, and the anticlockwise moment is the turning force due to the reaction r_B at support B. Since moment is given by force × distance from pivot, we can say:

Force due to mass × Distance from pivot = r_B × Length of beam
$$490.5 \times 0.4 = r_B \times 1$$

Therefore,
$$r_B = 196.2\,\text{N}$$

We have already shown that $r_A + r_B = 490.5\,\text{N}$, so
$$r_A = 490.5 - r_B$$
$$= 490.5 - 196.2$$
$$= 294.3\,\text{N}$$

Clockwise moment = $(4,000 \times 6) + (4,000 \times 1)$
$$= 24,000 + 4,000$$
$$= 28,000\,\text{N m}$$

Anticlockwise moment = $r_B \times 12$

Since clockwise and anticlockwise moments are equal:
$$r_B \times 12 = 28,000$$
$$r_B = 28,000/12$$
$$= 2,333\,\text{N}$$

Since the total reaction is equal to the total downward force $(4,000 + 4,000 = 8,000\,\text{N})$:
$$r_a = 8,000 - r_B$$
$$= 8,000 - 2,333$$
$$= 5,667\,\text{N}$$

Self-assessment tasks 7.2

1. Verify that the values of r_A and r_B obtained in Example 7.3 are correct, by taking support B as the pivot, and calculating clockwise and anticlockwise moments in the same way.
2. If the mass is moved again, so that it is now 30 cm from support B, what are the new values of r_A and r_B?

Self-assessment task 7.3

When designing equipment such as that used in the previous example, it is important to ensure that it is capable of bearing the maximum possible load. Assuming that only two parts are ever on the conveyor at one time, the maximum load on one support occurs when one part is exactly over Machine 2, and the next part is immediately behind it, as shown below. Calculate the reaction at each support.

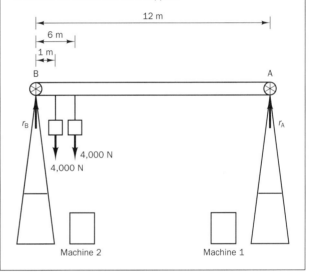

Shear force

When a beam is subjected to a *point load*, as in the above examples, it will be deformed to a greater or lesser extent, depending on the beam material. This is known as the *bending* of a beam.

To see the effect of this, we can think of the beam as being composed of a large number of thin 'slices' which can move relative to each other (Fig. 7.9).

When the beam is deformed we can think of the slices being displaced, with those slices closest to the point of application of the load being displaced the most. This sets up *shear forces* in the beam, which oppose the bending.

Shear forces are an important consideration in many engineering situations, and the ability to withstand shear forces without cracking is an important property of all engineering materials (see Unit 4).

Example 7.4

In a factory, parts are moved from one part of the assembly line to another using a conveyor, as shown in Fig. 7.8. Each part exerts a downward force of 4,000 N. The conveyor is designed to pick up parts from one machine and deposit them at the next, and it is only intended to carry a maximum of two parts at a time.

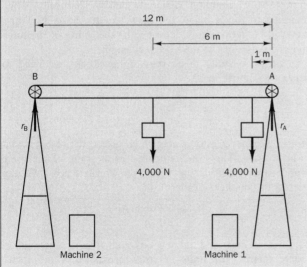

Figure 7.8 Conveyor – Example 7.4

We can calculate the reactions at A and B by the same method as we used in the previous example. The only difference is that we need to take account of two separate loads on the beam. For the situation shown in Fig. 7.8, where one part has just been collected from Machine 1 and another is halfway along the conveyor, we can calculate clockwise and anticlockwise moments about support A, as follows:

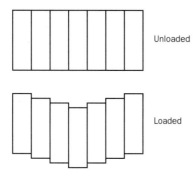

Figure 7.9 Deformation of a beam

We can calculate the shear force at a given point along a beam quite simply from the following:

The shear force at any section of a beam is equal to the *algebraic* sum of all the forces acting on the beam, to the left of the section.

'Algebraic' means we have to take into account the direction, as well as the magnitude of each force. Upward forces (reactions) are negative, and downward forces (loads) are positive.

Example 7.5

The beam in Fig. 7.10 is loaded with a single point load of 90 N, positioned two-thirds of the way along a 3 m beam, i.e. 2 m from one end and 1 m from the other end. In order to calculate the shear forces at x and y, we need first to calculate the reactions at A and B.

Using the method described above, we can calculate r_A and r_B by taking moments about r_A and r_B. At r_A,

Force due to mass × Distance from pivot = r_B × Length of beam
$$90 \times 2 = r_B \times 3$$
$$180 = r_B \times 3$$
$$r_B = 180/3 = 60 \text{ N}$$

Now $r_A + r_B = 90$ N, so

$$r_A = 90 - r_B$$
$$= 90 - 60 = 30 \text{ N}$$

So $r_A = 30$ N and $r_B = 60$ N.

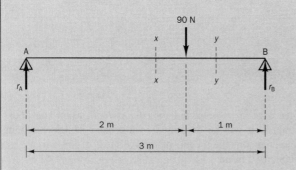

Figure 7.10 Example of points loads

The shear force at x is simply given by the value of the reaction at A:

$$SF_x = -r_A = -30 \text{ N}$$

The shear force at y is given by the sum of the reaction at A, which is negative, and the load, which is positive.

$$SF_y = -r_B + 70 = -30 + 90 = 60 \text{ N}$$

Self-assessment task 7.4

Using the example of the conveyor shown in task 7.3 calculate the shear force between the two parts being carried.

Wire in tension

The final example of static systems is one in which a mass is suspended from a wire or rope. In this case, the downward force exerted by the mass, due to gravity, is counteracted by an upward reaction in the wire (Fig. 7.11).

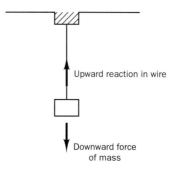

Figure 7.11 Mass suspended on a wire

Suspension of a mass from a wire in this way causes the length of the wire to increase slightly. *Hooke's law* states that this extension is *proportional* to the force exerted by the mass. For instance, if a 50 cm wire is stretched by 2 mm when subjected to a force of 100 N, it will be stretched by 4 mm when subjected to a force of 200 N.

If the mass is removed, the wire will generally return to its original length. A wire that behaves in this way is said to be *elastic*. Most common materials behave elastically when subjected to forces below the *elastic limit*. Above this limit, they will be permanently deformed by the action of the force and will not return to their original length.

In order to study this system more closely, we need to define two new terms:

- stress
- strain

Stress Whenever a force acts on an object, the object is said to be *stressed*. The stress is defined as the ratio of the force acting to the area of the object on which it acts. We can express this mathematically:

$$\text{Stress} = \text{Force}/\text{Area}$$
$$= F/A$$

As force is measured in newtons and area is measured in square metres, the units of stress are newtons per metre squared or N/m^2. Stress may also be given in pascals (Pa), where 1 Pa = 1 N/m^2.

Strain When a wire is stretched it is said to be under strain, and the strain is defined as the ratio of the extension of the wire to its original length. Again, we can express this mathematically:

$$\text{Strain} = \text{Extension}/\text{Original length}$$
$$= x/L$$

Since 'extension' and 'original length' have the same units (i.e. metres), the two units cancel out. Therefore strain has no units: it is a *dimensionless quantity*.

These two concepts are very important to engineers, and you will come across them frequently in your studies.

For a wire in tension, behaving elastically, we have seen that the extension of the wire is proportional to the applied force. However, the amount by which the length increases for a given increase in force (the *constant of proportionality*) has to be determined experimentally as it also depends on the original length of the wire, and its cross-sectional area.

If we look at the definitions of stress and strain, we can see that *stress* is related to the force, and *strain* is related to the extension. Therefore, the ratio of stress to strain is related to the ratio of force to extension:

$$\frac{\text{Stress}}{\text{Strain}} = \frac{F/A}{x/L} = (F/x) \times (L/A)$$

We know that, for a particular wire, the ratio of force to extension and the dimensions of the wire (L and A) are all constant, so the stress/strain ratio is also constant. The important thing to note is that this ratio is constant for any wire made of a particular material, i.e. the stress/strain ratio is a fundamental *property* of a material.

The stress/strain ratio is known as the *modulus of elasticity* of the material (symbol E). It is also known as Young's modulus. It is an important quantity in engineering because it allows us to predict the behaviour of a wire of a particular size and made of a particular material, without having to test it experimentally.

Since strain is dimensionless, the modulus of elasticity has the same units as the stress, i.e. N/m^2 or Pa.

Example 7.6

A copper wire of length 60 cm and cross-sectional area 0.8 mm^2 is suspended from a beam, and a series of weights are attached to the bottom. For each weight, the extension of the wire is measured. The results are shown in Table 7.1

Table 7.1 Results of an experiment to measure extension of a wire under load

Mass (kg)	Force* (N)	Extension (mm)
10	98.1	0.7
20	196.2	1.5
30	294.3	2.3
40	392.4	3.1
50	490.5	3.9
60	588.6	4.7

*Force = mass $\times$ g

If we know the original length and cross-sectional area of the wire, we can use these data to determine a value of the modulus of elasticity, E, for copper. From the definition of E given above we can write the following equation:

$$E = (F/x) \times (L/A)$$

We can rearrange this to give an equation for the relationship between applied force F and extension x:

$$x = \frac{FL}{EA} = \frac{L}{EA} \times F$$

From this we can see that a graph of x against F will be a straight line with a gradient equal to (L/EA). (If this is not clear to you, refer to Unit 8, Mathematics for Engineering.)

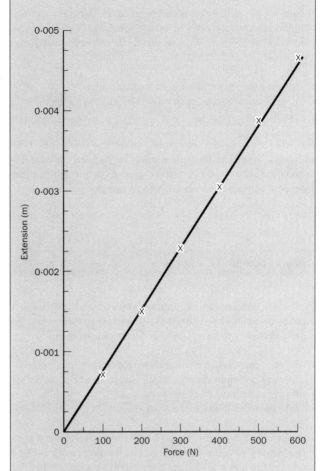

Figure 7.12 Graph of x against F

The data are plotted in Fig. 7.12. Note that the extension values have been converted into metres in order to keep the units correct. We can see that they do indeed form a straight line (allowing for experimental error in the measurements), and the gradient is given by:

$$\text{Gradient} = 4.7 \times 10^{-3}/600$$
$$= 7.8 \times 10^{-6}$$

We know the gradient is equal to L/EA, so:

$$E = L/(\text{Gradient} \times A)$$
$$= 0.6/(7.8 \times 10^{-6} \times 0.8 \times 10^{-6})$$
$$= 96 \times 10^9 \text{ Pa}$$

Self-assessment tasks 7.5

1. A simple crane has a hoisting cable made of steel rope. The rope has a cross-sectional area of 19.6 mm^2 and steel has a modulus of elasticity of 210 GPa. If there is 10 m of cable used to hoist a load off the ground, calculate the extension of the rope if a load of 100 kg is lifted.
2. Draw a graph of the predicted extension for loads varying from 0 to 500 kg in 50 kg steps.

7.2 Dynamic systems

In contrast to static systems described in the previous section, *dynamic systems* are ones in which some or all parts of the system are in *motion*. We can divide these systems into two categories:

- Linear motion, i.e. motion in a straight line.
- Rotational motion, i.e. circular motion about a fixed point.

We will look at the principles governing each of these categories, and see how they may be applied in real-life engineering situations, but first we need to define and explain three new quantities: *work*, *power* and *energy*.

Work, power and energy

These three concepts are all familiar from everyday life, but in a scientific or engineering context they have very specific meanings and it is necessary to have a good understanding of them.

- *Work* is done when a force is exerted against a resistance through a certain distance in the direction of the force.
- *Power* is the rate of doing work.
- *Energy* is the ability to do work.

The ultimate aim of any piece of engineering machinery is to perform work of some description. For instance, work is done when a crane lifts a load from the ground to a certain height, or when a car accelerates.

The SI unit of work is the joule (symbol J), and is defined as the work done when a force of 1 N is exerted through a distance of 1 m in the direction of the force.

One of the key parameters used to describe engineering machinery such as motors is the power. This is a measure of the speed at which the machine performs work.

The SI unit of power is the watt (symbol W), and 1 watt is defined as the power of a device that performs 1 J of work in 1 second.

In order to perform work, a device must contain energy. Energy exists in a number of different forms, and when work is done the energy is converted from one form to another. In dynamic systems we are concerned with two forms of energy:

- *Kinetic energy* is the energy possessed by a body by virtue of its motion.
- *Potential energy* is the energy possessed by a body by virtue of its position or state of strain.

Other important forms of energy are *heat* (p. 220), *chemical energy* and *electrical energy* (p. 225).

The units of energy are the same as those for work – joules. If work is done on one body by another, the first body gains an amount of energy equal to the amount of work done, and the second body loses an equal amount of energy. The total amount of energy in the system remains the same. This important principle is known as the *conservation of energy*.

Linear motion

When studying objects moving in a straight line, we need to consider five quantities or *variables*:

- initial velocity, u
- final velocity, v
- acceleration, a
- time, t
- distance travelled, s

These five variables are related to each other by four fairly simple equations as follows:

- $v = u + at$
- $s = \dfrac{v - u}{2} \times t$
- $s = ut + \tfrac{1}{2}at^2$
- $v^2 = u^2 + 2as$

Using these four equations, we can calculate the value of any one of the five variables, provided we know the values of at least three of the others. In practice, we usually need to know the final velocity v, or the distance travelled s, so the equations are usually written as above. The equations can be rearranged to find any of the other variables (see Unit 8 for more detail on rearranging equations).

Example 7.7

A train starts from a standstill and accelerates along a straight track, with an acceleration of 0.2 m/s². Using the equations above we can calculate: (a) the velocity after 12 seconds; (b) the distance travelled after 15 seconds; (c) the time taken to travel 50 m.

(a) We know $u = 0$ m/s, $a = 0.2$ m/s² and $t = 12$ s, and we need to know v. The relevant equation is therefore:

$$v = u + at = 0 + (0.2 \times 12) = 2.4 \text{ m/s}$$

(b) In this case, we know $u = 0$ m/s, $a = 0.2$ m/s² and $t = 15$ s, and we need to know s. The relevant equation is therefore:

$$s = ut + \tfrac{1}{2}at^2 = (0 \times 15) + \tfrac{1}{2}(0.2 \times 15^2) = 22.5 \text{ m}$$

(c) Here, we know $u = 0$ m/s, $a = 0.2$ m/s² and $s = 50$ m, and we need to know t. The relevant equation is the same as for part (b), but we need to rearrange it to make t the subject:

$$s = ut + \tfrac{1}{2}at^2$$

Because $u = 0$, this becomes:

$$s = \tfrac{1}{2}at^2$$
$$2s/a = t^2$$
$$t = \sqrt{(2s/a)} = \sqrt{(2 \times 50/0.2)} = \sqrt{500} = 22.4 \text{ s}$$

> **Self-assessment tasks 7.6**
>
> 1. A rock is dropped from a bridge into a river. If the parapet of the bridge is 16 m above the water, choose the relevant equations and calculate the time taken for the rock to reach the water, and its velocity at the moment of impact. In this case, the initial velocity is zero, and the acceleration is simply the acceleration due to gravity, 9.81 m/s^2.
> 2. Repeat the calculation for the case where the rock is thrown rather than dropped, giving it an initial velocity of 5 m/s.

Newton's laws of motion

Sir Isaac Newton proposed three fundamental laws related to motion. The three laws are stated below, and we will then examine each in more detail.

- *First law*: A body will continue in its state of rest or uniform motion in a straight line unless acted upon by an external force.
- *Second law*: When a body is acted upon by an external force, the rate of change of *momentum* is proportional to the magnitude of the force, and takes place in the direction of the force.
- *Third law*: To every action, there is an equal and opposite reaction.

The first law is fairly straightforward. A body that is accelerating, or decelerating, or changing direction, does so because it is being acted on by an external force. The force may not always be obvious, as when a bullet is gradually slowed by the effects of air resistance, but it is there.

In order to understand the second law, we need to define a new quantity: the *momentum* of a body is equal to the product of its mass and velocity.

If a body of mass m accelerates from an initial velocity u to a final velocity v under the influence of a force, the change in momentum is given by:

$$\text{Change in momentum} = mv - mu$$

If the acceleration takes place over a period of time t, the rate of change of momentum is given by:

$$\text{Rate of change} = \frac{mv - mu}{t}$$

By Newton's second law, this quantity is proportional to the applied force, F:

$$F = \frac{mv - mu}{t} \times \text{Constant (the } constant\ of\ proportionality\text{)}$$

$$= m \times \frac{v - u}{t} \times \text{Constant}$$

Since $(v - u)/t$ is equal to the acceleration, we can say that the applied force is proportional to the mass of the body, *and* proportional to its acceleration.

We saw earlier that the SI unit of force is the newton (N). The formal definition of the newton is as follows:

A force of 1 N is the force required to cause an object of mass 1 kg to change its velocity at a rate of 1 metre per second per second.

The effect of defining the unit in this way is that the *constant of proportionality* in the above equation is equal to 1, and force (in newtons) is equal to the product of mass (in kilograms) and acceleration (in metres per second per second):

$$F = ma$$

> **Example 7.8**
>
> (a) If an object with a mass of 1 kg is subjected to a constant force of 1 N, its velocity will increase by 1 metre per second (m/s) for each second that the force is applied. If it is stationary at the start, after 1 second it will be moving at 1 m/s, after 2 seconds it will be moving at 2 m/s and so on.
>
> If the object is twice as heavy, i.e. 2 kg, the acceleration caused by the force is halved, i.e. 0.5 m/s per second (m/s^2). After 1 second its velocity is 0.5 m/s, after 2 seconds it is 1 m/s and so on.
>
> We can say that acceleration is *proportional* to force, but *inversely proportional* to mass:
>
> $$a = F/m$$
>
> (b) A car moving at 15 m/s is subjected to a braking force of 600 N. If the car weighs 1,200 kg, how long does it take to come to a standstill?
>
> As we have seen, the acceleration or deceleration is found by dividing the force by the mass of the car:
>
> $$a = -600/1{,}200 = -0.5\,\text{m/s}^2$$
>
> The minus sign indicates that this is a braking or *retarding* force, and that the velocity is therefore decreasing.
>
> We can now find the time taken for the car to come to a stop. If the initial velocity is 15 m/s and the rate of deceleration is 0.5 m/s^2, the time taken is given by:
>
> $$\text{Time} = 15/0.5 = 30\,\text{s}$$

We have already come across Newton's third law, in a different context. Remember that in the case of a loaded beam we were able to calculate the *reactions* of the beam supports, counteracting the effect of the load. In this case the total reaction was equal to the total load, and the beam was therefore in static equilibrium.

Similarly, but perhaps less obviously, if a car is towing a trailer, the car exerts a force on the trailer, *and* the trailer exerts an equal but opposite force on the car.

> **Self-assessment tasks 7.7**
>
> 1. If the train mentioned on page 214 has a mass of 20 t (20,000 kg), what is its momentum after 12 s? If it has an acceleration of 0.2 ms^2 what is the force applied?
> 2. In the case of the dropped rock (tasks 7.6) what is its initial momentum if it has a mass of 0.5 kg?
> 3. How far will the car used in the example above have travelled in the 30 s it takes to come to a halt?

Work, power and energy in linear motion

We have seen from Newton's second law that the force required to accelerate a body is equal to the product of mass and acceleration. We also know that the work done is equal to the product of the applied force and the distance moved, so we can calculate the work done in accelerating a body.

Example 7.9

A mass of 3 kg is suspended from a steel wire, causing it to stretch by 12 mm. What is the work done?
First we need to calculate the force exerted by the load:

$$F = ma$$
$$= 3 \times 9.81 \text{ (acceleration due to gravity)}$$
$$= 29.43 \text{ N}$$

The work done, W, is equal to the force multiplied by the distance moved:

$$W = Fd$$
$$= 29.43 \times 0.012 \text{ (distance in metres to get the correct units)}$$
$$= 0.35 \text{ J}$$

Example 7.10

A hoist in a workshop is used to raise and lower equipment. Calculate the work done in raising a 500 kg lathe from the ground to a height of 7 m.
The lathe exerts a downward force of $500 \times 9.81 = 4{,}905$ N. From Newton's third law we know that the hoist exerts an equal force on the lathe, and it is this force that does the work. The work done is equal to the product of force and distance travelled, as before:

$$W = Fd = 4{,}905 \times 7 = 34{,}335 \text{ J} = 34.3 \text{ kJ}$$

If the hoist is powered by an electric motor, we can also calculate the power of the motor if it takes 9 s to raise the lathe through 7 m:

$$\text{Power, } P = W/t = 34{,}335/9 = 3{,}815 \text{ W}$$

Kinetic and potential energy

We saw earlier that energy can exist in different forms. A body in motion possesses *kinetic energy* and we can calculate this if we know the mass and velocity of the body:

$$\text{Kinetic energy} = \tfrac{1}{2}mv^2$$

Example 7.11

A lorry of mass 2,000 kg accelerates from rest at a rate of 2 m/s^2 for 12 s. Calculate the final velocity, and hence the kinetic energy of the lorry at this point.
First, we need to select the appropriate equation for calculating the final velocity. Since we know the initial velocity, the acceleration and the time taken, the equation we need is:

$$v = u + at = 0 + (2 \times 12) = 24 \text{ m/s}$$

The kinetic energy is given by the equation stated above:

$$\text{Kinetic energy} = \tfrac{1}{2}mv^2 = \tfrac{1}{2}(2{,}000 \times 24^2)$$
$$= 576{,}000 \text{ J or } 576 \text{ kJ}$$

Self-assessment tasks 7.8

1. What is the kinetic energy of the 20 t train travelling at 180 km/h?
2. What is the initial kinetic energy of the dropped rock (tasks 7.6) when it has an initial velocity of 5 ms^{-1}? What is its kinetic energy just before it hits the ground?
3. What is the kinetic energy of the car on p. 215 before it starts braking?

Potential energy is the energy possessed by a body by virtue of its position or state of stress. Two examples will illustrate this:

- An object raised above the ground possesses potential energy because its weight can do work as it returns to the ground (and because work has been done on it to raise it).
- A compressed spring has potential energy because it can do work as it expands to its original state.

In order to calculate the potential energy, we need to calculate the amount of work that can be done. For a body raised above the ground this is quite straightforward: the work done when the body returns to the ground is equal to the force on the body (its mass times the acceleration due to gravity) times the distance travelled.

$$\text{Potential energy} = mgh$$

Example 7.12

A crane on a building site raises a load of 50 bricks to a height of 20 m. (a) If each brick weighs 5 kg, what is the potential energy of the total load? (b) If one brick is dislodged and falls to the ground, what will be its velocity just before it hits the ground?

(a) The total mass of the bricks is $50 \times 5 = 250$ kg. Therefore the total potential energy is given by:

$$\text{Potential energy} = 250 \times 9.81 \times 20 = 49{,}050 \text{ J}$$

(b) The total potential energy calculated in (a) is divided equally between 50 bricks, therefore one brick has $49{,}050/50 = 981$ J. As the brick falls, by the principle of conservation of energy, this potential energy is converted into kinetic energy. Therefore, just before it hits the ground the brick has 981 J of kinetic energy. From the equation for kinetic energy, we can therefore calculate the final velocity.

$$\text{Kinetic energy} = \tfrac{1}{2}mv^2$$
$$v^2 = (2 \times \text{Kinetic energy})/m$$
$$= (2 \times 981)/5 = 392$$

Therefore,

$$v = \sqrt{392} = 19.8 \text{ m/s}$$

Self-assessment tasks 7.9

1. What is the potential energy gained by the lathe when it has been raised by 7 m?
2. A car is at the top of a hill which is 50 m high. The car freewheels down the road which is straight. The car does not brake as it goes down the hill. If the initial velocity is zero, what will be the velocity of the car by the time it reaches the bottom of the hill? The car has a mass of 900 kg. Assume that all the potential energy is converted into kinetic energy.
3. A maintenance person is servicing the air-conditioning machinery at the top of Canary Wharf Tower in London. He accidentally drops a spanner over the side of the building. What is the spanner's potential energy at the moment it is dropped? What is the spanner's velocity just before it hits the ground? What is the time taken for the spanner to reach the ground? What is the spanner's momentum just before it hits the ground? Assume that the tower is 250 m high and that the spanner has a mass of 750 g.

Rotational motion

The second type of dynamic system involves *rotational* motion. In these systems, an object moves in a circular path about a fixed point. Important examples include:

- Wheels of vehicles.
- The pedals used to power a bicycle.
- Flywheels used to drive mechanical devices.

We can represent a simple rotational system by a diagram such as Fig. 7.13.

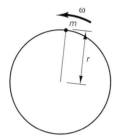

Figure 7.13 Rotational motion

The key quantities in this system are:

- The mass of the object, m.
- The distance of the object from the centre of rotation, r.
- The *angular velocity*, ω (Greek omega).

The angular velocity is similar to the linear velocity, but rather than measuring the distance covered in a given time interval, it measures the angle through which the body rotates in a given time.

You may be used to measuring angles in degrees, but the SI unit of angle is the *radian*, and the angular velocity is measured in radians per second (rad/s). Whereas the number of degrees in a complete circle is 360, the number of radians is 2π. The angular velocity of an object making one complete revolution every second is $\omega = 2\pi$.

The equation is

$$\omega = 2\pi N$$

where ω is the angular velocity in radians per second and N is the number or revolutions per second.

Example 7.13

A machine lathe is rotating at 3,600 revolutions per minute and we want to calculate its angular velocity.
First we need to work out the number of revolutions per second:

$$3{,}600 \text{ rev/min} = 3{,}600/600 \text{ rev/s} = 60 \text{ rev/s}$$
$$\text{Angular velocity} = 2\pi N = 2\pi \times 60 = 120\pi \text{ rad/s}$$

Self-assessment tasks 7.10

1. A flywheel is driven at a rate of 4,000 revolutions per minute. What is its angular velocity?
2. An engine crankshaft is roating at 2,500 revolutions per minute. What is its angular velocity?
3. A bicycle wheel has an angular velocity of 4π rad/s. How many revolutions per minute is the wheel doing?

Centripetal force

We have already seen Newton's first law which states that a body continues in its state of rest or uniform *linear* motion unless it is acted upon by an external force. Since rotating bodies are neither at rest nor in a state of linear motion, they must be acted upon by a force, and this force is known as the *centripetal* force.

The centripetal force acts towards the centre of rotation of the body, and causes the body to accelerate continuously towards the centre. The magnitude of the force is given by the following equation:

$$F = mr^2\omega$$

Example 7.14

A mass of 200 g is attached to a string of length 30 cm, and the string is spun round at a rate of 150 revolutions per minute. (a) Calculate the angular velocity of the mass, and the centripetal force. (b) If the string is replaced with one of length 60 cm, calculate the value of the centripetal force if the angular velocity is unchanged.

(a) If the mass completes 150 revolutions in a minute, it completes $150/60 = 2.5$ revolutions per second. Since each revolution is equivalent to 2π radians, the angular velocity is given by:

$$\omega = 2.5 \times 2\pi = 15.7 \text{ rad/s}$$

Centripetal force is given by:

$$F = mr^2\omega = 0.2 \times 0.3^2 \times 15.7 = 0.28 \text{ N}$$

(b) If the longer string is used, and the angular velocity remains the same, the equation becomes:

$$F = 0.2 \times 0.6^2 \times 15.7 = 1.13 \text{ N}$$

From the above example it is clear that the centripetal force required is dependent on the radius of the circle traced out by the mass, as well as its mass and angular velocity.

Linear velocity in a rotational system

If a body is rotating with a constant angular velocity, it also has a linear velocity. This velocity is of constant magnitude, but is constantly changing direction as the body rotates. The distance travelled by a rotating object in 1 (v) second is obtained by multiplying the number of revolutions per second (N) by the circumference of the circle (c) in which it moves:

$$v = c \times N$$

The circumference of a circle is given by the famous formula:

$$c = 2\pi r$$

The number of revolutions per second is obtained by dividing the angular velocity (in radians per second) by the number of radians in one complete revolution, 2π:

$$N = \omega/2\pi$$

Therefore, the linear velocity is given by:

$$v = c \times N = 2\pi r \times (\omega/2\pi) = r\omega$$

This velocity is always at right angles to a line joining the object to the centre of a circle, or at a *tangent* to the circle (Fig. 7.14).

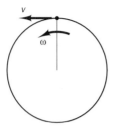

Figure 7.14 Velocity in rotational motion

For small objects rotating, as in the previous example, the linear velocity represents the velocity the object would have if the string were suddenly snapped. It would fly off in a straight line at a tangent to the circle with velocity v.

For a wheel on a vehicle, the linear velocity represents the speed of the vehicle if the wheels are rotating with a particular angular velocity.

Self-assessment tasks 7.11

1. A cyclist keeps the wheels of the bicycle turning at a steady angular velocity of 30 rad/s. If the wheels are 60 cm across, how fast is the cyclist going?
2. What is the linear velocity of the mass in the example above?
3. If a car is travelling at 30 m/s and the wheel has a radius of 0.3 m, what is the angular velocity of the wheel?

Measurement

Micrometer

To measure components to the accuracy required for engineering purposes, a micrometer caliper (commonly called a micrometer) is used. Micrometers have varying ranges and you will need to select a micrometer with the appropriate range for the component that you are measuring.

A typical micrometer is shown in Fig. 7.15. The two anvil faces are closed around the object to be measured so that they are touching the object. The measurement is then read off the scale as follows:

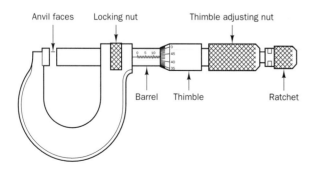

Figure 7.15 Typical micrometer

- Take the largest visible whole millimetre reading.
- If the next half millimetre is visible, add this.
- Note the thimble division that is closest to the datum line.
- Add all three together to give the measurement.

Example 7.15

The reading on the micrometer in Fig. 7.16 (a) is as follows

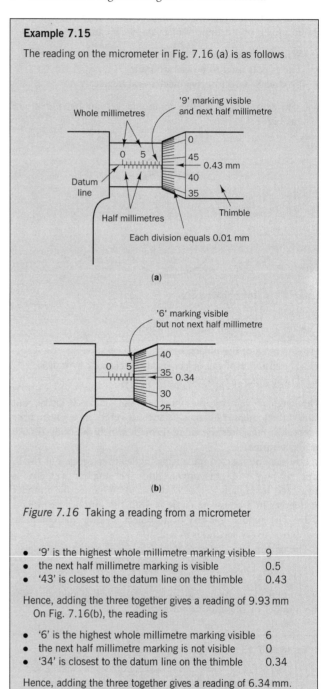

Figure 7.16 Taking a reading from a micrometer

- '9' is the highest whole millimetre marking visible   9
- the next half millimetre marking is visible   0.5
- '43' is closest to the datum line on the thimble   0.43

Hence, adding the three together gives a reading of 9.93 mm
On Fig. 7.16(b), the reading is

- '6' is the highest whole millimetre marking visible   6
- the next half millimetre marking is not visible   0
- '34' is closest to the datum line on the thimble   0.34

Hence, adding the three together gives a reading of 6.34 mm.

Sources of error

You should observe the following points to ensure that readings you take from a micrometer are accurate:

- Wipe the anvil faces of the micrometer clean before use. If they are dirty this could affect the reading.
- Do not use too much pressure: two clicks of the ratchet are enough.
- Do not leave the anvil faces touching when you are not using the micrometer.
- Check that the micrometer reads zero when the two anvil faces are touching.
- A micrometer is usually accurate to 0.01 mm.

Self-assessment task 7.12

What is the reading on the micrometer shown below?

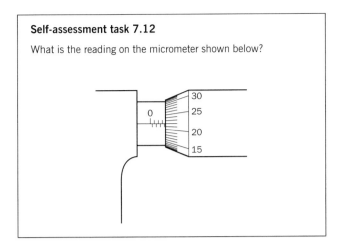

Spring balance

A typical spring balance being used to measure weight is shown in Fig. 7.17. Spring balances are used to measure force in newtons. When used as shown in Fig. 7.17, you are measuring weight. To determine mass, divide the weight by the acceleration due to gravity, g.

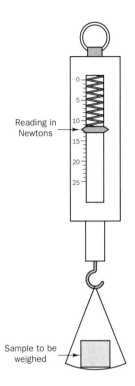

Figure 7.17 A spring balance

Sources of error
When using a spring balance, you should observe the following points to ensure that readings you take are as accurate as possible:

- Check that the balance reads zero when it is not loaded.
- Allow the balance to settle down before taking a reading so that it is not bouncing up and down.

7.3 Thermal systems

Thermal systems involve heat energy in some way. Examples of thermal systems are

- *Compressors* – in which heat is an unwanted by-product.
- *Internal combustion engines* – which convert heat to work.
- *Refrigerators* – which remove heat.
- *Central heating systems* – in which heat energy is produced from burning fuel.

In all thermal systems heat is transferred by a working fluid, i.e. a liquid, gas or mixture of the two. Referring to the previous examples, the working fluids in each case are

- Compressors: the gas being compressed.
- Internal combustion engines: air.
- Refrigerators: the refrigerant, e.g. CFCs, ammonia.
- Central heating systems: water.

In this section we will examine

- How we can quantify the state of a gas.
- What happens when a substance (solid, liquid or gas) is heated.

Quantifying the state of a gas: the gas law

The state of a gas can be quantified by

- Its absolute pressure, P, measured in Pascals (Pa).
- Its absolute temperature, T, measured in Kelvins (K).
- Its volume, V, measured in cubic metres (m^3).

Provided the temperature and pressure are not too high, a gas can be treated as perfect, i.e. the molecules of the gas do not interact. This being the case, the gas law can be applied to determine the state of the gas when the conditions are changed.

The gas law

The gas law states that for a perfect gas

$$\frac{\text{Absolute pressure} \times \text{Volume}}{\text{Absolute temperature}} = \text{Constant}$$

This is only true if

- The gas can be considered as perfect.
- The gas is in a state of equilibrium, i.e. the pressure, volume and temperature are not changing rapidly.

Note that both pressure and temperature must be absolute

Absolute pressure = Gauge pressure + Atmospheric pressure

Absolute temperature = temperature (°C) + 273

i.e. temperature above absolute zero measured in Kelvins. The gauge pressure is the pressure you measure using, for example, a manometer or Bourdon gauge relative to atmospheric pressure.

To illustrate the gas law, we will consider a simple experiment in which we have a given mass of gas in a cylinder. The cylinder is fitted with a piston to change the volume of the gas and a means of heating to change the temperature of the gas (see Fig. 7.18).

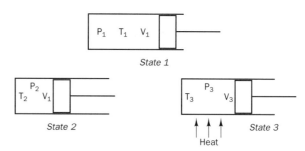

Figure 7.18 Changing the state of a gas

The diagrams show the gas inside the cylinder in three different states. According to the gas law, provided the mass of gas is constant, i.e. no leaks, and the gas can be considered as perfect, then

$$\frac{P_1 V_1}{T_1} = \frac{P_2 V_2}{T_2} = \frac{P_3 V_3}{T_3}$$

No matter how much heat is supplied or where the piston is positioned PV/T is constant.

We will now put the gas law into action in an engineering situation.

Example 7.16

If you have ever watched a grand prix on television, you will know that tyres, and in particular the temperature of the tyres used on Formula One cars are very important. Tyre warmers are used to preheat the rubber before new wheels and tyres are fitted. This is because the rubber only works properly (i.e. gives maximum grip) at about 60 °C.

One of the potential problems which technicians have to overcome is that cold air is put into the tyres which then heats up to operating temperature on the track, thereby changing the tyre pressure (Fig. 7.19). The internal volume of the tyre remains more or less constant, so, as you will see, its value is important.

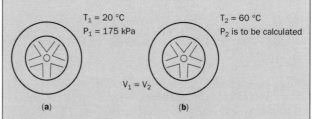

Figure 7.19 Formula One car tyre before fitting and at operating temperature

Before we apply the gas law, the units must be sorted out, i.e. converted into absolute form.

Absolute temperature: $T_1 = 20 + 273 = 293$ K
$T_2 = 60 + 273 = 333$ K

Absolute pressure: Atmospheric pressure $\cong 1$ bar $= 1 \times 10^5$ Pa
$P_1 = 1.75 \times 10^5 + 1 \times 10^5 = 2.75 \times 10^5$ Pa

The gas law states

$$\frac{P_1 V_1}{T_1} = \frac{P_2 V_2}{T_2}$$

As we are going to determine P_2, the equation must be transposed, i.e.

$$P_2 = \frac{P_1 V_1}{T_1} \times \frac{T_2}{V_2}$$

As $V_1 = V_2$ they can be cancelled, so

$$P_2 = \frac{P_1 T_2}{T_1} = \frac{2.75 \times 10^5 \times 333}{293} = 3.13 \times 10^5 \text{ Pa}$$

This is a pressure increase of nearly 14 per cent, which will increase the rigidity of the inflated tyre considerably. Tyre technicians would allow for this increase when inflating the tyre, but they would really be using the gas law.

Self-assessment tasks 7.13

1. When, in an internal combustion engine, a cylinder has filled with petrol vapour and closed before combustion, the piston head moves up the cylinder and compresses the petrol/air mixture. If the volume of the cylinder before compression is 500 cm³ (i.e. 0.5 litre), and 50 cm³ after compression and the pressure before compression is atmospheric, what is the pressure in the cylinder just before combustion? Assume that the temperature remains the same.
2. If the tyres in example above are cooled to 0 °C, what is the pressure in the tyre?

Applying heat to a substance: specific and latent heat

If heat is applied to a solid or liquid, one of two things can occur:

- The temperature of the substance will increase.
- The substance could melt, i.e. change its phase.

The process of heating a solid so that it melts and then further heating the liquid so that it evaporates can be illustrated with a graph (see Fig. 7.20). Similarly when the gas is cooled, the process is illustrated in Fig. 7.21.

Water is perhaps the most easily understood substance. At standard atmospheric pressure (1.013 bar or 1.013×10^5 Pa)

Melting or fusion temperature = 0 °C (273 K)
Evaporating or condensing temperature = 100 °C (373 K)

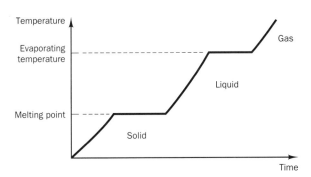

Figure 7.20 Temperature change over time when a solid is heated

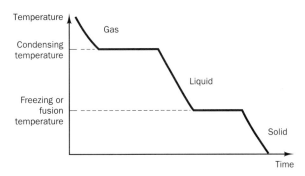

Figure 7.21 Temperature change over time when a gas is cooled

The important thing to note in these graphs is that while a substance is changing phase from

- solid to liquid
- liquid to gas
- gas to liquid or
- liquid to solid

the temperature remains constant even though heat is still being applied.

When the phase of the substance is not changing, the temperature will increase as heat is applied and fall when heat is removed.

Increasing temperature

Heat which causes a temperature increase is referred to as **sensible heat**, i.e. the heat can be sensed by a change in temperature.

The amount of heat required to raise the temperature of a substance (Q) depends on

- the mass of the substance (m) in kg
- the temperature increase required (T_2, T_1) in °C
- the specific heat of the substance (c) in $\text{J kg}^{-1}\,°\text{C}^{-1}$

The specific heat of a substance is the amount of heat energy required to raise the temperature of 1 kg of the substance by 1 °C or 1 K. Written as an equation this is:

Heat required $Q = mc(T_2 - T_1)$ Joules

Table 7.2 Specific heat capacities for a range of materials

Substance	Specific heat capacity (J/kg K)
Aluminium	920
Iron	460
Water	4,200
Steam	2,100
Oil	2,000
Mercury	136
Rubber	1,800

Example 7.17

To fit a threaded insert, made from steel, in an aluminium component, the aluminium is heated to 120 °C from a room temperature of 20 °C so that the hole into which the insert is to be fitted expands. The mass of the aluminium component is 0.5 kg and that of the insert 0.1 kg.

We will first determine the heat energy required to heat the aluminium. From Table 7.2, the specific heat capacity of aluminium is 0.92 kJ/kg/°C = 920 J/kg/°C.

So the heat energy required is

$$Q = mc(T_2 - T_1)$$
$$= 0.5 \times 920 \times (120 - 20) = 46,000 \text{ J}$$

Before the insert is fitted, it is cooled to 5 °C so that it contracts. When it is placed in the aluminium component, the temperature of the aluminium will fall and that of the steel will rise. If we assume that the operation is carried out in insulated surroundings so that rapid cooling does not occur due to the air temperature, then we can determine the temperature at which the complete component will settle.

The calculation is based on the fact that

Heat gained by steel = Heat lost by aluminium
$$Q_S = Q_A$$

Now, referring to the temperatures on Fig. 7.22:

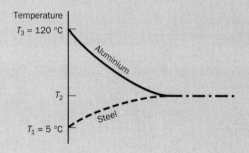

Figure 7.22 Temperature change of the steel insert and aluminium component

and
$$Q_S = m_S c_S (T_2 - T_1)$$
$$Q_A = m_A c_A (T_3 - T_2)$$
but as
$$Q_S = Q_A,$$
$$m_S c_S (T_2 - T_1) = m_A c_A (T_3 - T_2)$$

Putting in the values:

$$0.1 \times 460(T_2 - 5) = 0.5 \times 920(120 - T_2)$$
$$46(T_2 - 5) = 460(120 - T_2)$$
$$T_2 - 5 = \frac{460}{46}(120 - T_2)$$
$$= 10(120 - T_2)$$
$$= 1,200 - 10T_2$$
$$T_2 + 10T_2 = 1,200 + 5$$
$$11T_2 = 1,205$$

which gives

$$T_2 = 1,205/11 = 109.5 \text{ °C}$$

i.e. the temperature at which the component will settle is about 110 °C.

Remember: we do not need to convert the temperature to Kelvins because we are dealing with temperature *differences*, rather than an absolute value of temperature.

Self-assessment tasks 7.14

1. Calculate the energy required to heat the Formula One racing tyres from 20 °C to 60 °C. Assume that the tyre has a mass of 10 kg. Use the values for specific heat capacity given in Table 7.2.
2. A steel cold chisel is to be tempered by heating it up and quenching it in oil. Calculate the amount of energy required to heat the tool from room temperature to 280 °C. Then find the resultant temperature of the oil bath and the tool after quenching. Assume that the tool has a mass of 2 kg and the oil bath contains 20 kg of oil. Use the figures given in Table 7.2.

Changing Phase

The heat required to change the phase of a substance is known as **latent heat**.

- Latent heat of fusion is the heat required to melt (or solidify) 1 kg of a substance.
- Latent heat of vaporisation is the heat required to evaporate (or condense) 1 kg of a substance.

Example 7.18

This scenario is probably more applicable to the house than any manufacturing process in industry, other than making the tea!

A kettle is rated at 0.75 kW. This means that it can supply energy at the rate of 0.75 kJ/s. In this scenario the cut-out switch on a kettle fails and the kettle boils dry. The kettle initially contains 1.5 litres of water. We are going to calculate

- The heat required to evaporate all the water.
- The time taken for the water to evaporate.

Volume of water $= 1.5 \text{ l} = 1.5 \times 10^{-3} \text{ m}^3$

Mass of water $=$ Volume $\times$ Density
$$= 1.5 \times 10^{-3} \times 1,000 = 1.5 \text{ kg}$$

Latent heat of vaporisation for water $= 2,257$ kJ/kg

So the heat required to evaporate all the water is:

$$Q = \text{Mass} \times \text{Latent heat}$$
$$= 1.5 \times 2,257 = 3,385.5 \text{ kJ}$$

Now we know that the kettle supplies heat at a rate of 0.75 kJ s, so that the time taken for the kettle to boil dry is:

$$3,385.5/0.75 = 4,514 \text{ seconds}$$

This is one and a quarter hours, so it is quite likely that someone will have noticed the steam by then.

The next example combines the two concepts of sensible and latent heat.

Example 7.19

A steam boiler contains 250 litres of water initially at 25 °C. To find out the correct size of gas burner used to heat the water we need to calculate the amount of heat required to convert all the water to steam at 150 °C, at atmospheric pressure.

To start the problem, we will sketch a graph to show the heating process (Fig. 7.23). The calculation must be carried out in three stages:

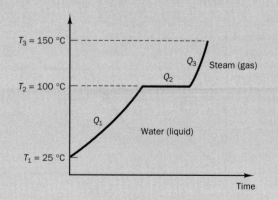

Figure 7.23 Heating process in Example 7.19

- Calculate the heat required to raise the water temperature from 20 °C to 100 °C (Q_1).
- Calculate the heat required to convert 250 litres of water to steam at 100 °C (Q_2).
- Calculate the heat required to raise the steam temperature from 100 °C to 150 °C (Q_3).

We will also have to look up the specific heat capacities of water and steam (they are very different). If you refer to Table 7.2 on page 221, you will see that the values are

$$c_W = 4.2 \text{ kJ/kg °C for water}$$
$$c_S = 2.1 \text{ kJ/kg °C for steam}$$

The latent heat of evaporation for water will also be required (Table 7.3). The symbol h is used:

$$h = 2{,}257 \text{ kJ kg}$$

We can now start the calculation.

Volume of water = 250 litres = 0.25 m³

Mass of water = volume × density
$$= 0.25 \times 1{,}000 \text{ kg} = 250 \text{ kg}$$

(It is convenient to remember that 1 litre of water has a mass of 1 kg.)

Heat required to raise water temperature to 100 °C:

$$Q_1 = mc_W(T_2 - T_1) = 250 \times 4{,}200(100 - 25)$$
$$= 7.875 \times 10^7 \text{ J} = 78.75 \text{ MJ}$$

Heat required to evaporate 250 kg of water:

$$Q_2 = mh = 250 \times 2{,}257 \times 10^3$$
$$= 5.6425 \times 10^8 \text{ J} = 564.25 \text{ MJ}$$

Heat required to raise steam temperature to 150 °C:

$$Q_3 = mc_S(T_3 - T_2) = 250 \times 2{,}100(150 - 100)$$
$$= 2.625 \times 10^7 \text{ J} = 26.25 \text{ MJ}$$

Total heat required is:

$$Q = Q_1 + Q_2 + Q_3$$
$$= 78.75 + 564.25 + 26.25 = 669.25 \text{ MJ}$$

Table 7.3 Latent heats for a range of materials

Substance	Melting or vaporisation	Temperature (K)	Latent heat (kJ/Kg)
Water	Vaporisation	373	2,257
Aluminium	Melting	933	396
Mercury	Melting	234	11.5

Self-assessment tasks 7.15

Use the values given in Table 7.3

1. A part is to be cast in aluminium and 3 kg of the metal will be needed. Assuming that the aluminium is initially at room temperature (20 °C), calculate the amount of energy needed to melt the aluminium.
2. How much energy would need to be taken away from 1 kg of mercury at room temperature (20 °C) to solidify it?

Measuring heat

Heat cannot be measured directly, but its effects can. Temperature is the most obvious effect of heat and you should be careful not to confuse the two.

- Heat is a form of energy.
- Temperature is a measure of the molecular motion in a substance. As energy is supplied, so the motion increases speed.

Thermometers

The thermometer is the most common and simplest instrument for measuring temperature. Most thermometers are filled with mercury but at lower temperatures (below −30 °C) alcohol is used. The alcohol is dyed, often red, so that the thermometer can be easily read.

For best accuracy, the following points should be observed:

- Use a long thermometer.
- Use a thermometer with a small range, e.g. if you want to measure a temperature of about 60 °C do not use an instrument with a range of 0 to 200 °C; choose one which goes up to 100 °C.
- Be careful to follow the maker's instructions with regard to the depth to which the thermometer should be immersed.
- When using a mercury thermometer, always read from the top of the meniscus, and ensure that your sight line is at right angles to the thermometer (Fig. 7.24).

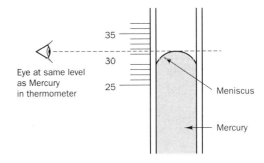

Figure 7.24 Reading the temperature off a thermometer

Thermocouples

These are widely available and are often sold as **digital thermometers**. Laboratory standard instruments usually come with a probe remote from the readout device.

Thermocouples of different types are produced for different temperature ranges and operating environments. If you refer to an electronics components catalogue, you will see the products available.

For best accuracy the following points should be observed:

- Use a thermocouple with an appropriate temperature range.
- Check that the connections between the probe and the meter are good.

Measuring pressure

There are two main types of instrument (or pressure gauge) which are used to measure pressure. They are:

- a manometer
- a Bourdon gauge

Both measure gauge pressure, that is the pressure relative to atmospheric pressure.

Manometer

A manometer is shown in Fig. 7.25. It is used for measuring low pressures at low temperatures. The difference in height of the levels of the liquid in each arm is used to calculate the pressure. If the liquid in each arm is at the same level, the gas is at atmospheric pressure.

For best accuracy the following points should be observed:

- Ensure that the manometer is upright, otherwise you will not get a true reading.
- Do not use a manometer at high temperatures and pressures.

Bourdon gauge

A bourdon gauge is used to measure high pressures which can also be at high temperatures, e.g. steam pressure in a boiler. It is compact and easy to use.

For best accuracy the following points should be observed:

- Ensure that the needle on the meter is reading zero at atmospheric pressure.
- Do not use for low pressures as it is not as sensitive or as accurate as the U-tube manometer.

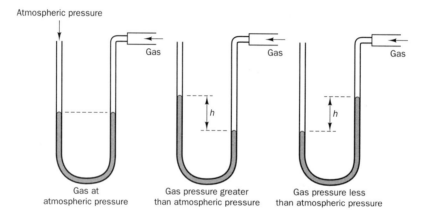

Figure 7.25 A manometer

7.4 Electrical systems

Electrical systems involve electrical energy in some way. The system is usually in the form of a circuit of some kind. Examples of electrical systems are:

- *Lights* – where an electrical system is used to provide lighting.
- *Heating* – such as a cooker or kettle where an electrical system is used to provide heat.
- *Motors* – when an electrical system is used to provide mechanical energy.

Electrical systems make use of a flow of electrons (current) to transfer energy. Electrons are negatively charged subatomic particles which will flow through a conductor such as copper or aluminium when provided with a potential difference or voltage.

Voltage can be thought of as the pressure behind the electron flow. This is also known as the **electromotive force** or **e.m.f.** The e.m.f. is the voltage measured across the terminals of a battery or (cell) when no load is connected, i.e. no current is flowing.

The **potential difference** is the difference in voltage or potential measured across any two points in a circuit when a current is flowing.

As the electrons pass through a conductor they interact with other particles and in this way the flow is resisted to some extent, i.e. there is resistance.

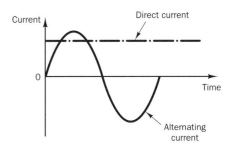

Figure 7.26 Alternating and direct current

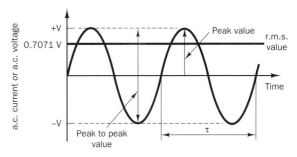

Figure 7.27 Peak value, period and frequency

Direct and alternating current

When a conductor is connected to a voltage source such as a battery, the electron flow is in one direction, from negative to positive. This is known as direct current (d.c.). It is conventional to show current flow from positive to negative. This makes no difference to calculations. Examples of items that use direct current are:

- a torch
- the electrical system of a car
- a pocket calculator

If the conductor is connected to the mains supply, the electrons 'shuffle' backwards and forwards – that is, the electrons oscillate about a fixed position. The motion is sinusoidal. This is known as an alternating current – that is, it alternates between positive and negative (see Fig. 7.26). If there is an alternating current, there must be an alternating voltage to provide the pressure. Sine waves are explained in Unit 8.

R.M.S. values

When considering the effect of passing a current through a resistance, it is useful to be able to compare d.c. directly with a.c. To do this we compare the r.m.s. (root mean square) value of the alternating current to the direct current (see Fig. 7.27)

For example 3 A r.m.s. a.c. is equivalent to 3 A d.c. as it has the same heating effect when it passes through a resistance, and for the same reason 12 V r.m.s. a.c. is equivalent to 12 V d.c.

The r.m.s. value is defined as

$$\text{r.m.s. value} = \text{Peak voltage}/0.7071$$

When we refer to the mains voltage as 230 V a.c., we are talking about an r.m.s. value. The peak voltage would be

$$\text{r.m.s. value}/0.7071 = 230/0.7071 = 325.3\,\text{V}$$

Frequency

The freqency of an alternating current or an alternating voltage is the number of cycles of the wave that there are in one second. For example, if the time taken for one cycle of the sinewave in Fig. 7.22 is τ seconds, the frequency is $1/\tau$ cycles per second. The unit of measurement is hertz (Hz).

In the UK, the mains voltage is 230 V a.c. and the frequency of the mains supply is 50 cycles per second or 50 Hz. This means that the time taken for one complete cycle of the sine wave is 1/50th of a second or 20 milliseconds (ms).

Self-assessment tasks 7.16

1. What is the peak-to-peak value of the mains voltage?
2. What is the r.m.s. value of a peak voltage of 34 V?

Ohm's law

Ohm's law describes the relationship between voltage, current and resistance in a circuit. The law states that

Voltage (V) = Current (I) × Resistance (R)

where the voltage is measured in volts (V), the current is measured in amperes (A) and the resistance is measured in ohms (Ω).

The following example should help you to understand Ohm's law.

Example 7.20

A d.c. motor is used to drive a cooling fan in a car. The resistance of the motor is $2.4\,\Omega$. We need to know the size of current which will flow to provide the correct gauge of wire when connecting the motor. Assume that the car has a 12 V battery.
Using Ohm's law

$$V = IR$$

We need I, so transposing the equation we get

$$I = V/R = 12/4.8 = 2.5\,A$$

i.e. when the motor is running a current of 2.5 A will flow. In reality, the motor would warm up which would have the effect of increasing the resistance, thereby decreasing the current.

Conversely the starter motor of a car is often required to work at low temperatures in the winter so that its resistance will decrease and a higher current can be drawn.

Other factors which might affect the size of current drawn by the motor are:

- The load put on the motor by the device being driven.
- Friction in the motor bearings.
- The condition of the motor brushes.

These are all beyond the scope of this section but may be worth considering if you ever use a d.c. motor in a design project.

Note that Ohm's law can also be applied to alternating current systems provided:

- There is only resistance in the circuit.
- The r.m.s. values of voltage and current are used.

Self-assessment tasks 7.17

1. A one-bar (nominally rated at 1 kW) electric fire has a resistance of $53\,\Omega$. What is the current flowing through the heating element? Assume that the mains supply is 230 V r.m.s.
2. A light bulb on a Christmas tree has a resistance of $27\,\Omega$ and the current flowing through it is 0.425 A. What is the potential difference (voltage) across the lamp?
3. A kettle has a current of 8.7 A flowing through the heating element. What is the resistance of the heating element?

Resistance in a circuit

Although all conductors offer some resistance to electron flow, it is often assumed that the resistance of connecting wires is negligible compared to the resistance of other components. This assumption may not be valid in systems such as automotive wiring looms where connectors particularly may offer a significant resistance, especially when corrosion has taken place.

In electrical and electronic devices resistances of known values (resistors) are used to control voltage and current. The symbol for a resistor is given in Fig. 7.28.

Figure 7.28 Symbol for a resistor

Resistors can be combined in one of two ways:

- in series, i.e. one after the other
- in parallel, i.e. side by side

Combinations of series and parallel resistances are also used, but this will not be dealt with in this section.

Series resistors

A group of resistors connected in series is shown in Fig. 7.29.

Figure 7.29 Resistors in series

- The total resistance is $R = R_1 + R_2 + R_3$.
- Current (I) through each resistance is the same.
- Remember that the total resistance of resistors conected in series is always greater than that of any resistance in the series.

Parallel resistors

A group of resistors connected in parallel is shown in Fig. 7.30.

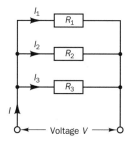

Figure 7.30 Resistors in parallel

- The total resistance R is found from the following equation:

$$\frac{1}{R} = \frac{1}{R_1} + \frac{1}{R_2} + \frac{1}{R_3}$$

- Total current $I = I_1 + I_2 + I_3$.
- The voltage V across each resistance is the same.
- Remember that the total resistance of resistors connected in parallel is always less than that of any resistor in the group.

To illustrate the two types of circuit we will look at some engineering examples.

Example 7.21

The headlight on a motorcycle fails in such a way that it just glows rather than shining brightly. The circuit is shown in Fig. 7.31.

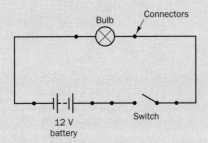

Figure 7.31 Motorcycle headlamp circuit

When the bulb is tested off the bike, it is found to be satisfactory; however, when the battery is checked it is found to be fully charged. The bulb is rated at 55 watts power consumption. Using the formula for electrical power:

Power (P) = Voltage (V) × Current (I)

but we need current, so

$$I = P/V$$

Assuming that the bike has a 12 V d.c. power supply, the current required to illuminate the bulb fully is:

$$I = 55/12 = 4.58\,\text{A}$$

Clearly the current is less than this, as the bulb is not shining brightly. So there must be extra resistance in the circuit. When the connectors are checked, it is found that, due to corrosion, two of them have significant resistances, one being 2.5 Ω and the other 1.5 Ω. Using Ohm's law, the bulb resistance can be calculated.

$$V = IR$$

But we need R, so by transposing the equation,

$$R = V/I = 12/4.58 = 2.62\,\Omega$$

The circuit can now be regarded as having three resistances in series, as shown in Fig. 7.32. The total resistance of the complete circuit is

$$R = R_1 + R_2 + R_3 = 1.5 + 2.5 + 2.6 = 6.6\,\Omega$$

i.e. the total resistance is 6.6 Ω.

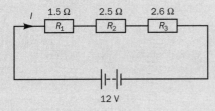

Figure 7.32 Motorcycle headlamp circuit with resistances of connectors shown

Using Ohm's law, we can now determine the current which is actually flowing in the circuit.

$$I = V/R = 12/6.6 = 1.82\,\text{A}$$

As the current flowing is only 1.82 A, it is not surprising that the bulb is only glowing dimly.

Example 7.22

The loudspeaker outputs from amplifiers usually state the minimum impedance of the speakers that can be connected. Impedance, in this sense, can be regarded as the resistance of the speakers. Impedance actually takes account of other factors in an a.c. circuit component which are beyond the scope of this section.

An amplifier has four speaker outputs, with the *minimum* total impedance (i.e. resistance) being specified as 4 Ω.

Initially two speakers are connected, each of impedance 8 Ω. These are connected in parallel (see Fig. 7.33). The circuit can be simplified to Fig. 7.34.

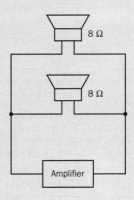

Figure 7.33 Loudspeaker circuit with two loudspeakers

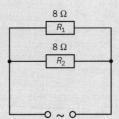

Figure 7.34 Simplified loudspeaker circuit

The total resistance is given by

$$\frac{1}{R} = \frac{1}{R_1} + \frac{1}{R_2}$$

Putting in the resistance values we get

$$\frac{1}{R} = \frac{1}{8} + \frac{1}{8} = \frac{2}{8}$$

Hence

$$R = 8/2 = 4\,\Omega$$

In this setup then, the combined impedance (resistance) of the two speakers is equal to the minimum specified, so the system will function correctly without risk of damage.

Now we will look at the situation in which four speakers of impedance 8 Ω are connected (see Fig. 7.35).

In this case, the total impedance is given by

$$\frac{1}{R} = \frac{1}{R_1} + \frac{1}{R_2} + \frac{1}{R_3} + \frac{1}{R_4} = \frac{1}{8} + \frac{1}{8} + \frac{1}{8} + \frac{1}{8} = \frac{4}{8}$$

Hence

$$R = 8/4 = 2\,\Omega$$

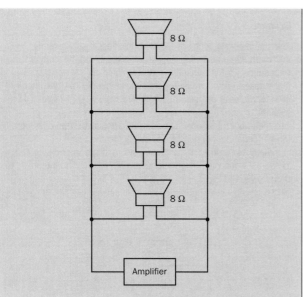

Figure 7.35 Loudspeaker circuit with four loudspeakers

So in this arrangement the combined impedance is too low. The result of this would be that at high volume the current flowing to the speakers would be too high for the output stage of the amplifier to cope with and, if a protection device such as a fuse is not fitted, damage would result.

Self-assessment tasks 7.18

1. A string of 20 Christmas tree lights is connected in series. If the mains supply is 230 V a.c. and the resistance of each lamp is 25 Ω, what is the size of the current flowing through each lamp?
2. What would be the combined resistance of (a) two and (b) four of the loudspeakers used in Example 7.22 above if they were connected in series?
3. What would be the combined resistance of the motorcycle lamp and the connectors if they were in parallel? What would be the current flowing in the circuit?

Kirchhoff's laws

When electrical circuits become more complex, Ohm's law is not sufficient to allow a complete analysis of circuit currents and voltages. Gustav Kirchhoff, a German physicist, made observations which now enable us to develop simultaneous equations to solve for the unknown values in a circuit.

Kirchhoff's current law

Kirchhoff's current law states that at any moment in time, the algebraic sum of all currents at a junction is zero. In mathematical terms this is expressed as:

$$\sum I = 0$$

'Algebraic' means allowing for the direction of the current. Look at the junction shown in Fig. 7.36. According to Kirchhoff's first law

$$I_1 - I_2 + I_3 + I_4 - I_5 = 0$$

Note that I_1, I_3 and I_4 are flowing into the junction whereas I_2 and I_5 are flowing away from the junction and so I_2 and I_5 are negative. It would also be correct to write

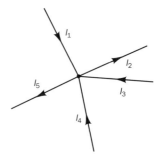

Figure 7.36 Kirchhoff's first law

$$-I_1 + I_2 - I_3 - I_4 + I_5 = 0$$

or even

$$I_1 + I_3 + I_4 = I_2 + I_5.$$

Example 7.23

Look at the circuit in Fig. 7.37 which is the example of the four loudspeakers connected in parallel. At node A, I_1 is flowing into the junction and I_2 and I_3 are flowing away from the junction. Hence, according to Kirchhoff's current law,

$$I_1 - I_2 - I_3 = 0$$

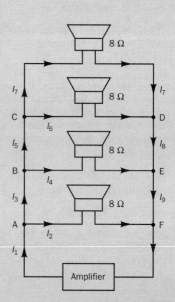

Figure 7.37 Loudspeaker circuit: Kirchhoff's current law

From earlier, we know that the combined impedance (resistance) of the combination is 2 Ω. If the voltage is 2 V r.m.s., then using Ohm's law the total current is

$$I_1 = V/R = 2/2 = 1\,A$$

Hence, $1 - I_2 - I_3 = 0$ and rearranging gives $I_2 + I_3 = 1\,A$.
As the loudspeakers all have the same impedance, i.e each is 8 Ω, the current through each of them will be the same. Hence $I_2 = I_4 = I_6 = I_7$.
Looking at node B, $I_3 = I_4 + I_5$. But $I_4 = I_2$, so $I_3 = I_2 + I_5$.
At node C, $I_5 = I_6 + I_7$. But $I_6 = I_7 = I_2$, so $I_5 = I_2 + I_2 = 2I_2$.
Substituting for I_5 in the equation for I_3 gives

$$I_3 = I_2 + 2I_2 = 3I_2$$

From earlier, $I_2 + I_3 = 1\,A$, so

$$I_2 + 3I_2 = 1\,A, \quad 4I_2 = 1\,A, \quad \text{and} \quad I_2 = 0.25\,A$$

Self-assessment tasks 7.19

1. Work out the values of I_3, I_4, I_5, I_6 and I_7 in Fig. 7.37.
2. Work out the currents flowing in and out of nodes D, E and F in Fig. 7.37.

Kirchhoff's voltage law

Kirchhoff's voltage law states that, in any closed circuit, the algebraic sum of the products of the current and the resistance of each part of the circuit is equal to the resultant e.m.f. in the circuit. In mathematical form this is

$$\sum E = \sum IR$$

Example 7.24

Consider the example of the motorcycle lamp circuit with corroded connectors. In this case, the e.m.f. is the e.m.f. of the battery, which is 12 V. From before we know that the current is 1.82 A. Applying Kirchhoff's voltage law:

$$12 = (1.82 \times 1.5) + (1.82 \times 2.5) + (1.82 \times 2.6)$$
$$= 2.73 + 4.55 + 4.72 = 12$$

as expected.
In general terms this is

$$V(\text{or } E) = IR_1 + IR_2 + IR_3 = I(R_1 + R_2 + R_3)$$

which is a restatement of Ohm's law.

If we also consider the circuit shown in Fig. 7.38 (the loudspeaker circuit again), we can apply Kirchhoff's voltage law to this circuit. Look at loop 1. There is one resistance and one voltage source. According to the voltage law

$$E = I_2 R$$

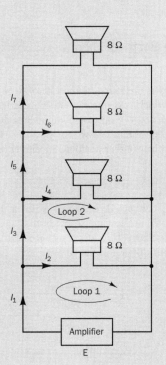

Figure 7.38 Loudspeaker circuit: Kirchhoff's voltage law

We already know that the resistance is $8\,\Omega$ and I_2 is 0.25 A. Putting these values in the equation gives

$$E = 0.25 \times 8 = 2\,\text{V}$$

The voltage is what we should have expected, as we already know that the voltage is 2 V. Substituting values that you have calculated into an equation to calculate one of the original values is a good way of checking if you have worked out the answer correctly.

Self-assessment tasks 7.20

1. Use Kirchhoff's current law to calculate the unknown currents shown in the circuit below.

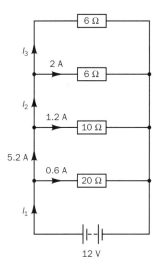

2. Use Kirchoff's voltage law to calculate the unknown currents shown in the circuit below.

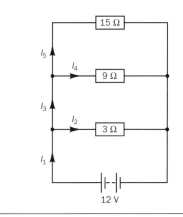

Capacitors

Before finishing this section on electrical systems, we will look at another common electronic component which is widely used in electronic circuits. The capacitor is a device for storing electrical charge temporarily.

To understand what is meant by the term 'charge', we will look at the behaviour of a capacitor in a d.c. circuit, after first describing the construction of a simple capacitor. A simple capacitor is shown in Fig. 7.39.

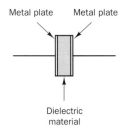

Figure 7.39 Simple capacitor

A capacitor is simply two metal plates between which is sandwiched insulating material called the *dielectric*. As this method of construction would make capacitors very large they are made commercially by sandwiching paper between sheets of foil and then rolling them up like a Swiss roll.

When the capacitor is connected as shown in Fig. 7.40(a), with switch B closed, electrons will flow from the negative side of the battery to the right-hand plate and electrons on the left-hand plate will flow to the battery. This results in the left plate being positively (+ve) charged and the right-hand plate being negatively (−ve) charged, i.e. the capacitor is charged.

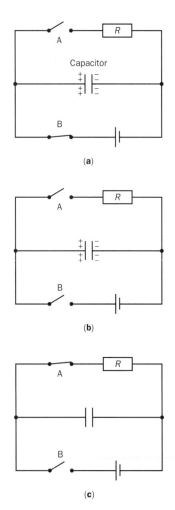

Figure 7.40 Simple circuit with a capacitor

If switch B is now opened the capacitor will remain in a charged state (Fig. 7.40(b)). If switch A is closed (Fig. 7.40(c)), the charged capacitor will be in a circuit with the resistor R and electrons will flow from the right plate through the resistor to the left plate. This will continue until no charge remains on the plates, i.e. the capacitor has discharged.

When a capacitor discharges, it does so according to an exponential law. A graph of charge against time is shown in Fig. 7.41.

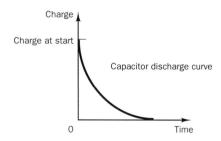

Figure 7.41 Capacitor discharge curve

Measuring charge and capacitance

When a capacitor charges it does so according to a simple law:

$$\text{Charge} = \text{Capacitance} \times \text{Voltage}$$
$$Q = CV$$

where the charge Q is measured in coulombs (C) and the capacitance C is measured in farads (F). The farad is a very large unit and capacitances are usually given in microfarads (μF).

Self-assessment task 7.21

If the circuit shown in Fig. 7.40 has the following values, calculate the charge stored on the capacitor:

(a) $V = 6\,\text{V}, C = 6\,\mu\text{F}$
(b) $V = 12\,\text{V}, C = 5\,\mu\text{F}$
(c) $V = 1.5\,\text{V}, C = 2\,\mu\text{F}$

Measuring electrical energy

Electrical energy is measured by the rate of flow of electrons in an instrument which is connected either in series or in parallel with a component or group of components. Instruments use the principles of Ohm's law to measure the quantities. Instruments can have either analogue or digital readouts.

Ammeters

Ammeters are connected in series with a component to measure the current flowing through that component (Fig. 7.42). They must have very low resistance so that they do not affect the current flowing in the circuit.

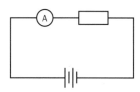

Figure 7.42 An ammeter in a circuit

Voltmeters

Voltmeters are connected in parallel with the component to measure the potential difference across the component (Fig. 7.43). They must have very high resistance so that only a very small current flows through the meter, otherwise it will affect the current flowing through the component and hence the voltage across the component.

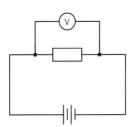

Figure 7.43 A voltmeter in a circuit

Multimeters

Multimeters can be either analogue or digital. Usually you will be able to measure current and voltage using a multimeter. Very often you can also use the meter to measure resistance.

Multimeters are connected up in the same way as an ammeter or a voltmeter, depending on whether you want to measure voltage or current. For measuring resistance, the multimeter is connected to both terminals of the component, when the component has been removed from the circuit.

Oscilloscopes

Oscilloscopes consist of a cathode-ray tube with a single beam of light. It uses the same principle as a television set, but is much simpler. The beam of light can be deflected by a current or a voltage to the inputs. The beam will move upwards or downwards depending on whether the current or voltage is positive or negative.

There will be a grid on the screen of an oscilloscope marked in centimetres. The input will have a switchable scale marked as, for example, '1 V/cm', '10 V/cm', '100 V/cm'. If the input is switched to 1 V/cm, this means that a voltage of 1 V will deflect the beam of light by 1 cm on the screen. This can be either the vertical or horizontal scale.

In addition, on the horizontal scale, the oscilloscope has a **timebase**. This moves the beam at a constant rate across the screen. The scale is switchable and will be marked, for example, '1 s/cm' '100 ms/cm'. Time can be measured by measuring the horizontal length of the trace on the screen, i.e. across the screen. If the timebase is switched to 1 s/cm, then 1 cm will represent 1 s.

The timebase is used when you want to study the shape of an alternating current or voltage. If the mains voltage is applied to the input, you will see a sinewave (Fig. 7.44). The inputs here have been set to 50 V/cm and 5 ms/cm.

The length of one complete cycle of the sinewave τ is 4 cm. As the scale is 5 ms/cm, this means that the period is $4 \times 5 \, \text{ms} = 20 \, \text{ms}$. The frequency is $1/\tau = 1/20 \, \text{Hz} = 50 \, \text{Hz}$.

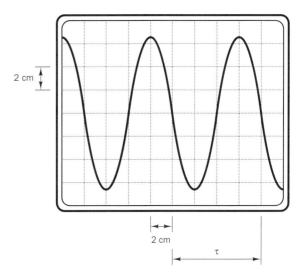

Figure 7.44 Oscilloscope screen

Sources of error

Possible sources of error when using the instruments described above are as follows:

- When reading an analogue meter, always make sure that your line of sight is at right angles to the needle of the meter that you are reading. Some meters have a mirror by the scale so that if you can see a reflection of the needle, you know that your line of sight is not at right angles to the needle.
- Always check that meters are connected properly. Bad connections can mean either that nothing is measured, or that no current flows in the circuit.
- Always check that you are using an instrument with an appropriate range. For example, if the current you will be measuring is in the range 0–10 A, use a meter which has this range, or the closest to it. Do not use a meter with a range 0–100 A.
- If you are unsure of the range that you will need for measuring the current or voltage, start with a meter with a much higher range. If the range is too high, try a meter with a smaller range. Keep on doing this until the current or voltage you are measuring makes full use of the range of the meter. If you use a meter with a range which is too low, the pointer of an analogue meter will be deflected past the top of the scale and it could well damage the meter.
- With digital meters beware of taking too many decimal places. If the meter has four or five decimal places, it is usually only worth going to two, perhaps three, unless you are measuring something very accurately.
- Always check that the meter reads zero when it is not connected to the circuit. If it does not, adjust it to read zero.
- When using an oscilloscope, check which switchable scale you are using.

Self-assessment task 7.22

What is the frequency of the sinewave shown on the oscilloscope screen below? What is the peak voltage? What is the peak-to-peak voltage? Calculate the r.m.s. voltage. The vertical scale has been set to 1 V/cm and the timebase to 0.1 ms/cm.

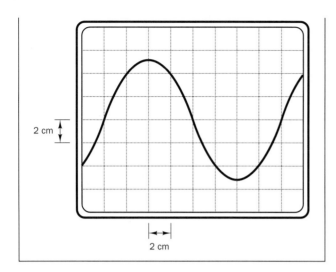

7.5 Electromagnetic systems

Electromagnetic systems involve the interaction of

- electrical energy (an electric current)
- magnetic energy (a magnetic field) and
- mechanical energy (motion)

Examples of electromagnetic systems are:

- An electric motor where the interaction of an electric current with a magnetic field produces motion.
- A generator where the interaction of motion and a magnetic field produces an electric current.
- A loudspeaker where the interaction of a current and a magnetic field produces motion, which generates the sound.

The principles govern the operation of many devices which are in everyday use, including all electric motors, loudspeakers and microphones and the generation of electricity to power these devices.

Magnetic fields

Magnetic fields exist naturally. Some materials can be permanently magnetic, for example iron, nickel or cobalt. A permanent magnet has its own magnetic field with a north pole and a south pole. The earth can be thought of as a very large permanent magnet.

Magnetic fields are also created by a current flowing in a wire. Look at Fig. 7.45. It shows the magnetic field produced by a current flowing through a wire.

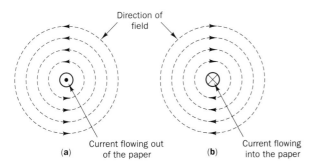

Figure 7.45 Magnetic field generated by a current in a straight conductor

Figure 7.45(a) shows the field when the current is flowing towards you. Think of the direction of the current as an arrowhead coming towards you. Figure 7.45(b) shows the field when the current is flowing away from you. In this case think of the current direction as the tails of an arrow flying away from you. The field lines always go in closed loops and never cross each other.

If you know the direction of the current, the following is an easy way to work out the direction of the field. Hold your right hand in a fist but with the thumb sticking out at right angles to it, as shown in Fig. 7.46. If the thumb is pointing in the direction of the current, the fingers give the direction of the field.

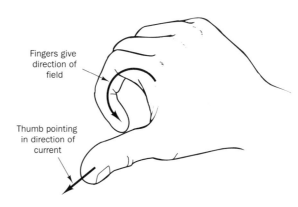

Figure 7.46 Working out the direction of the field

If you have a wire wound in a coil similar to a spring, as shown in Fig. 7.47, you have a solenoid. The magnetic field produced by a solenoid is shown in Fig. 7.48. The field can be made stronger by putting an iron core in the coil, as shown in Fig. 7.49.

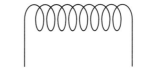

Figure 7.47 A simple solenoid

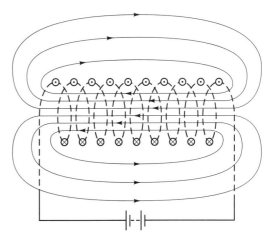

Figure 7.48 Magnetic field produced by a solenoid

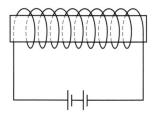

Figure 7.49 Solenoid with an iron core

Force on a conductor

If you have a wire (conductor) in a magnetic field which has a current flowing through it, there is a force on the wire which is proportional to the strength of the magnetic field, the magnitude of the current and the length of the conductor which is in the magnetic field (Fig. 7.50). The force is given by the equation

$$F = BIl$$

where F is the force in newtons, B is the magnetic flux density in teslas, I is the current in amps and l is the length of wire in the magnetic field.

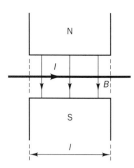

Figure 7.50 Force on a conductor in a magnetic field

Look at Fig. 7.51(a) which shows the field between the poles of a permanent magnet. If a conductor carrying a current is put between the poles of the magnet, the field is disturbed as shown in Fig. 7.51(b). To the left of the conductor you can see that as the lines of force are going in the same direction; the field is stronger to the left of the conductor. To the right of the conductor, the lines are opposing each other and so the field is weaker. The stronger field forces the conductor away.

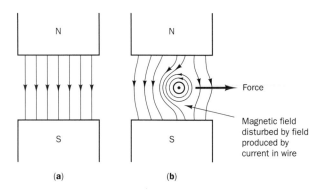

Figure 7.51 Magnetic field between the poles of a permanent magnet

Fleming's left-hand rule

If you know the direction of the magnetic field and current, the direction of the force can be predicted by Fleming's left-hand rule. It is also known as the motor rule.

The rule states that if you hold the thumb and the first and second fingers of the left hand at right angles to each other (Fig. 7.52), and

- the first finger is pointing in the direction of the field

and

- the second finger is pointing in the direction of the current

then

- the thumb will be pointing in the direction of the motion

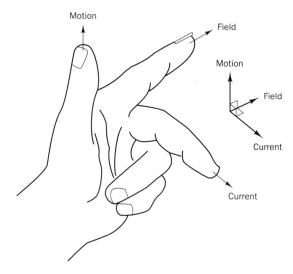

Figure 7.52 Fleming's left-hand rule

Magnetic flux density

The magnetic flux density is defined as the density of a magnetic flux such that a conductor carrying 1 ampere at right angles to the flux has a force of 1 newton acting on it. Its unit is the tesla (T).

Magnetic flux

If the magnetic field has a cross-sectional area a and a uniform flux density, the total flux is given by

$$\Phi = B \times a \text{ webers}$$

where Φ is the magnetic flux in webers, B is the magnetic flux density in teslas and a is the cross-sectional area of the magnetic field in square metres.

Example 7.25

Look at the arrangement shown in Fig 7.53. The disc is free to rotate and has a connection in the spindle and another connection at the edge of the disc, which can be assumed to be frictionless. Part of the disc cuts through the magnetic field.

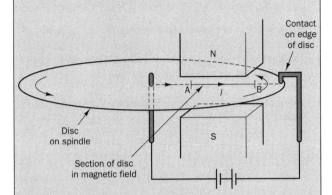

Figure 7.53 A rotating disc in a magnetic field

The disc has a current of 2 A flowing through it. The length of the disc that is in the field is 5 cm and the magnetic flux density is 2 T.

The force on the disc along line AB is given by

$$F = BIl = 2 \times 2 \times 0.05\,\text{N} = 0.2\,\text{N}$$

Using Fleming's left-hand rule, we can see that the direction of the force on the disc is into the page. From the information on page 215 we know that $F = ma$. So if we know that the mass of the disc is 100 g, then

$$a = F/m = 0.2/0.1 = 2\,\text{m/s}^2$$

As the disc is free to rotate, it will move. So a new piece of the disc comes into the field. This part also has a current flowing through it. There will be a force on the new part of the disc which also tries to push the disc in the same direction. So the disc continues to move.

This is the principle behind the electricity meter that you will see in your own home. The meter is used to measure the amount of electrical energy that you have used. This is also the principle on which electric motors work, but the actual construction of a motor will be much more complex.

Self assessment tasks 7.23

1. If a conductor is in a uniform magnetic field, calculate the force using the following values for the current, magnetic flux density and length of conductor in the field:
   (a) $B = 1\,\text{T}$, $I = 1\,\text{A}$, $l = 10\,\text{cm}$
   (b) $B = 3\,\text{T}$, $I = 5\,\text{A}$, $l = 10\,\text{cm}$
   (c) $B = 10\,\text{T}$, $I = 2\,\text{A}$, $l = 25\,\text{cm}$

2. Look at this circuit for a simple doorbell. Explain how you think it works.

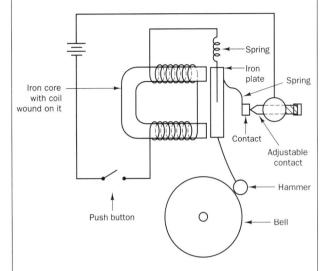

3. Look at this circuit for a simple telephone receiver. Explain how you think it converts electrical energy to sound.

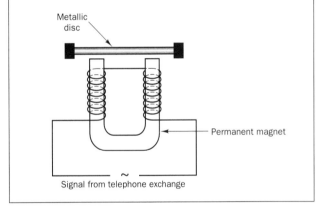

Faraday's law of electromagnetic induction

Faraday's law of electromagnetic induction states that

- An e.m.f. is induced (set up) whenever the magnetic field changes.
- The magnitude of the e.m.f. is proportional to the rate of change of magnetic flux.

Let us now look at the practical effects and applications of these laws.

Induced e.m.f.

If a conductor is moved in a magnetic field, so that it cuts through the flux of a magnetic field, an e.m.f. is induced in the conductor. If the conductor is part of a circuit, a current will flow in the circuit. This e.m.f. is said to be induced and the effect is known as electromagnetic induction.

Example 7.26

Consider the car shown in Fig. 7.54. When the car moves, it is cutting through the earth's magnetic field. The car has components which are conductors, for example, the engine block and the axles. According to Faraday's law of electromagnetic induction, an e.m.f. will be induced in the metallic parts of the car.

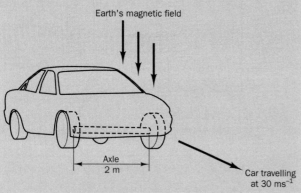

Figure 7.54 Car axle in the earth's magnetic field

The induced e.m.f. is given by

$$E = BA/t$$

where E is the induced e.m.f. in volts, B is the magnetic flux density in teslas, A is the area of the magnetic field cut by the conductor and t is the time.

We will now calculate the e.m.f. induced in one of the car's axles. The area is given by the length of the axle (l) multiplied by the distance the car has gone in a certain time t. Now the distance divided by time is velocity (or speed)(v). Hence

$$E = Blv$$

If the length of the axle is 2 m, the car is travelling at 30 m/s (108 km/h) and the vertical component of the earth's magnetic field is 40 μT,

$$E = 40 \times 10^{-6} \times 2 \times 30 = 2{,}400 \times 10^{-6}\,V$$
$$= 2{,}400\,\mu V = 2.4\,mV$$

If the circuit is completed, a current will flow in the circuit. From p. 234, we know that $B = \Phi/a$; hence

$$E = \Phi l v/a = \Phi/t$$

This gives us the definition of the weber, which is the magnetic flux which, when cut by a conductor in 1 second, generates an e.m.f. of 1 volt.

This effect of induced e.m.f. is fundamental to the generation of electricity. It is the principle behind both the generation of electricity and the alternator in a car engine.

A further practical application of this is in transporting flammable liquids. If the voltage cannot be dissipated by completing a circuit, a charge will build up. Eventually the charge becomes so great that if another conductor comes near, the charge sparks between the conductors. If you have flammable liquids this could ignite the liquid.

Fleming's right-hand rule

If you know the direction of motion and the magnetic field, the direction of the e.m.f. can be predicted using Fleming's right-hand rule. It is also known as the generator rule.

The rule states that if you hold the thumb and the first and second fingers of the right hand at right angles to each other (as in Fig. 7.55), and

- the first finger is pointing in the direction of the field *and*
- the thumb is pointing in the direction of the motion of the conductor relative to the magnetic field *then*
- the second finger is pointing in the direction of the induced e.m.f.

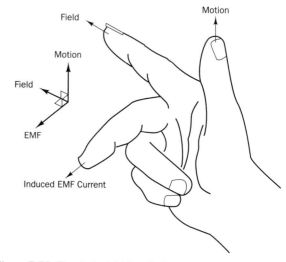

Figure 7.55 Fleming's right-hand rule

Magnetically coupled systems (transformer)

Transformers are used for stepping voltages (and currents) up or down. Electricity is generated at 230 V a.c. and then stepped up to 132 kV or higher for transmission over the National Grid. It is then stepped back down again to 230 V a.c. for delivery to houses, shops and offices.

A transformer, in its simplest form, is two coils which have been wound round an iron core, as shown in Fig. 7.56. If an alternating current is applied to the primary coil, the current flowing in the coil generates an alternating magnetic field in the iron core.

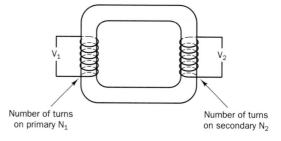

Figure 7.56 A simple transformer

As a result of Faraday's law of electromagnetic induction, the alternating magnetic field generates an e.m.f. in the secondary coil. The magnitude of the e.m.f. generated in the secondary coil is related to

- The number of turns on the primary N_1.
- The number of turns on the secondary N_2.

The relationship is

$$\frac{V_1}{V_2} = \frac{N_1}{N_2}$$

If we wish to calculate V_2, rearranging the equation gives

$$V_2 = V_1 \frac{N_2}{N_1}$$

N_2/N_1 is also known as the *turns ratio*.

If there are 100 turns on the primary, 1,000 turns on the secondary and V_1 is 230 V a.c.,

$$V_2 = 230 \times (1{,}000/100) = 230 \times 10 = 2{,}300 \text{ V a.c.}$$

Self-assessment task 7.24

If V_1 is 230 V a.c. and there are 1,000 turns on the primary and 50 turns on the secondary, what is V_2?

7.6 Unit test

Test yourself on this unit with these multiple-choice questions.

1. Which of the following measurement devices makes use of the principle of moments?

    (a) simple balance
    (b) spring balance
    (c) thermocouple
    (d) manometer

2. A material that obeys Hooke's law is said to be:

    (a) plastic
    (b) elastic
    (c) static
    (d) rigid

3. The relationship between force and momentum is defined by:

    (a) principle of moments
    (b) Newton's second law
    (c) Hooke's law
    (d) Ohm's law

4. Which of the following statements describes Kirchhoff's current law?

    (a) current is proportional to voltage
    (b) current flows from negative to positive
    (c) the algebraic sum of currents at a junction is zero
    (d) current is the same throughout a circuit

5. Which of the following best describes the concept of power?

    (a) the rate of doing work
    (b) the ability to do work
    (c) the force required to perform a task
    (d) the heat emitted by a machine

6. What name is used for the force that holds an object in a state of rotational motion?

    (a) gravity
    (b) friction
    (c) centripetal force
    (d) centrifugal force

7. The working fluid in an internal combustion engine is:

    (a) petrol
    (b) oil
    (c) water
    (d) air

8. Which of the following is a dimensionless quantity?

    (a) work
    (b) stress
    (c) modulus of elasticity
    (d) strain

9. What is the downward force exerted by a mass of 14 kg?

    (a) 14 N
    (b) 140 N
    (c) 137.3 N
    (d) 68.6 N

10. A 2 m steel wire of Young's modulus 190×10^9 Pa is stretched by a mass of 100 kg. If the resulting extension of the wire is 3 mm, calculate the cross-sectional area of the wire.

    (a) 0.34 mm^2
    (b) $3.4 \times 10^{-6} \text{ mm}^2$
    (c) $1.7 \times 10^{-6} \text{ m}^2$
    (d) 3.4 mm^2

11. A satellite in orbit around the earth completes exactly two orbits every day. If the distance from the satellite to the centre of the earth is 15,000 km, what is its linear velocity?

    (a) 2,182 m/s
    (b) 13,710 m/s
    (c) 347 m/s
    (d) 4,364 m/s

12. A car, initially stationary, accelerates smoothly for 20 s with an acceleration of 0.8 m/s^2, and then decelerates smoothly at a rate of 0.5 m/s^2. How long does it take to become stationary once again?

    (a) 20 s
    (b) 8 s
    (c) 40 s
    (d) 32 s

13. A water heater heats a bath containing 15 kg of water from room temperature to 40 °C in 6 minutes. What is the power of the heater? (Assume there are no heat losses from the bath.)

    (a) 7 kW
    (b) 3.5 kW
    (c) 210 kW
    (d) 35 kW

14. In the circuit shown in Fig. 7.T1, the ammeter shows a current of 2 A. What is the value of the unknown resistance R?

    (a) 2 Ω
    (b) 0.2 Ω
    (c) 20 Ω
    (d) 4 Ω

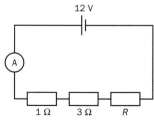

Figure 7.T1

15. A transformer is used to reduce the mains voltage for use on a small piece of machinery. If the transformer has 3,000 turns on the input coil, and 150 on the output coil, what is the r.m.s. output voltage?

    (a) 230 V
    (b) 1.5 V
    (c) 23 V
    (d) 11.5 V

16. A 50 cm beam pivoted at its midpoint has a 500 g mass suspended from its left end. How far from the right end of the beam would you have to place a 1 kg mass for the beam to balance?

    (a) 10 cm
    (b) 15 cm
    (c) 2.5 cm
    (d) 12.5 cm

17. A 2-litre flask is sealed at room temperature (20 °C) and atmospheric pressure. Air is then pumped out until the pressure reaches 65 kPa. What is the temperature inside the flask?

    (a) 190 K
    (b) 451 K
    (c) 13 °C
    (d) 20 °C

18. The tesla (T) is the unit of:

    (a) magnetic field strength
    (b) magnetic flux density
    (c) electromotive force
    (d) electric field strength

19. What would you measure with a Bourdon gauge?

    (a) temperature
    (b) current
    (c) shear force
    (d) pressure

20. Multimeters can be used to measure:

    (a) voltage and current at the same time
    (b) voltage or current, but not both
    (c) voltage and capacitance
    (d) resistance and impedance

Mathematics for Engineering

Engineering is concerned with manufacturing, designing and repairing. It involves making measurements and combining a large number of measurements to make things work. Decisions have to be taken based on these complex measurements and calculations have to be made to assist in making these decisions. Mathematics is a tool to describe and model engineering processes and, of greater importance, to assist in developing a product from the initial design through to the manufacturing stage.

This unit provides a study of mathematics which underpins the other mandatory units of the GNVQ (Advanced) in Engineering.

In this unit the student will learn how to:

- Use number and algebra to solve engineering problems.
- Use trigonometry to solve engineering problems.
- Use of functions and graphs to model engineering situations and solve engineering problems.

After reading this unit the student will be able to:

- Represent and calculate numbers appropriate to engineering.
- Use and manipulate algebraic expressions.
- Express solutions to engineering problems in algebraic form.
- Select trigonometric ratios and use Pythagoras's theorem to solve engineering problems.
- Solve problems involving triangles using trigonometry.
- Solve engineering problems using triangles of forces.
- Represent data graphically.
- Use functions and graphs to model and solve engineering problems.
- Represent trigonometric functions by a rotating vector.
- Reduce data to linear form.
- Interpret rates of change of functions to solve engineering problems.
- Interpret areas under graphs to solve engineering problems.
- Use differentiation to solve engineering problems.

8.1 Use numbers and algebra to solve engineering problems

Negative numbers and modulus

Signs are used with numbers to denote direction or to denote opposites, as shown in Fig. 8.1.

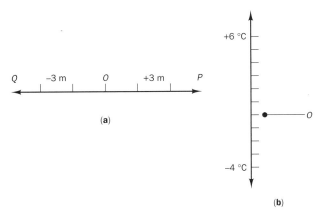

Figure 8.1

For example, in Fig. 8.1(a), $+3$ m measures 3 metres from left to right, that is OP; -3 m measures 3 metres from right to left, that is, OQ. Figure 8.1(b) shows positive and negative temperatures.

The **modulus** of a number refers to negative numbers and is the magnitude or size of such a number. The modulus of -3, written as $|-3|$ is 3.

Multiplication and division of numbers

The following examples show the rule:

$$(-3) \times (-4) = +12; \quad (-3) \times (+4) = -12;$$
$$(+20) \div (-5) = -4; \quad (-20) \div (-5) = +4$$

Rule 1

- Numbers with like signs when multiplied or divided give a positive (+) answer.
- Numbers with unlike signs when multiplied or divided give a negative (−) answer.

Addition and subtraction of positive numbers

We see that

$$-3 + 4 = +1; \quad -7 + 3 = -4; \quad -4 - 5 = -9$$

Rule 2

- Numbers with the same sign are added and the answer takes the common sign.
- Numbers with unlike signs are subtracted and the answer takes the sign of the larger number.

Addition and subtraction of negative numbers

Using Fig. 8.2 the following addition and subtraction can be worked out:

$$-4 - (-6) \quad \text{and} \quad 5 + (-6)$$

$L \bullet \longrightarrow M$		$LM = +6$
$L \longleftarrow \bullet M$		$ML = -6$

Figure 8.2

In Fig. 8.2,

$$-6 = ML, \quad -(-6) = -ML = LM = +6,$$
$$+(-6) = +ML = -6$$

Therefore,

$$-4 - (-6) = -4 + 6 = +2$$
$$+5 + (-6) = 5 - 6 = -1$$

Rule 3

- Addition of a negative number remains a negative number.
- Subtraction of a negative number converts it to a positive number.

Example 8.1

Evaluate the following:

$-8 - 5$; $\quad -7 + 2$; $\quad -3 + 8$; $\quad -4 - (-7)$; $\quad +6 + (-8)$;

$(-6) \times (-10)$; $\quad (-8) \times (2)$; $\quad (-15) \div (-3)$.

Using the above three rules we obtain

$-8 - 5 = -13$	$-4 - (-7) = -4 + 7 = +3$	$-6 \times (-10) = +60$
$-7 + 2 = -5$	$+6 + (-8) = 6 - 8 = -2$	$(-8) \times (2) = -16$
$-3 + 8 = +5$		$(-15) \div (-3) = +5$

Self-assessment tasks 8.1

Work out the following:

1. $-6 + 7$, $-3 - 9$, $4 - 9$, $-7 + 18$, $-3 + 7 - 2 + 1$
2. $-8 - (-7)$, $-3 + (-6)$, $-10 - (-8)$
3. $-6 \times (-10)$, $-4 \times (6)$, $+2 \times (-4)$
4. $-21 \div (-7)$, $42 \div (-6)$, $-24 \div 3$

Indices

A number such as 64 can be written in its factors $2 \times 2 \times 2 \times 2 \times 2 \times 2$. Such a number can be written in short form, called the index form, as 2^6. The number 6 is the **index** or **power** and 2 is the **base**. Similarly:

$$1{,}000 = 10 \times 10 \times 10 = 10^3$$

Multiplication of numbers in index form

Consider $(2 \times 2 \times 2 \times 2) \times (2 \times 2 \times 2)$ which is equal to $(2 \times 2 \times 2 \times 2 \times 2 \times 2 \times 2)$:

$$2^4 \times 2^3 = 2^7 \quad \text{that is} \quad 2^{4+3}$$

Rule 4

- Multiplication of numbers in index form on the same base is carried out by adding the indices:

$$a^n \times a^m = a^{n+m} \qquad (1)$$

Division of numbers in index form

Consider $(6 \times 6 \times 6 \times 6 \times 6) \div (6 \times 6 \times 6)$ which is equal to 6×6:

$$6^5 \div 6^3 = 6^2 \quad \text{that is} \quad 6^{5-3}$$

Rule 5

- Division of numbers in index form, on the same base, is carried out by subtracting the indices:

$$a^n \div a^m = a^{n-m} \qquad (2)$$

Numbers with an index of 0

Using rule 5 we have $7^3 \div 7^3 = 7^0$; i.e.

$$1 = 7^0$$

Rule 6

Any number raised to the index 0 has a value 1.

$$a^0 = 1 \qquad (3)$$

Negative indices

Using rule 5 we have $9^2 \div 9^5 = 9^{-3}$. But

$$9^2 \div 9^5 = \frac{9 \times 9}{9 \times 9 \times 9 \times 9 \times 9} = \frac{1}{9^3}$$

Rule 7

- A negative index signifies a reciprocal. Changing the sign to $+$ will invert the number:

$$a^{-n} = \frac{1}{a^n} \qquad (4)$$

Numbers in index form raised to another index

For example,

$$(7^2)^3 = 7^2 \times 7^2 \times 7^2 = 7^{2+2+2} = 7^{3 \times 2} = 7^6$$

Rule 8

- The value of a number in the form of an index raised to another index is found by multiplying the two indices:

$$(a^n)^m = a^{n \times m} \qquad (5)$$

Fractional indices

Using Rule 8:

$$81^{\frac{1}{2}} = (3^4)^{\frac{1}{2}} = 3^{4 \times \frac{1}{2}} = 3^2 = 9$$

Therefore the index $\frac{1}{2}$ is the square root.

Rule 9

- A fractional index denotes a root.

Example 8.2

Evaluate the following:

1. $2^7 \times 2^4$, $4^3 \times 4$, $9^7 - 9^2$, $10^8 - 10^3$
2. Write with $+$ indices: 8^{-3}, $\dfrac{1}{9^{-1}}$, $3^{-\frac{1}{2}}$
3. $(5^2)^3$, $(2^3)^7$, $(5^6)^0$
4. $(32)^{\frac{1}{5}}$, $(125)^{\frac{2}{3}}$, $(81)^{\frac{3}{4}}$

1. Using rules 4 and 5:
   $2^7 \times 2^4 = 2^{11}$, $4^3 \times 4^1 = 4^4$,
   $9^7 \div 9^2 = 9^5$, $10^8 \div 10^3 = 10^5$,
2. From rule 7:
   $8^{-3} = \dfrac{1}{8^3}$, $\dfrac{1}{9^{-1}} = 9$, $3^{-\frac{1}{2}} = \dfrac{1}{3^{\frac{1}{2}}}$
3. From rules 6/8:
   $(5^2)^3 = 5^6$, $(2^3)^7 = 2^{21}$, $(5^6)^0 = 5^0 = 1$.
4. From rule 8:
   $(32)^{\frac{1}{5}} = (2^5)^{\frac{1}{5}} = 2^1 = 2$, $125^{\frac{2}{3}} = (5^3)^{\frac{2}{3}} = 5^2$,
   $81^{\frac{3}{4}} = (3^4)^{\frac{3}{4}} = 3^3$

Self-assessment tasks 8.2

1. Work out the following multiplications and divisions, leaving your answer in index form:
   $8^3 \times 8$, $5^4 \times 5^7$, $3^8 \times 3^{-4}$, $9^8 \div 9$, $6^5 \div 6^2$, $8^2 \div 8^{-7}$
2. Write the following with positive indices:
   8^{-1}, 7^{-4}, $\dfrac{1}{5^{-3}}$, $\dfrac{3}{2^{-4}}$, $\left(\dfrac{3}{4}\right)^{-4}$.

Numbers in standard form

Numbers in standard form are written as a decimal number between 1 and 10, multiplied by multiples or submultiples of 10. For example,

$$637.5 = 6.375 \times 10^2$$
$$7,391 = 7.391 \times 10^3$$
$$0.0572 = 5.72 \times 10^{-2}$$

Self-assessment tasks 8.3

1. Write the following in standard form: 43.5, 756, 300.5, 0.0097, 0.002, 0.000572.
2. Evaluate: $\dfrac{3.3 \times 10^{-4} \times 4.2 \times 10^8}{1.1 \times 10^{-3} \times 2.1 \times 10^2}$
3. A grinding wheel makes 2,455 rev/min. Express this in standard form and calculate the revolutions per second.
4. The current in an operational amplifier is 0.0000856 A. What is this in standard form? Find the value in standard form if the current is trebled.

Binary numbers

In our normal counting system, called the **denary system**, there are 10 figures: 0, 1, 2, 3, 4, ..., 9. This system is on a base of 10. A number such as 429 can be considered as

$$4 \text{ sets of } 100 + 2 \text{ sets of } 10 + 9 \text{ sets of } 1$$

that is

$$4 \times 100 + 2 \times 10 + 9 \times 1$$
$$4 \times 10^2 + 2 \times 10 + 9 \times 1$$

It is seen that each digit in the number moving right to left signifies units, tens, hundreds, thousands, etc.; that is 10^0, 10^1, 10^2, 10^3, etc. Therefore, for example,

$$2,536 = 2 \times 10^3 + 5 \times 10^2 + 3 \times 10^1 + 6 \times 10^0$$

The **binary system** has a base of 2 and possesses two figures only, 0 and 1. Binary numbers can be expressed in the same format as denary numbers but with the base number of 2 rather than 10. Using the same procedure with binary numbers:

$$10011 = 1 \times 2^4 + 0 \times 2^3 + 0 \times 2^2 + 1 \times 2^1 + 1 \times 2^0 \quad (6)$$

Binary counting

In denary, $9 + 1 = 10$, $19 + 1 = 20$, etc., where 9 is the highest figure. In binary the highest figure is 1, as shown in Table 8.1.

Table 8.1 Binary counting and denary equivalents

Binary	Denary equivalent
$1 + 1 = 10$	2
$10 + 1 = 11$	3
$11 + 1 = 100$	4
$100 + 1 = 101$	5
$101 + 1 = 110$	6
$110 + 1 = 111$	7
$111 + 1 = 1000$	8

Conversion of binary numbers to denary numbers

Using the procedure shown in equation (6) a binary number can be converted to denary as follows

$$111 = (1 \times 2^2) + (1 \times 2^1) + (1 \times 2^0)$$
$$= 4 + 2 + 1 = 7$$
$$10011 = (1 \times 2^4) + (0 \times 2^3) + (0 \times 2^2) + (1 \times 2^1) + (1 \times 2^0)$$
$$= 16 + 0 + 0 + 2 + 1 = 19$$

Conversion of denary numbers to binary

The denary number is separated out into multiples of 2, in descending order: ..., 2^5, 2^4, 2^3, 2^2, 2^1, 2^0; that is, ..., 32, 16, 8, 4, 2, 1.

Consider

$$59 = 32 + 16 + 8 + 2 + 1$$

Comparing this with the above sequence we see that 4 is missing. Therefore

$$59 = 32 + 16 + 8 + 0 + 2 + 1$$

and writing in multiples of 2

$$59 = 2^5 + 2^4 + 2^3 + 0 + 2^1 + 2^0$$
$$= 1 \times 2^5 + 1 \times 2^4 + 1 \times 2^3 + 0 \times 2^2 + 1 \times 2^1 + 1 \times 2^0$$

giving a binary number 111011.

Addition of binary numbers

Add together the two numbers 10101 and 11101 using the result that $1 + 0 = 1$ and $1 + 1 = 10$:

```
  10101
  11101
 ------
 110010
  1111    carried over from the previous column
```

Self-assessment tasks 8.4

1. Convert the following binary numbers into denary: 110111, 100001, 11011.
2. Convert the following denary numbers into binary: 10, 29, 45, 167.
3. Add the following binary numbers: $101 + 10$, $1110 + 1011$, $110111 + 10110$, $11111 + 1$

Errors and significant figures

Significant figures

The significant figure of a number is the number of working digits it contains, and indicates the level of accuracy of the number. Thus,

- 761.43   is correct to 5 significant figures
- 0.00347   is correct to 3 significant figures because the zeros after the decimal point are not working figures contributing to the overall accuracy
- 4.00347   is correct to 6 significant figures because of the whole number 4 makes the zeros working figures
- 0.3007   is correct to 4 significant figures

If we correct 760.43 to 2 significant figures we obtain 760. However, a number such as this with a zero at the end produces an uncertainty. Consider the number 760.43 corrected back to 3 significant figures; it becomes 760. Therefore, 760 may be correct to 2 or 3 significant figures.

Accuracy of a number

The number of significant figures indicates the level of accuracy of the number. Consider a length of wood measured and found to be 27.4 cm, which is correct to 3 significant figures. This means that the wood has a true length between 27.35 cm and 27.45 cm as shown in Fig. 8.3(a).

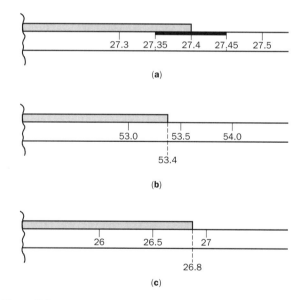

Figure 8.3

Rounding off a number

This is the term for the reduction of the significant figures of a number. The rule for doing so can be deduced from Fig. 8.3(b) and (c). In (b) it is seen that 53.4 is closer to 53 than to 54, so that 53.4 rounded to 2 significant figures becomes 53. In (c) 26.8 is nearer to 27 than it is to 26 so that 26.8 rounded to 2 significant figures becomes 27. By rounding numbers in this way the error introduced as a result of reducing the significant figures is minimised.

Rule 10

- If, in any number, the significant figures are to be reduced, and if the figure to be removed is greater than 5, the figure immediately preceding it is increased by 1. If the figures to be removed is less than 5, the figure preceding it is unchanged.

A difficulty arises when the figure to be removed is exactly 5. In order to reduce the error the rule is

Rule 11

- If the figure to be removed is 5 and the figure immediately preceding it is odd, this figure is increased by 1. If the figure preceding it is even, it remains unchanged.

Example 8.3

Correct the following to the respective significant figure:

(a)   375.47   to 3 significant figures
(b)   0.00687 to 2 significant figures
(c)   1.00911 to 3 significant figures
(d)   478.5   to 3 significant figures
(e)   1975   to 3 significant figures

Using the two rules stated above we obtain (a) 375, (b) 0.0069, (c) 1.01, (d) 478, (e) 1980.

Computational accuracy

If a quantity used in engineering is the result of arithmetic operations on a number of measurements it is important to express the result to the correct level of accuracy. Consider the area of a metal plate of length 44 cm and breadth 36 cm. Multiplying out gives the area as 1,584 if the two measures were exact. If, however, the two measures are correct to 2 significant figures it is necessary to examine the level of accuracy which is reasonable to expect in the value of the area. The length lies between 43.5 cm and 44.5 cm; the breadth lies between 35.5 cm and 36.5 cm.

The minimum possible area is, therefore,

$$43.5 \times 35.5 = 1,544.25 \, cm^2$$

The maximum possible area is, therefore,

$$44.5 \times 36.5 = 1,624.25 \, cm^2$$

The true area lies between these two values. This means that the area 1,584 is only accurate within ±40, that is, to 2 significant figures. It is seen that the accuracy of the result is only the same as the number of significant figures in the original measurements.

Rule 12

- The result of any computation should be expressed to the same number of significant figures as that of the least accurate data. However, if the result of this computation is to be used in a further computation, temporarily it is expressed to one more significant figure.

(Note: These rules will be used throughout the remainder of the mathematics section when carrying out any computation.)

Units of measurement

In almost all instances in engineering measurements are made in SI (System Internationale) units, which is essentially a metric system based on 10s. Every quantity must have a basic measure in a similar way that the metre or the foot is a basic measure of length. This basic measure is called a *unit* of measure.

Multiples and sub-multiples of basic units

The multiples of basic units are generally expressed in factors of 1,000 or 10^3. Sub-multiples are expressed as divisors of 1,000 or 10^{-3}.

Table 8.2 Multiples and submultiples

Factor	Name	Symbol	Example unit
10^9	giga	G	GV (gigavolt)
10^6	mega	M	MW (megawatt)
10^3	kilo	k	kW (kilowatt)
10^{-2}	centi	c	cm (centimetre)
10^{-3}	milli	m	ms (millisecond)
10^{-6}	micro	μ	μA (microampere)
10^{-9}	nano	n	ns (nanosecond)

Table 8.3 Basic and derived units

Quantity	Basic unit	Other units	Conversions
Length	metre (m)	kilometre (km)	1 km $= 10^3$ m
		centimetre (cm)	1 cm $= 10^{-2}$ m
		millimetre (mm)	1 mm $= 10^{-3}$ m
		micrometre (μm)	1 μm $= 10^{-6}$ m
Area	square metre (m^2)	square kilometre (km^2)	1 $km^2 = 10^6\ m^2$
		square centimetre (cm^2)	1 $cm^2 = 10^{-4}\ m^2$
		square millimetre (mm^2)	1 $mm^2 = 10^{-6}\ m^2$
Volume	cubic metre (m^3)	litre (l)	1 l $= 10^{-3}\ m^3$
		cubic centimetre (cm^3)	1 $cm^3 = 10^{-6}\ m^3$
Mass	kilogram (kg)	tonne (t)	1 t $= 10^3$ kg
		gram (g)	1 g $= 10^{-3}$ kg
		milligram (mg)	1 mg $= 10^{-6}$ kg
Time	second (s)	minute (min)	1 min $= 60$ s
		millisecond (ms)	1 ms $= 10^{-3}$ s
		microsecond (μs)	1 μs $= 10^{-6}$ s
		nanosecond (ns)	1 ns $= 10^{-9}$ s

Table 8.4 Units used in engineering

Quantity	Measurement	Unit symbol
Velocity	Distance moved per second	m/s
Acceleration	Velocity change per second	m/s^2
Force	Mass $\times$ acceleration	N (newton) (1 N $= 1$ kg $\times 1\ m/s^2$)
Work	Force $\times$ distance	J (joule) (1 J $= 1$ N $\times 1$ m)
Energy	Work done equals energy used	J
Power	Energy produced or used per second	W (watt) (1 W $= 1$ J/s)
Current	Ampere	A
Voltage	Volt	V
Temperature	degree Celsius	°C
Density	mass per cubic metre	kg/m^3 (which is the same as g/l)
Stress	force per square metre	N/m^2

Volumes of engineering objects

Most objects in engineering are machined or fabricated into or from regular objects. These regular objects are

- rectangular prism (oblong box)
- other prisms
- cylinder
- cone
- sphere
- pyramid

These are shown in Fig. 8.4.

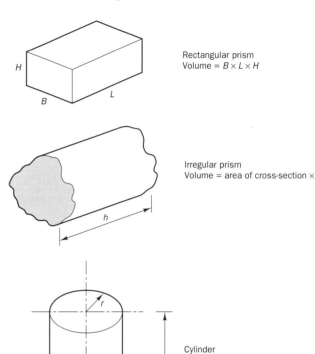

Rectangular prism
Volume $= B \times L \times H$

Irregular prism
Volume $=$ area of cross-section $\times$

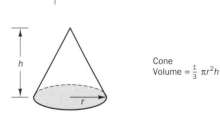

Cylinder
Volume $= \pi r^2 h$

Cone
Volume $= \frac{1}{3} \pi r^2 h$

Sphere
Volume $= \frac{4}{3} \pi r^3$

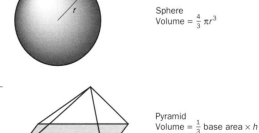

Pyramid
Volume $= \frac{1}{3}$ base area $\times h$

Figure 8.4

Example 8.4

Calculate the volume of the V-block shown in Fig. 8.5

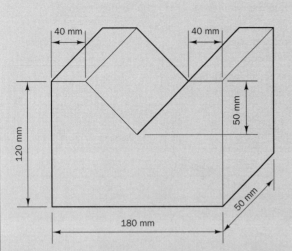

Figure 8.5

The block is composed of a rectangular prism from which a triangular prism has been machined out.

Volume of rectangular prism:

$$180 \times 120 \times 50 \, \text{mm}^3 = 1,080,000 \, \text{mm}^3$$

Volume of triangular prism:

$$(\tfrac{1}{2} \times \text{base} \times \text{height of triangle}) \times 50 = \tfrac{1}{2} \times 100 \times 50 \times 50$$
$$= 125,000 \, \text{mm}^3$$

Therefore, volume of block:

$$1,080,000 - 125,000 = 955,000 \, \text{mm}^3$$
$$= 960,000 \, \text{mm}^3 \text{ correct to 2 significant figures}$$
$$= 9.6 \times 10^5 \, \text{mm}^3 \text{ in standard form}$$

Example 8.5

Calculate the volume of a sphere of radius 3.4 cm.
Using the formula for the volume of a sphere, and substituting $r = 3.4$.

$$V = \tfrac{4}{3} \times \pi \times r^3 = \tfrac{4}{3} \times \pi \times (3.4)^3$$
$$= 160 \, \text{cm}^3 \text{ correct to 2 significant figures}$$

Self-assessment tasks 8.5

1. Calculate the volumes of the following figures
   (a) a cylinder, height 50 mm, base radius 25 mm
   (b) a cone of height 13 cm and base radius 7.2 cm
   (c) a sphere of radius 4.12 m
   Express the result to the appropriate significant figure and in standard form.

2. In order to balance a high-speed disc a hole is drilled near its rim, as shown below. Find the volume of the metal removed.

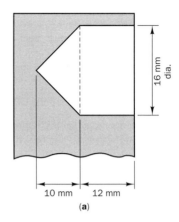

(a)

3. Calculate the volume of the lathe centre shown below.

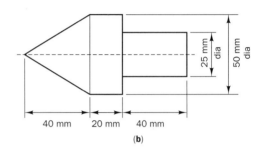

(b)

4. Calculate the volume of the casting shown below.

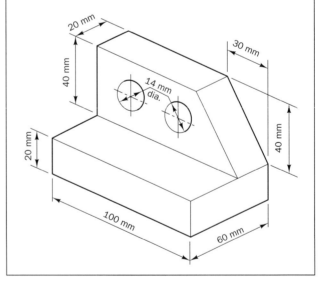

Density and mass

In any object, composed of the same material throughout, the mass of 1 cm³ in one part of the object will be the same as the mass of 1 cm³ in another part of the object. This mass of 1 cm³ is called the *density*. Therefore, if the mass of an object is M and its volume is V then each $1\,\text{cm}^3$ will have a mass of M/V, that is

$$\text{Density } D = \frac{M}{V} \quad (7)$$

From this formula the unit of density is **units of mass/units of volume**, g/cm^3 and kg/m^3.

To convert g/cm^3 to kg/m^3 and vice versa

The density of lead is $11.4\,\text{g/cm}^3$, which means that $1\,\text{cm}^3$ of lead has a mass of $11.4\,\text{g}$. A volume of $1\,\text{m}^3 = 10^6\,\text{cm}^3$ has a mass of $11.4 \times 10^6\,\text{g} = 11.4 \times 10^3\,\text{kg}$. Hence density of any material is D (g/cm^3) or $D \times 10^3$ (kg/m^3).

Example 8.6

An aluminium piston head has a volume of $96\,\text{cm}^3$ and a mass $260\,\text{g}$. Find its density in g/cm^3 and convert to kg/m^3.
Using formula (7)

$$\text{Density} = \frac{\text{Mass}}{\text{Volume}} = \frac{260}{96} = 2.71\,\text{g/cm}^3$$

Converting:

$$\text{Density} = 2.71 \times 10^3\,\text{kg/m}^3$$

Example 8.7

A steel ball bearing has a density of $7.9\,\text{g/cm}^3$. Its diameter in $2.7\,\text{cm}$. Calculate its volume and hence its mass.

- Volume of sphere $= \frac{4}{3} \cdot \pi r^3 = 10.3\,\text{cm}^3$ correct to 3 significant figures because of ongoing calculation.
- Each $1\,\text{cm}^3$ weighs $7.9\,\text{g}$.
- Mass of ball bearing $= 7.9 \times 10.3\,\text{g} = 81\,\text{g}$ correct to 2 significant figures.

Self-assessment tasks 8.6

1. In Q1, of Self-assessment tasks 8.5, each of the objects is made of duralumin, of density $2.88\,\text{g/cm}^3$. Calculate the mass of each object.
2. A steel chuck has a volume of $220\,\text{cm}^3$ and mass of $1.72\,\text{kg}$. Calculate the density in g/cm^3 and in kg/m^3.
3. If a spanner, made of steel, has a mass of $860\,\text{g}$, find its volume. (Density of steel $= 7,800\,\text{kg/m}^3$.)

Algebraic numbers

In algebra letters are used to represent general numbers. They are general numbers in that they can take variable values. For example, as in

$$6 + 6 + 6 + 6 + 6 = 5 \times 6$$

so also

$$z + z + z + z + z = 5 \times z$$

Generally the multiplication sign is omitted and written as $5z$. The algebraic number $5z$ is called a **term**; the number in front of the letter is called the **coefficient**. A number of terms can be added and subtracted provided the algebraic part (the letters) is the same. Thus, we have

$$5m + 7m = 12m$$

that is, the coefficients are added. Two numbers $5m$ and $7c$, cannot be added. Similarly, $8a - 5a = 3a$, while $8x - 5y$ cannot be subtracted.

Rule 13

- Addition and subtraction of algebraic numbers is carried out on the coefficients, provided the letters are the same.

Note: The rules for positive and negative numbers given previously apply to these coefficients as well.

Example 8.8

In the following we add and subtract the left-hand side to give

(a) $14x - 9x = 5x$
(b) $7z - 13z = -6z$
(c) $-4p - 2p - 3p = -9p$
(d) $5s - 6t + 3s - 5t$

First of all the terms in s and the terms in t are collected together, i.e.

$$5s + 3s - 6t - 5t = 8s - 11t$$

This last expression cannot be simplified further.

Self-assessment tasks 8.7

1. Reduce the following to the simplest expression:
   $7x + 5x$; $2m - 7m$; $-8t - 7t$; $9u - 5v - 11u + 7v$
2. A bar is $12x$ long. Pieces are cut off on the guillotine of lengths $2x$, $5x$, $3x$. What is the length of the remaining off-cut?
3. An ammeter has divisions on it. It is used to measure the current through each of three resistors connected in parallel. It records $4d$, $3d$, d. If the total current is $16\,\text{A}$ what value of current does d represent?

Multiplication and indices

Multiplication

When two algebraic numbers are multiplied together the coefficients are multiplied, using the rule for multiplication of signs, with the letters arranged in the order of the alphabet. Thus

$-6b \times 5a = -30ab$ (the '$\times$' is understood to be included).

> **Example 8.9**
>
> $$6 \times 5p = 30p$$
>
> $$-8m \times 5a = -40am$$
>
> Brackets are used to denote multiplication of more complex expressions. For example, $4(2x + 7)$ means that $2x$ is multiplied by 4 and the 7 multiplied by 4, that is
>
> $$4(2x + 7) = 8x + 28$$
>
> This is further illustrated (using the rules for negative numbers) by
>
> $$-2(4x - 3y) = -8x + 6y$$

Indices

In algebra numbers are generally expressed in index form. Multiplication and division are carried out according to the rules established in the section on 'Indices' (p. 000).

Note the difference between the following:

$$a^4 = a \times a \times a \times a \quad \text{and} \quad 4a = a + a + a + a$$

Rule 4: $a^n \times a^m = a^{n+m}$

Examples: $y^2 \times y^7 = y^9$

$$3z^2 \times -5z^3 = -15x^5$$

Rule 5: $a^n \div a^m = a^{n-m}$

Examples: $t^7 \div t^2 = t^5$

$$18z^5 \div -2z = -9z^4$$

Rule 6: $a^0 = 1$

Rule 7: $a^{-n} = \dfrac{1}{a^n}$

Examples: $s^{-6} = \dfrac{1}{s^6}$

$$4q^{-7} = \dfrac{4}{q^7}$$

$$\left(\tfrac{3}{4}\right)^{-6} = \left(\tfrac{4}{3}\right)^6$$

Rule 8: $(a^n)^m = a^{n \times m}$

Examples: $(p^3)^4 = p^{12}$

$$(3t^2)^4 = 3^4 t^8 = 81 t^8$$

$$-4(t^2)^3 = -4t^6$$

Rule 9: $a^{\frac{1}{m}} = \sqrt[m]{a}$

Examples: $(4x^6)^{\frac{1}{2}} = 2x^3$

$$\left(\frac{81}{256}\right)^{\frac{1}{2}} = \frac{9}{16}$$

$$\left(\frac{125}{64}\right)^{-\frac{1}{3}} = \left(\frac{64}{125}\right)^{\frac{1}{3}} = \frac{4}{5}$$

> **Self-assessment tasks 8.8**
>
> Simplify the following, leaving the result in index form:
>
> 1. $-3t^2 \times 6t^4$, $\;-5y^2 \times -4y^2$, $\;-3m \times -4m^2 \times -5m^3$
> 2. $-27f^4 \div 3f$, $\;18q^4 \div -2q^4$
> 3. $(7w^3)^2$, $\;7(w^3)^2$, $\;(5t^0)^7$
> 4. The radius of a ball bearing is $2y$. Express the formula for the volume in terms of y.
> 5. An axle is $4t$ long. Its radius is to be machined to be $\tfrac{1}{20}$ of its length. Find the volume of the finished axle.

Logarithms

Although the electronic calculator has replaced logarithm tables for carrying out calculations, nevertheless the basic theory of logarithms is still important in engineering.

The meaning of a logarithm

Consider the number $64 = 2^6$. The index 6 is said to be the logarithm of 64 on the base of 2. This definition is written as

$$6 = \log_2 64$$

Further examples:

$$1{,}000 = 10^3 \;:\; 3 = \log_{10} 1{,}000$$
$$25 = 5^2 \;:\; 2 = \log_5 25$$
$$N = a^t \;:\; t = \log_a N$$

Two important logarithms

- Logarithms on the base of 10, that is $\log_{10} N$ but written as $\log N$. Its value for any number, say 410, may be obtained on the electronic calculator as follows:

  enter 410 *press* $\boxed{\log}$ key *displays* 2.61

  that is $\log 410 = 2.61$ correct to 3 significant figures

- Logarithms on the base of e, that is $\log_e N$ but written as $\ln N$. The number 'e' is called the *exponential* and is an important number in engineering. Its value is the sum of the series

$$e = 1 + 1 + \frac{1}{1 \times 2} + \frac{1}{1 \times 2 \times 3} + \frac{1}{1 \times 2 \times 3 \times 4} + \ldots$$

$$= 2.718 \text{ correct to 4 significant figures}$$

The ln value for any number, say 39.6, may be obtained on the electronic calculator as follows:

enter 39.6 *press* $\boxed{\ln}$ key *displays* 3.68

that is $\ln 39.6 = 3.68$ correct to 3 significant figures

Changing a logarithm back to the initial equation is as follows:

$$\ln y = x$$
$$y = e^x$$

It is found that

$$e^x = 1 + x + \frac{x^2}{1 \cdot 2} + \frac{x^3}{1 \cdot 2 \cdot 3} + \frac{x^4}{1 \cdot 2 \cdot 3 \cdot 4} + \ldots$$

The laws of logarithms

Law 1: $\log M \times N = \log M + \log N$

e.g. $\log 5 \times 10 = \log 5 + \log 10$

Law 2: $\log \dfrac{M}{N} = \log M - \log N$

e.g. $\log \dfrac{12}{2} = \log 12 - \log 2$

Law 3: $\log M^n = n \cdot \log M$

e.g. $\log 8^3 = 3\log 8$

These three laws are true for any base.

Example 8.10

Simplify the following logarithmic expressions:

(a) $\log 20 + \log 30 - \log 25$
(b) $3 \log 2a - 2 \log 4a$
(c) $\log 3x^3 + \log 4x - \log 2x^2$

(a) $\log 20 + \log 30 - \log 25 = \log 20 \times 30/25 = \log 24$
(b) Laws 1 and 2 cannot be used when the log terms have coefficients other than 1. First of all we use law 3 to take the coefficient into the number, as follows:

$$3 \log 2a - 2 \log 4a = \log (2a)^3 - \log (4a)^2$$
$$= \log \frac{(2a)^3}{(4a)^2}$$
$$= \log \frac{2a \cdot 2a \cdot 2a}{4a \cdot 4a} = \log \frac{a}{2}$$

(c) $\log 3x^3 + \log 4x - \log 2x^2 = \log \dfrac{3x^3 \cdot 4x}{2x^2} = \log 6x^2$

Example 8.11

Simplify the following:

(a) $\dfrac{\log 64}{\log 32}$ (b) $\dfrac{\log 64 + \log 4 - \log 8}{\log 1{,}024}$

(a) The laws of logarithms do not apply to dividing one log term with another. First of all we write 64 and 32 as indexed numbers. Therefore we have

$$\frac{\log 64}{\log 32} = \frac{\log 2^6}{\log 2^5} = \frac{6 \log 2}{5 \log 2} = \frac{6}{5}$$

(b) We now have:

$$\frac{\log 64 + \log 4 - \log 8}{\log 1{,}024} = \frac{\log 2^6 + \log 2^2 - \log 2^3}{\log 2^{10}}$$

$$= \frac{6 \log 2 + 2 \log 2 - 3 \log 2}{10 \log 2} = \frac{5 \log 2}{10 \log 2} = \frac{1}{2}$$

These examples involve logarithms on the base of 10, but the same working out would apply to any base, provided they were the same throughout the example.

Properties of logarithms

(a) $\log_a 1$. Let $\log_a 1 = s$. Therefore $a^s = 1$. From the rules of indices this can only be true if $s = 0$:

$$\log_a 1 = 0$$

(b) $\log_a a$. Let $\log_a a = t$. Therefore $a^t = a$. From the rules of indices this can only be true if $t = 1$:

$$\log_a a = 1$$

Logarithms and exponentials in engineering

Throughout engineering logarithms and exponentials are used to solve problems.

Current through an inductance

When the switch in Fig. 8.6 is opened the electric current does not cease immediately but decays over a short period caused by the inductive effect. If the original current flowing is I_0 and the current i at any later time t after the switch has been opened, then

$$i = I_0 \, e^{-kt} \quad \text{where } k \text{ is the time constant}$$

Convert this to the logarithm

$$\frac{i}{I_0} = e^{-kt} \quad \text{so that} \quad -kt = \ln \frac{i}{I_0}$$

from which k or t may be calculated.

A similar result is obtained for the discharge of a capacitor across a resistor.

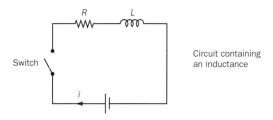

Circuit containing an inductance

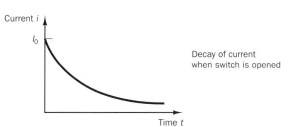

Decay of current when switch is opened

Figure 8.6

Newton's law of cooling

If an object is at a temperature θ above its surroundings after cooling from a temperature θ_0 above its surroundings, in a time t, then the law states that

$$\theta = \theta_0 e^{-kt}$$

This reduces to the logarithmic form to give

$$-kt = \ln \frac{\theta}{\theta_0}$$

from which k or t can be determined.

Indicial equations

Logarithms are used to solve equations of the type $x^4 = 20$ and $6.2^x = 39$.

- $x^4 = 20$: Take logarithms on both sides of the equation:
  $\log x^4 = \log 20$
  $4 \log x = 1.301$
  $\log x = 0.325$
  Using the calculator:

  $\quad$ enter 0.325 $\quad$ press $\boxed{\text{INV} \mid \log}$ $\quad$ keys $\quad$ displays 2.11

  $x = 2.11$ to 3 significant figures

- $6.2^x = 39$: Take logarithms on both sides of the equation:
  $\log 6.2^x = \log 39$
  $x \log 6.2 = 1.591$
  $x \cdot 0.792 = 1.591$
  $\quad x = 1.591/0.792 = 2.0$ to 2 significant figures

Self-assessment tasks 8.9

1. (a) Find, using a calculator, (i) $\log 372.1$, (ii) $\ln 42.55$, (iii) $\log 7.2 - \ln 0.63$, (iv) $\log 36/\ln 36$.
   (b) (i) If $\log x = 5.6$ find x, (ii) If $\ln y = 4.32$ find y.
2. A metal rod, after grinding, cools from $300\,°C$ above room temperature to $100\,°C$ above room temperature in $90\,s$. Find k if
   $$\theta = \theta_0 e^{-kt}$$
3. Simplify, without using the calculator,
   (a) $\log 81 + \log 3 - \log 27$
   (b) $\log 625 - 5 \log 5 + \log 25$
   (c) $\dfrac{\frac{1}{2} \log 81 + \frac{1}{3} \log 27}{\log 9}$
   (d) $\log 3a^3 b^2 - \log 3a^2 b^3$
4. Solve (a) $5^x = 37$, (b) $x^{2.3} = 16$, (c) $6^{2x} = 74$

Algebraic operations 1: addition and subtraction of fractions

For simple fractions the addition and subtraction is carried out as in arithmetic by finding the LCM.

Example 8.12

Simplify the following expression

$$\frac{3t}{4} - \frac{2t}{3} + \frac{5t}{8}$$

The LCM of 4, 3 and 8 is 24, therefore the expression becomes

$$\frac{6 \times 3t - 8 \times 2t + 3 \times 5t}{24} = \frac{18t - 16t + 15t}{24} = \frac{17t}{24}$$

For more complex fractions with letters in the denominator, the same process is used. In the case of

$$\frac{3}{x} + \frac{4}{y}$$

the LCM is the product xy, so that the expression becomes

$$\frac{3y + 4x}{xy}$$

A variation of this is an expression of the type

$$\frac{4}{x} + \frac{2}{x^2}$$

The LCM in this case is always the highest power of x, that is, x^2, giving

$$\frac{4x + 2}{x^2}$$

Example 8.13

Simplify the following:

(a) $\dfrac{y}{5} + \dfrac{3y}{10} - \dfrac{3y}{20}$ $\quad$ (b) $\dfrac{5}{t} + \dfrac{3}{2t^2} - \dfrac{4}{3t^3}$

(c) $\dfrac{2t-3}{4} - \dfrac{3t-5}{3}$

(a) The LCM of 5, 10 and 20 is 20 so that the expression is

$$\frac{4y + 6y - 3y}{20} = \frac{7y}{20}$$

(b) The LCM of t, $2t^2$, $3t^3$ is $6t^3$ to give

$$\frac{30t^2 + 9t - 8}{6t^3}$$

(c) The LCM is 12:

$$\frac{2t-3}{4} - \frac{3t-5}{3} = \frac{3(2t-3) - 4(3t-5)}{12}$$
$$= \frac{6t - 9 - 12t + 20}{12}$$
$$= \frac{-6t + 11}{12}$$

Example 8.14

In an electric circuit two resistors R_1, R_2 are connected in parallel. The equivalent resistance is

$$\frac{1}{R} = \frac{1}{R_1} + \frac{1}{R_2}$$

Rearrange and find the formula for R.

Adding two fractions the LCM is $R_1 R_2$ so that

$$\frac{1}{R} = \frac{R_2 + R_1}{R_1 \cdot R_2}$$

Turning the fractions over

$$R = \frac{R_1 \cdot R_2}{R_1 + R_2}$$

Self-assessment tasks 8.10

1. Simplify the following:
   (a) $\dfrac{2a}{5} + \dfrac{3a}{4}$ (b) $\dfrac{2t}{3} - \dfrac{5t}{7} + \dfrac{9t}{21}$
   (c) $\dfrac{4(2z+5)}{5} - \dfrac{2(4-2z)}{7}$

2. The equivalent resistance of three resistors connected in parallel is
   $$\dfrac{1}{R} = \dfrac{1}{3r} + \dfrac{1}{4r} + \dfrac{1}{6r}$$
   Find the formula for R and determine its value if $r = 4\,\Omega$.

Algebraic operations 2: simplification of fractions

In simplifying fractions the laws of indices are used, as shown in the following example.

Example 8.15

Simplify the following:

(a) $\dfrac{21x^6 y^5}{14x^7 y^4}$ (b) $\dfrac{-15u^2(3w-5)^3}{-25u^{0.5}(3w-5)^2}$

(a) As in arithmetic, 7 will divide into both coefficients. The algebraic terms are simplified using the rules of indices:

$$\dfrac{21x^6 y^5}{14x^7 y^4} = \dfrac{3}{2} x^{6-7} y^{5-4} = \dfrac{3}{2} x^{-1} y = \dfrac{3y}{2x}$$

(b) Both coefficients are divisible by -5. The algebraic terms are simplified using the rules of indices, which also applies to the composite term $(3w - 5)$.

$$\dfrac{-15u^2(3w-5)^3}{-25u^{0.5}(3w-5)^2} = \dfrac{3}{5} u^{2-0.5}(3w-5)^{3-2}$$

$$= \dfrac{3}{5} u^{1.5}(3w-5)$$

Self-assessment tasks 8.11

Simplify the following:

1. $\dfrac{35y^4}{-15y^2}$, $\dfrac{27s^{1.5}}{-15s}$, $\dfrac{22t^{2.5}}{12t^{-0.5}}$

2. $\dfrac{5y^4(7t-3)^5}{15y^3(7t-3)^2}$

3. $\dfrac{6ab}{5cd^2} \times \dfrac{15c^2 d}{2a^2 b^2}$

Expansion of brackets

We have already seen that a term such as $6(2x - 5)$ means that both numbers inside the bracket are multiplied by 6. We now extend this to the multiplication of two bracketed terms, such as, for example $(2x - 5)(3x + 1)$. This means that both terms in the second bracket are multiplied by both terms in the first bracket. It is expanded as follows:

$$(2x - 5)(3x + 1) = 2x(3x + 1) - 5(3x + 1)$$
$$= 6x^2 + 2x - 15x - 5$$

remembering the rules for multiplying by a negative number

$$= 6x^2 - 13x - 5$$

Example 8.16

Multiply out the following brackets and simplify:

(a) $4(3a - 2b + 5c) - 2(a - 5b + 6c)$
(b) $(3c + 5)(3c - 1)(7c - 2)$

(a) We multiply the first bracket by 4 and the second by -2:

$12a - 8b + 20c - 2a + 10b - 12c = 10a + 2b + 8c$

(b) The multiplication is carried out first of all with the second and third pair of brackets:

$(3c + 5)(3c - 1)(7c - 2)$
$= (3c + 5)[3c(7c - 2) - 1(7c - 2)]$
$= (3c + 5)[21c^2 - 6c - 7c + 2]$
$= (3c + 5)(21c^2 - 13c + 2)$
$= 3c(21c^2 - 13c + 2) + 5(21c^2 - 13c + 2)$
$= 63c^3 - 39c^2 + 6c + 105c^2 - 65c + 10$
$= 63c^3 + 66c^2 - 59c + 10$

Self-assessment tasks 8.12

Multiply the following brackets:

1. $(2t - 1)(4t + 7)$, $(7a + 2)(a - 3)$, $(s - 2)(s + 4)$, $(r - 3)(r + 3)$.

2. $(u + 2)(u + 3)(u + 4)$, $(v - 2)(v + 2)(2v - 1)$

Factorisation

Linear expressions

In a linear expression all the algebraic terms have powers of 1.

On page 247 (Multiplication and indices) we showed that an expression such as $3(2x + 5)$ multiplies out as $3 \times 2x + 3 \times 5$. Factorisation is the reverse of this. The common number 3 is divided into both numbers and placed as a common multiplier outside a bracket. Consider the expression $8px + 8pz$. Common to both terms is $8p$, which is called a common factor. We can divide both terms by $8p$ to give $x + z$, and place it outside the bracket as a multiplier, to keep the overall expression the same, that is

$$8px + 8pz = 8p(x + z)$$

> **Example 8.17**
>
> Factorise the following expressions:
>
> (a) $4x + 6y + 10z$ (b) $px + qx + rx$
> (c) $x(t - 2) + 2y(t - 2)$
>
> (a) The common factor to the three terms is 2. We divide each term by 2 and place it outside as a multiplier:
>
> $$4x + 6y + 10z = 2(2x + 3y + 5z)$$
>
> (b) The common factor is x so that it factorises to $x(p + q + r)$.
> (c) The common factor is $(t - 2)$, and the procedure is exactly the same, giving
>
> $$(t - 2)(x + 2y)$$

> **Example 8.18**
>
> Factorise $ab - ac + eb - ec$.
> In this example the common factors occur in pairs:
>
> $$(ab - ac) + (eb - ec) = a(b - c) + e(b - c)$$
> $$= (b - c)(a + e)$$

> **Self-assessment tasks 8.13**
>
> Factorise the following:
>
> 1. $6a + 9b - 12c$, $4ap - 20aq - 10ar$
> 2. $ut - vt - us + vs$, $5ay + 25az - 5by - 25bz$

Quadratic expressions

A quadratic expression contains a term with an index of 2.

Difference of two squares

Multiplying out the brackets $(x - y)(x + y)$ we obtain:

$$(x - y)(x + y) = x(x + y) - y(x + y)$$
$$= x^2 + xy - xy - y^2 = x^2 - y^2$$

that is, it is the difference of two squares. Reversing this process gives a pair of factors:

$$x^2 - y^2 = (x - y)(x + y)$$

> **Example 8.19**
>
> Factorise the following using the difference of two squares:
>
> (a) $4t^2 - 25s^2$ (b) $100z^2 - 81$
>
> (a) Both terms are squares, $(2t)^2$ and $(5s)^2$, so that
>
> $$4t^2 - 25s^2 = (2t - 5s)(2t + 5s)$$
>
> (b) Similarly,
>
> $$100z^2 - 81 = (10z - 9)(10z + 9)$$

> **Self-assessment tasks 8.14**
>
> 1. Factorise: $p^2 - q^2$, $m^2 - 16n^2$, $(300t^2 - 12)$.
> 2. An annulus has internal and external radii r and R. The area is $\pi R^2 - \pi r^2$. Factorise the expression.

Trinomial expressions

> **Example 8.20**
>
> Factorise the trinomial: $6x^2 + 7x + 2$.
> In order to avoid the trial and error method the following steps make the process easy.
>
> - Multiply the coefficient of x^2 and the constant, that is, $6 \times 2 = 12$.
> - Write out the factor pairs of 12: (1,12) (2,6) (3,4)
> - **Since the constant term is positive find the factor pair which adds to make 7 the coefficient of x. It is (3,4).**
> - Write $7x$ as $3x + 4x$.
> - The trinomial becomes
>
> $$6x^2 + 3x + 4x + 2$$
>
> which we can now factorise in pairs:
>
> $$3x(2x + 1) + 2(2x + 1) = (2x + 1)(3x + 2)$$

> **Example 8.21**
>
> Factorise: $4x^2 - x - 5$.
>
> - Multiply the coefficient 4 by the constant term 5: 20
> - Write out the factor pairs of 20: (1,20) (2,10) (4,5).
> - **Since the constant term is negative find the factor pair which subtracts to give -1 the coefficient of x. It is (4,5).**
> - Write $-x$ as $4x - 5x$.
> - The trinomial becomes
>
> $$4x^2 + 4x - 5x - 5$$
> $$4x(x + 1) - 5(x + 1) = (x + 1)(4x - 5)$$

> **Self-assessment tasks 8.15**
>
> Factorise the trinomials:
>
> 1. $x^2 - 2x - 24$ 2. $x^2 - 5x + 6$
> 3. $6x^2 + x - 15$ 4. $3a^2 - 7a + 2$
> 5. $4t^2 - 16t + 15$ 6. $8m^2 + 6m - 9$

Transposition of formulae

In engineering use is made of formulae and algebraic equations, such as the equations of motion $v = u + at$, or $V = IR$ for Ohm's law. These formulae may have to be rearranged to obtain a formula for one of the other quantities, say t, thus making t the new subject instead of v in the above example. This operation of changing the formula around is called *transposition of formulae*.

The process is to ensure that any changes made on one side of the equation are repeated on the other side in order to keep both sides equal.

Linear formulae

Consider the formulae $v = u + at$ and make t the subject.

- Step 1 – Subtract u from both sides:

$$v - u = at$$

- Step 2 – Divide both sides by a:

$$\frac{v - u}{a} = \frac{at}{a} = t$$

$$t = \frac{v - u}{a}$$

Example 8.22

A pulley is in equilibrium under the action of a number of forces, the tensions T_1 and T_2 in the ropes acting upwards, two loads, 50 and M, acting downwards, and a lifting force P pushing upwards on the pulley. Find the formula for the load M.

The pulley is in equilibrium so that

$$T_1 + T_2 = 50 + M - P$$

Subtract 50 from both sides of the equation:

$$T_1 + T_2 - 50 = 50 + M - P - 50$$
$$T_1 + T_2 - 50 = M - P$$

Add P to both sides of the equation:

$$T_1 + T_2 - 50 + P = M - P + P$$
$$M = T_1 + T_2 - 50 + P$$

Formulae where the variables are raised to a power

Consider the formula for the volume of a cylinder, $V = \pi r^2 h$. The expression for r is found as follows

- Step 1 – Divide both sides by πh:

$$\frac{V}{\pi h} = r^2$$

- Step 2 – Take square roots on both sides:

$$r = \sqrt{\frac{V}{\pi h}}$$

Example 8.23

The equation of a circle is $x^2 + y^2 = 100$. Find the equation for y.

Using the same procedure as above, subtract x^2 from both sides:

$$x^2 + y^2 - x^2 = 100 - x^2$$
$$y^2 = 100 - x^2$$

Take square roots on both sides of the equation, remembering that square root has an index of $\frac{1}{2}$

$$y = (100 - x^2)^{\frac{1}{2}} = \sqrt{(100 - x^2)}$$

Formulae involving fractions

Consider the formula for the equivalent resistance of two resistors in parallel:

$$\frac{1}{R} = \frac{1}{10} + \frac{1}{r}$$

In order to find the formula for r, the first step, as before, is to subtract 1/10 from both sides of the equation:

$$\frac{1}{R} - \frac{1}{10} = \frac{1}{10} + \frac{1}{r} - \frac{1}{10} = \frac{1}{r}$$

Take the LCM of the left-hand side:

$$\frac{10 - R}{10R} = \frac{1}{r}$$

Invert the fraction on both sides of the equation:

$$\frac{10R}{10 - R} = r$$

Note: **This procedure of inverting fractions can only be done if there is only one fraction on either side.**

Self-assessment tasks 8.16

1. The length of a rod, L, after heating through $t\,°C$ is given by $L = l(1 + \alpha t)$, where l and α are constants. Find the equation for α.
2. The area of a flat ring is given by $40 = \pi R^2 - \pi r^2$. Find the formula for r.
3. The equivalent resistance is given by $1/R = 2 + (1/r)$. Find the equation for r.
4. The period of a simple pendulum is given by $T = 2\pi\sqrt{(l/g)}$. Find the formula for g.

Simple equations

In the beginning of this unit it was stated that letters are used to denote general numbers, in that they can take a variable number of values. The relationship between two or more of these general numbers is called an *equation*. For example, $y = 3x + 2$ states that the value of y will always be dependent upon the value of x. If $x = 5$, for instance, then $y = 3 \times 5 + 2 = 17$. In such an equation x is called the **independent** variable and y is called the **dependent** variable.

A simple equation such as $4x + 5 = 17$ is one which contains one variable only, and is linear; that is, the variable has an index of 1. Finding the value of the variable x is called finding the solution. In the case of a simple equation the variable can only have one value.

The first step is to get the term containing x on its own on the left-hand side of the equation.

- Step 1 – Subtract 5 from both sides:

$$4x + 5 - 5 = 17 - 5$$
$$4x = 12$$

- Step 2 – Divide by the coefficient of x:

$$4x/4 = 12/4$$
$$x = 3$$

As a check if we substitute this back into the original equation we have $12 + 5 = 17$; that is, it balances.

Example 8.24

Solve the equation: $7y - 14 = 2y - 6 - 3y + 16$.

- Step 1 – Simplify both sides of the equation:

$$7y - 14 = -y + 10$$

- Step 2 – Add y to both sides of the equation:

$$7y - 14 + y = -y + 10 + y$$
$$8y - 14 = 10$$

- Step 3 – Add 14 to both sides of the equation:

$$8y - 14 + 14 = 10 + 14$$
$$8y = 24$$

- Step 4 – Divide both sides by 8:

$$8y/8 = 24/8$$
$$y = 3$$

Slightly more complicated equations involve brackets or fractions, but the operations discussed earlier will reduce the equations to the form in Example 8.25.

Example 8.25

Solve the equation: $3(x-4) + 4x = 2(2x+5) - 1$.

We remove the brackets first and then proceed as in Example 8.24.

$$3x - 12 + 4x = 4x + 10 - 1$$
$$7x - 12 = 4x + 9$$

Subtract $4x$ from both sides:

$$7x - 12 - 4x = 4x + 9 - 4x$$
$$3x - 12 = 9$$

Add 12 to both sides:

$$3x - 12 + 12 = 9 + 12$$
$$3x = 21$$

Divide both sides by 3:

$$x = 21/3 = 7.$$

Example 8.26

Solve the equation: $\dfrac{3x-4}{3} - \dfrac{3x}{4} = \dfrac{1}{6}$

- Step 1 – The LCM of 3, 4, 6 is 12.
- Step 2 – Multiply each term by the LCM:

$$\frac{12(3x-4)}{3} - \frac{12 \times 3x}{4} = 12 \times \frac{1}{6}$$

- Step 3 – Cancel out the denominator in each term:

$$4(3x-4) - 3 \times 3x = 2$$
$$12x - 16 - 9x = 2$$
$$3x - 16 = 2$$
$$3x - 16 + 16 = 2 + 16$$
$$3x = 18$$
$$x = 6$$

Example 8.27

A rectangular plate has to be cut so that its length is 3 m longer than its width. Its perimeter is 22 m. Find its length and breadth.

Let the width of the plate be x (m); then its length is $(x+3)$. Its perimeter is

$$2x + 2(x+3) = 22$$

This is a simple equation. Multiply out the bracket:

$$2x + 2x + 6 = 22$$
$$4x + 6 - 6 = 22 - 6$$
$$4x = 16$$
$$x = 4$$

Therefore the width is 4 m and the length is 7 m.

Self-assessment tasks 8.17

Solve the following equations:

1. $3y - 7 = 17$
2. $20 - 3s + 3 = 2s - 12$
3. $-3x + 21 + 6x + 14 = -2x - 5$
4. $7(4t - 5) - 5t = 5(3t + 1)$
5. $3(4z - 7) + 40 = 10z$
6. $\dfrac{2x-1}{4} + \dfrac{4x+5}{5} = \dfrac{x}{20} + 1$
7. A wire, 32 cm long is to be bent into a rectangular frame, whose length and breadth are in the ratio 5 : 3. Find the area of the frame.
8. An electric current of 16 A divides into three branches of resistors in parallel. The current in the first branch is 4 times that in the second branch; the current in the third branch is 6 A. Find the value of the current in each of the first two branches.

Simultaneous equations

In a simple equation with only one variable there is only one possible solution, that is only one value of the variable. For an equation such as $y = -3x + 7$ there is an infinite number of solutions. For every possible value of x there will be a corresponding value of y. The same will be true of any other equation in x and y, such as $x + y = 3$. However, there will only be one pair of values which will be a solution to both equations. This pair of x and y values will be the solution of the pair of equations. The pair of equations is called **simultaneous equations**.

The solutions can be found by eliminating one of the variables. If we consider the two equations already noted above:

$$x + 3y = 7 \quad \text{(i)}$$
$$x + y = 3 \quad \text{(ii)}$$

Subtract the two equations
$$2y = 4$$
$$y = 2$$

Substitute $y = 2$ into either of the original equations.

In equation (ii) $\qquad x + 2 = 3$
$$x = 1$$

The solution of the simultaneous pair is $x = 1, x = 2$. This result can be checked by substituting into equation (i), that is $1 + 3 \times 2 = 7$, which is seen to balance.

Example 8.28

Solve the following pair of simultaneous equations:

$$2x + 3y = 5$$
$$3x - 4y = -1$$

To solve the equations we eliminate either variable. To eliminate x we multiply equation (i) by 3, the coefficient of x in equation (ii), and multiply equation (ii) by 2, the coefficient of x in equation (i):

$$\times 3 \quad 2x + 3y = 5 \quad \text{(i)}$$
$$\times 2 \quad 3x - 4y = -1 \quad \text{(ii)}$$

That is:
$$6x + 9y = 15$$
$$6x - 8y = -2$$

Subtract to eliminate x:
$$17y = 17$$
$$y = 1$$

Substitute into (i):
$$2x + 3 \times 1 = 5$$
$$2x = 2$$
$$x = 1$$

The solution is $x = 1$ and $y = 1$, which may be checked by substituting into equation (ii).

> **Self-assessment tasks 8.18**
>
> Solve the following simultaneous equations:
>
> 1. $3x + 2y = 7$
> $4x - y = 2$
> 2. $x + 2y = -4$
> $2x - 3y = 13$
> 3. $3E + 4I = -8$
> $-2E - 3I = 7$
> 4. 4 lathes of type A and 10 lathes of type B turn out 450 components each week. 2 lathes of type A and 8 of type B turn out 270 components each week. Find the number of components turned out by each lathe of type A and each of type B each week.

Quadratic equations

A quadratic equation contains only one variable, but in one of the terms the variable will have an index of 2. A typical equation is

$$2x^2 - 3x + 4 = 0$$

There are two values of x which are solutions of a quadratic equation.

Solutions of quadratic equations by factorisation

In finding the solution to a quadratic equation by factorisation it is important to note that if two numbers $X \times Y = 0$ then either $X = 0$ or $Y = 0$.

The procedure is to factorise the quadratic expression using the method described in the section on Factorisation (p. 251).

> **Example 8.29**
>
> Solve the equation: $6x^2 - x - 2 = 0$.
>
> - Step 1 – Multiply the coefficient of x^2 by the constant: $6 \times 2 = 12$.
> - Step 2 – Factorise the expression on the left-hand side by using factor pairs of 12, (1,12) (2,6) (3,4). Since the constant term is negative we need the pair which subtracts to give -1. It is (3, 4). Therefore the equation is rewritten as:
>
> $$6x^2 + 3x - 4x - 2 = 0$$
>
> - Step 3 – Factorise in pairs:
>
> $$3x(2x + 1) - 2(2x + 1) = 0$$
> $$(2x + 1)(3x - 2) = 0$$
>
> This has the form
>
> $$X \times Y = 0$$
>
> Using the conclusion above, then
>
> $$2x + 1 = 0 \quad \text{or} \quad 3x - 2 = 0$$
>
> The solution is:
>
> $$x = -\tfrac{1}{2} \quad \text{or} \quad x = \tfrac{2}{3}$$

It is seen that quadratic equations have two solutions. However, when the quadratic is a perfect square, the two solutions will be identical. An example of a perfect square is

$$x^2 - 6x + 9 = 0$$

This factorises to $(x - 3)/(x - 3) = 0$, so that $x = 3$ is the single solution. The perfect square can be identified because half the coefficient of x, -3, when squared gives the constant term, 9, which will always be $+$. Another example is

$$x^2 + 10x + 25 = 0$$

Solution of quadratic equations by formula

Note: Since $3^2 = +9$, and $(-3)^2 = +9$

$$\sqrt{9} = 3 \quad \text{or} \quad -3$$

that is, a square root of a number will have two values, a positive and a negative one.

Not all quadratic expressions will factorise and then it is necessary to resort to a formula. To obtain the formula we use the general form of the equation

$$ax^2 + bx + c = 0$$

First of all we make the left-hand side a perfect square.

- Step 1 – Divide throughout by a:

$$x^2 + \frac{b}{a}x + \frac{c}{a} = 0$$

- Step 2 – Subtract c/a from both sides:

$$x^2 + \frac{b}{a}x = -\frac{c}{a}$$

- Step 3 – Halve b/a, square it, and add to both sides:

$$x^2 + \frac{b}{a}x + \left(\frac{b}{2a}\right)^2 = \left(\frac{b}{2a}\right)^2 - \frac{c}{a}$$

- Step 4 – The left is now a perfect square:

$$\left(x + \frac{b}{2a}\right)^2 = \frac{b^2}{4a^2} - \frac{c}{a}$$

- Step 5 – Take the LCM of the right:

$$\left(x + \frac{b}{2a}\right)^2 = \frac{b^2 - 4ac}{4a^2}$$

- Step 6 – Take the square root of both sides, remembering that the square root will have two values, '+' and '−':

$$x + \frac{b}{2a} = \frac{\pm\sqrt{(b^2 - 4ac)}}{2a}$$

The two solutions are:

$$x = \frac{-b + \sqrt{(b^2 - 4ac)}}{2a} \quad \text{or} \quad x = \frac{-b - \sqrt{(b^2 - 4ac)}}{2a}$$

> **Example 8.30**
>
> Solve the quadratic using the formula $2x^2 + 3x - 4 = 0$.
> Comparing this with the general equation then $a = 2$, $b = 3$, $c = -4$. Substitute these values into the equation for x to obtain
>
> $$x = \frac{-3 + \sqrt{(3^2 - 4 \times 2 \times -4)}}{2 \times 2} \quad \text{or} \quad x = \frac{-3 - \sqrt{(3^2 - 4 \times 2 \times -4)}}{2 \times 2}$$
>
> $$= \frac{-3 + \sqrt{(9 + 32)}}{4} \quad \text{or} \quad = \frac{-3 - \sqrt{(9 + 32)}}{4}$$
>
> $$= \frac{-3 + 6.40}{4} \quad \quad = \frac{-3 - 6.40}{4}$$
>
> $$= 0.85 \quad \quad = -2.35$$
>
> correct to 2 decimal places

Example 8.31

The length of a rectangular plate is 10 cm more than its width. Its area is 264 cm². Find the length and width.

Let the length be x cm; then the width is $x - 10$. Therefore, the area is:

$$x(x - 10) = 264$$
$$x^2 - 10x - 264 = 0$$
$$(x - 22)(x + 12) = 0$$

Therefore $x = 22$ or $x = -12$, which is inadmissible because it is negative. The rectangle is 22 cm long and 12 cm wide.

Self-assessment tasks 8.19

1. Solve the following equations by factorisation
   (a) $x^2 - 2x + 1 = 0$
   (b) $3x^2 + x - 10 = 0$
   (c) $2x^2 - 11x + 5 = 0$
   (d) $4x^2 + 12x - 7 = 0$
   (e) $3x^2 - 7x + 2 = 0$

2. Solve the following equations using the formula, correct to 2 decimal places:
   (a) $x^2 - x - 1 = 0$
   (b) $3x^2 - 7x + 3 = 0$
   (c) $5x^2 + 4x - 3 = 0$
   (d) $4x^2 - 2x - 7 = 0$
   (e) $2x^2 - 7x - 6 = 0$

8.2 Use trigonometry to solve engineering problems

Introduction

This is the branch of mathematics that deals with the relationships between the sides and the angles of a triangle. Provided that a minimum of (a) 3 sides, (b) 2 sides and 1 angle or (c) 1 side and two angles are known about any triangle, the other unknown sides or angles can be calculated using trigonometrical methods.

Trigonometry is based on the trigonometrical ratios of *sine*, *cosine* and *tangent*, which are the ratios between the sides of a right-angled triangle. The values of these ratios depend upon the size of the angles, and do not depend on the size of the triangle.

Trigonometric ratios

In Fig. 8.7 sides are given names with respect to the angle θ. The side opposite the right angle (90°) is called the *hypotenuse*. The other two sides are named according to their position relative to the angle. The ratios are defined as:

$$\text{sine of } \theta = \frac{\text{opposite}}{\text{hypotenuse}} \quad \text{or} \quad \sin \theta = o/h$$

$$\text{cosine of } \theta = \frac{\text{adjacent}}{\text{hypotenuse}} \quad \text{or} \quad \cos \theta = a/h \qquad (8)$$

$$\text{tangent of } \theta = \frac{\text{opposite}}{\text{adjacent}} \quad \text{or} \quad \tan \theta = o/a$$

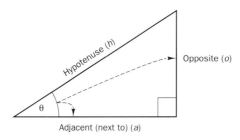

Figure 8.7

For a known angle the values of these ratios can be obtained from an electronic calculator. For example, $\sin 40° = 0.6428$ can be obtained by entering 40 and pressing the $\boxed{\sin}$ key.

Example 8.32

Find the lengths of the unknown sides in the steel bracket *ABC* shown in Fig. 8.8

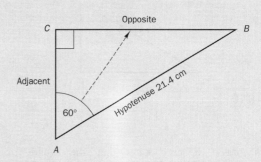

Figure 8.8

Using $\sin \theta = o/h$ with $\theta = 60°$ and $h = 21.4$ cm:

$$\sin 60° = o/21.4$$

Multiply both sides by 21.4:

$$o = 21.4 \times \sin 60 = 21.4 \times 0.866$$
$$BC = 18.5 \text{ cm}$$

Using $\cos \theta = a/h$ again with $\theta = 60°$ and $h = 21.4$ cm:

$$\cos \theta = a/21.4$$
$$a = 21.4 \times \cos 60 = 21.4 \times 0.5$$
$$AC = 10.7 \text{ cm}$$

Example 8.33

Figure 8.9 shows a voltage diagram for an electronic circuit. Calculate the phase angle ϕ between the two voltages.

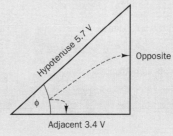

Figure 8.9

Using $\cos \phi = a/h$ with $a = 3.4$ V and $h = 5.7$ V:

$$\phi = 3.4/5.7 = 0.5965$$
$$= \cos^{-1} 0.5965 = 53.4°$$

Note: $\phi = \cos^{-1} 0.5965$ means that ϕ is an angle which has a cos of 0.5965. The value of ϕ can be obtained from the calculator by entering 0.5965 and pressing the $\boxed{\cos^{-1}}$ key.

Theorem of Pythagoras

This is an important theorem and is a useful alternative method of finding the third side of a right-angled triangle when the other two sides are known. In the right-angled triangle in Fig. 8.10 the sides are labelled a, b, c according to the angles that they are opposite.

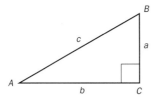

Figure 8.10

Pythagoras's theorem states that the square of the hypotenuse is equal to the sum of the squares of the other two sides, that is

$$c^2 = a^2 + b^2 \qquad (9)$$

Example 8.34

Find the unknown force F in the force diagram shown in Fig. 8.11.

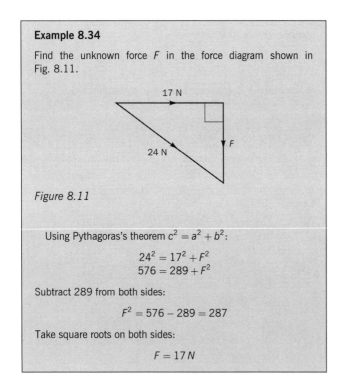

Figure 8.11

Using Pythagoras's theorem $c^2 = a^2 + b^2$:

$$24^2 = 17^2 + F^2$$
$$576 = 289 + F^2$$

Subtract 289 from both sides:

$$F^2 = 576 - 289 = 287$$

Take square roots on both sides:

$$F = 17\,N$$

Trigonometric ratios of angles greater than 90°

Since angles greater than 90° do not exist in a right-angled triangle a method must be devised for finding the trigonometric ratios for the angles greater than 90°. In Fig. 8.12(a)–(d) angles θ are generated by a line OP rotating in an anticlockwise direction, θ being the angle between the positive x-axis OX and the line OP. Since OP is a rotating line it is always positive. **The trigonometric ratios for θ are always**

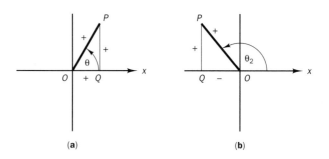

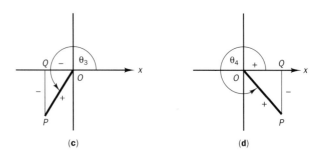

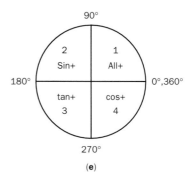

Figure 8.12

taken to be numerically equal to those for the angle POQ in the right-angled triangle OPQ. These ratios may be positive or negative depending upon the signs of the sides PQ and OQ in each quadrant, as explained below.

Angles between 0 and 90°

$$\theta = \text{angle } POQ \quad (\text{written } \angle POQ)$$

As shown in Fig. 8.12(a) the ratios are all positive, that is,

$$\sin \theta = PQ/OP = \frac{+}{+} = +\text{ value}$$

$$\cos \theta = OQ/OP = \frac{+}{+} = +\text{ value}$$

$$\tan \theta = PQ/OQ = \frac{+}{+} = +\text{ value}$$

Angles between 90 and 180°

$$\theta_2 = 180° - \angle POQ$$

The ratios are defined as above but the side OQ (Fig. 8.12(b)) is now negative:

$$\sin\theta_2 = PQ/OP = \frac{+}{+} = +\text{ value}$$

$$\cos\theta_2 = OQ/OP = \frac{-}{+} = -\text{ value}$$

$$\tan\theta_2 = PQ/OQ = \frac{+}{-} = -\text{ value}$$

Angles between 180 and 270°

$$\theta_3 = 180° + \angle POQ$$

The ratios are defined as above but the sides OQ and PQ (Fig. 8.12(c)) are both negative in the ratios, so that:

$$\sin\theta_3 = PQ/OP = \frac{-}{+} = -\text{ value}$$

$$\cos\theta_3 = OQ/OP = \frac{-}{+} = -\text{ value}$$

$$\tan\theta_3 = PQ/OQ = \frac{-}{-} = +\text{ value}$$

Angles between 270 and 360°

$$\theta_4 = 180° + \angle POQ$$

The ratios are defined as above but the side PQ (Fig. 8.12(d)) is now negative in the ratios, so that:

$$\sin\theta_4 = PQ/OP = \frac{-}{+} = -\text{ value}$$

$$\cos\theta_4 = OQ/OP = \frac{+}{+} = +\text{ value}$$

$$\tan\theta_4 = PQ/OQ = \frac{-}{+} = -\text{ value}$$

Figure 8.12(e) identifies which trigonometric ratios are positive in each quadrant. Example 8.35 illustrates the theory for finding the trigonometric ratio of an angle greater than 90°.

Example 8.35

Determine the value of sin 225°.
In Fig. 8.13 the angle 225° is in quadrant 3, where the sine is negative (see Fig. 8.12(e)).

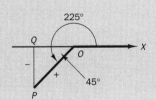

Figure 8.13

Since $\theta_3 = 180 + \angle POQ$:

$$225 = 180 + \angle POQ$$
$$\angle POQ = 45°,$$

so that

$$\sin 225 = -\sin 45 = -0.7071.$$

However, the values of the trigonometric ratios can be determined directly using the electronic calculator. In the above example

enter 225° *press* $\boxed{\sin}$ *key to display* $= -0.7071$

Note: If OP rotates in a clockwise direction the angle moved through is a negative angle. In Fig. 8.14 the rotation of OP in a clockwise direction from OX produces an angle of −50°, which is seen to be the same as a rotation in an anticlockwise direction of 310°.

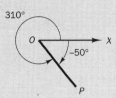

Figure 8.14

To determine the angle from a given trigonometrical ratio

There are always two values of θ in the range 0 to 360° with the same trigonometrical ratio. Using the calculator to determine the angle will only give one of these values. To find both values of θ first of all enter the **positive** value of the trigonometric ratio into the calculator to find $\angle POQ$. The two values of θ are then found using the method given above, as shown by the following examples.

Example 8.36

Find the values of θ in the range 0 to 360° for which $\tan\theta = -1.4$.
Using the calculator enter +1.4 and press the $\boxed{\tan^{-1}}$ key to display $\angle POQ = 54.5°$. Figure 8.12(e) shows that the tangent is negative in the second and fourth quadrants. Therefore the angles θ_2 and θ_4 are required, as shown in Fig. 8.15.

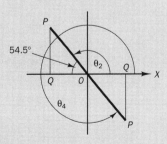

Figure 8.15

$$\begin{aligned}\theta_2 &= 180 - \angle POQ &\text{and}&& \theta_4 &= 360 - \angle POQ \\ &= 180 - 54.5 &\text{and}&& &= 360 - 54.5 \\ &= 125.5° &\text{and}&& &= 305.5°\end{aligned}$$

Example 8.37

Find the values of θ in the range 0 to 360° for which $\sin\theta = 0.722$.

Using the calculator we obtain $\angle POQ = 46.2°$ (the ratio is positive). Figure 8.12(e) shows that sine is positive in quadrants 1 and 2. Therefore, the two angles $\theta_1 \,(= \angle POQ)$ and θ_2 are required, as shown in Fig. 8.16:

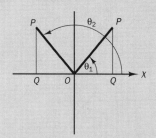

Figure 8.16

$$\angle POQ = 46.2° \quad \text{and} \quad \theta_2 = 180 - \angle POQ$$
$$= 180 - 46.2$$
$$= 133.8°$$

Example 8.38

Find the values of θ in the range 0 to 360° for which $\cos\theta = -0.61$.

From the calculator we find the angle corresponding to $\cos\theta = +0.61$, which is $POQ = 52.4°$. Figure 8.12(e) shows that the cosine is negative in the second and third quadrants. Therefore the angles θ_2 and θ_3 are required, as shown in Fig. 8.17:

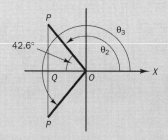

Figure 8.17

$$\theta_2 = 180 - \angle POQ \quad \text{and} \quad \theta_3 = 180 + \angle POQ$$
$$= 180 - 52.4 \quad \text{and} \quad = 180 + 52.4$$
$$= 127.6° \quad \text{and} \quad = 232.4°$$

Self-assessment tasks 8.20

Determine the two values of θ between 0 and 360° in each of the following ratios:

1. $\sin\theta = 0.64$ 2. $\sin\theta = -0.317$ 3. $\cos\theta = 0.3$
4. $\cos\theta = -0.542$ 5. $\tan\theta = 0.21$ 6. $\tan\theta = -2.1$

The sine and cosine rules

The trigonometrical ratios defined above are based on a right-angled triangle. Consequently they cannot be used for triangles without a right angle. In order to calculate sides and angles in such triangles the sine and cosine rules are used. The two rules may be derived by dividing any triangle, such as ABC in Fig. 8.18, into two right-angled triangles.

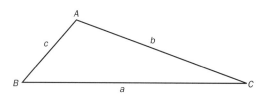

Figure 8.18

The sine rule states that

$$a/\sin A = b/\sin B = c/\sin C$$

The cosine rule states that

$$a^2 = b^2 + c^2 - 2bc \cos A$$
$$b^2 = a^2 + c^2 - 2ac \cos B$$
$$c^2 = a^2 + b^2 - 2ab \cos C$$

In any triangle there are 6 unknowns: 3 sides and 3 angles. In order to be able to determine all six values at least three of these must be known.

The cosine rule is used if the following are known: 3 sides or 2 sides and the angle in between them.

The sine rule is used if the following are known: 1 side and 2 angles or 2 sides and an angle which is **not** in between.

Note: At least one side must always be specified. The application of the two rules is shown in Examples 8.39 to 8.42.

Example 8.39

Find the length of the base of the template shown in Fig. 8.19.

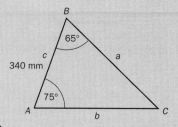

Figure 8.19

Since one side and two angles are known, the sine rule is used. Now,

$$\angle C = 180 - (65 + 75) = 40°$$

(because the angles of a triangle always total 180°)

$$\frac{b}{\sin B} = \frac{c}{\sin C} \quad \text{therefore} \quad \frac{b}{\sin 65} = \frac{340}{\sin 40}$$

Multiply by sin 65:

$$b = \frac{340 \sin 65}{\sin 40} = 340 \times \left(\frac{0.9063}{0.6428}\right)$$

$$= 479.4 \, \text{mm}$$

$$= 480 \, \text{mm} \quad \text{(correct to 2 significant figures)}$$

Example 8.40

Figure 8.20 shows an assembly for a crane. Find the angle between the girders AB and BC.

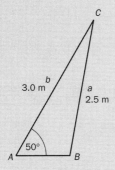

Figure 8.20

Two sides and an angle **not** between them are known. Therefore the sine rule is used. For ease of working we use the inverted form of the sine rule:

$$\frac{\sin A}{a} = \frac{\sin B}{b} \quad \text{therefore} \quad \frac{\sin 50}{2.5} = \frac{\sin B}{3.0}$$

Multiply by 3.0:

$$\sin B = \frac{3 \sin 50}{2.5} = 0.9193$$

$$B = \sin^{-1} 0.9193 = 66.8°$$

Referring to Fig. 8.20, the angle ABC is greater than 90°. The other solution, following Example 8.37, is in the second quadrant, that is $180 - 66.8 = 113.2°$, which is 110° correct to 2 significant figures.

Example 8.41

The total current I taken by an a.c. circuit is given by the line BC in Fig. 8.21. Calculate the value of this total current and the phase angle ϕ.

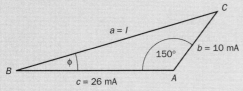

Figure 8.21

Two sides and the angle in between are known, so the cosine rule is used:

$$a^2 = b^2 + c^2 - 2bc \cos A$$

$$= 26^2 + 10^2 - 2 \times 25 \times 10 \cos 120$$

$$= 676 + 100 - 500(-0.866)$$

$$= 1,209$$

$$a = \sqrt{1,209} = 34.8$$

$$= 35 \, \text{mA} \quad \text{(correct to 2 significant figures)}$$

Now that four quantities are known, either rule may be used to find ϕ. Using the inverted sine rule:

$$\frac{\sin B}{b} = \frac{\sin A}{a} \quad \text{therefore} \quad \frac{\sin \phi}{10} = \frac{\sin 120}{34.8}$$

(using the value of one more significant figure for a further calculation as stated in Rule 12 on p. 244)

$$\sin \phi = \frac{10 \sin 150}{34.8} = 0.1437$$

$$\phi = \sin^{-1} 0.1437 = 8.26°$$

$$= 8.3° \quad \text{(correct to 2 significant figures)}$$

Example 8.42

Figure 8.22(a) shows three forces acting at a point. Figure 8.23(b) is the triangle of forces for this loading arrangement. Calculate the angle θ between the forces F_1 and F_2.

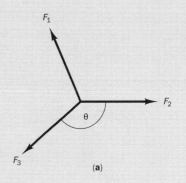

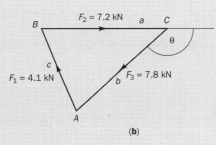

Figure 8.22

Since the three sides are known, the cosine rule is used to find the angle θ at C.

$$c^2 = a^2 + b^2 - 2ab \cos C$$

Transposing:

$$2ab \cos C = a^2 + b^2 - c^2$$

Dividing by $2ab$:

$$\cos C = \frac{a^2 + b^2 - c^2}{2ab}$$
$$= \frac{7.2^2 + 7.8^2 - 4.1^2}{2 \times 7.2 \times 7.8} = 0.8535$$
$$C = \cos^{-1} 0.8535$$
$$= 31.4°$$

Therefore
$$\theta = 180 - 31.4 = 148.6$$
$$= 150° \text{ (correct to 2 significant figures)}$$

Self-assessment tasks 8.21

1. A crank-connecting rod mechanism is shown below. Calculate, for the position shown, the distance AC and $\angle BAC$.

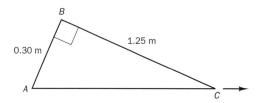

2. A tie wire 6.0 m long is fixed at one end to the top of a vertical radio mast, 4.2 m tall, and the other end is bolted to the horizontal ground. Calculate the angle of inclination of the wire and the distance of the bolted joint from the foot of the mast.
3. In this vector diagram calculate the voltage V_t across the resistance and the phase angle ϕ.

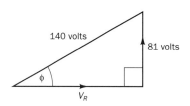

4. For this taper roller calculate the taper angle θ and the length of the bearing surface AB.

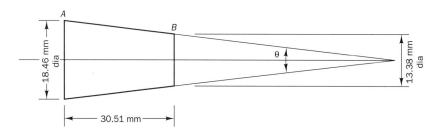

5. A wall crane is shown below. Find the length of the tie and the angle the tie makes with the wall.

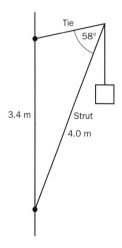

6. An electric cable is to run due east from a substation A for 800 m, and then north-east to a factory B. If the factory B is on a bearing of 055° from A, calculate the total length of cable required and the direct distance from A to B.
7. The vector diagram for an electrical circuit is given below. Calculate the voltage V and the angle θ it makes with the 3.2 V vector.

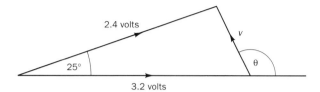

8. A parallelogram of vectors is shown below. Calculate the resultant vector R and the inclination ϕ it makes with the horizontal vector.

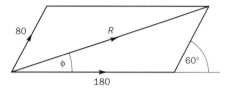

9. The installation of a new motor involves suspending it by two wire ropes as shown below. If the angle that the longer rope makes with the ceiling must not be less than 30, what is the minimum length of the smaller rope? For this arrangement what will be the distance of the point A below the ceiling?

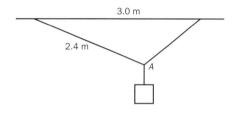

Area of a triangle

In Fig. 8.23 the area of a triangle is

$$\tfrac{1}{2} \text{ base} \times \text{height} = \tfrac{1}{2}ah$$

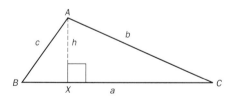

Figure 8.23

In triangle AXC

$$\frac{h}{b} = \sin C \quad \text{so that} \quad h = b \sin C$$

Area of $\triangle ABC = \tfrac{1}{2}ab \sin C$

$$= \tfrac{1}{2}ac \sin B \qquad (10)$$

$$= \tfrac{1}{2}bc \sin A \quad \text{by a similar manner}$$

In order to use these formulae two sides and the angle in between must be known.

If 3 sides are known and no angles are known then the following formula is suitable.

$$\text{Area of triangle} = \sqrt{[s(s-a)(s-b)(s-c)]}$$

where $\qquad 2s = a + b + c$

If one side and two angles are known the sine rule is used to find a second side; then the first equation (10) may be used to find the area.

Example 8.43

Calculate the area of sheet metal required to make the machine guard shown in Fig. 8.24.

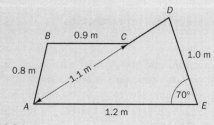

Figure 8.24

Area of $\triangle ADE = \frac{1}{2} ad \sin E = \frac{1}{2} \times 1.0 \times 1.2 \times \sin 70$
$= 0.564 \, m^2$

Area of $\triangle ABC = \sqrt{[s \cdot (s-a) \cdot (s-b) \cdot (s-c)]}$
where $2s = a + b + c = 0.9 + 1.1 + 0.8 = 2.8$
$s = 1.4$

Area of $\triangle ABC = \sqrt{[1.4(1.4 - 0.9) \cdot (1.4 - 1.1) \cdot (1.4 - 0.8)]}$
$= \sqrt{0.126}$
$= 0.355 \, m^2$

Therefore,

Area of guard $= 0.564 + 0.355$
$= 0.919$
$= 0.92 \, m^2$ (correct to 2 significant figures)

Self-assessment tasks 8.22

1. A support plate is in the form of $\triangle ABC$ with $AC = 10.4$ cm, $BC = 13.1$ cm, and $\angle C = 60°$. Calculate the area of the plate.
2. An electric heat sink is triangular in shape and to lose the required amount of heat it must have an area of at least 2,400 mm^2. If the two sides of this triangle must have lengths of 80 mm and 160 mm, what is the minimum value of the angle between these two sides?
3. A mesh guard for a robot is to be in the form of the trapezium shown below. Calculate the area of metal required for the guard.

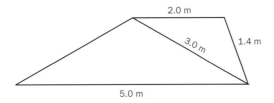

4. An instrument panel must have the dimensions shown below. Find the area of the panel.

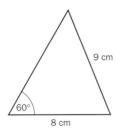

Cartesian and polar co-ordinates

To produce graphs and engineering drawings either by hand or by computer it is necessary to have a method of locating the positions of points on paper or on screen. At least two numbers, called **co-ordinates**, are required to locate a point in a plane. Two systems are used.

Cartesian co-ordinates (x, y)

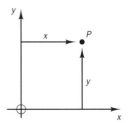

Figure 8.25

This is the most commonly used system. Two perpendicular datum lines are used; the horizontal line is called the *x axis*, the vertical line is called the *y axis*, as shown in Fig. 8.25. The point of intersection of the two axes is called the **origin**. Any point P is located by its perpendicular distances from the two axes, these distances being called the *x* and *y* co-ordinates.

Polar co-ordinates (r, θ)

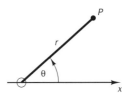

Figure 8.26

In this system a point is located at a distance r along a line OP from a fixed point O, called the **pole**, as shown in Fig. 8.26. θ is the angle that the line OP makes with the reference $+x$ axis. It is important to remember that θ is positive when OP rotates anticlockwise.

Figure 8.27 shows the relationship between both systems. Pythagoras's theorem and trigonometry can be used to change from one system to another.

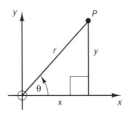

Figure 8.27

Conversion from Cartesian to polar co-ordinates

From Pythagoras: $r = \sqrt{(x^2 + y^2)}$

and $\theta = \tan^{-1}(y/x)$

The smallest value of θ is usually quoted and can be positive or negative. The value of θ obtained must be checked so that it places P in the correct quadrant. This can be done by using a sketch to check the results, as shown in Example 8.44.

Conversion from polar to Cartesian co-ordinates

From Fig. 8.27 using trigonometrical ratios in a right-angled triangle

$$x = r\cos\theta, \quad y = r\sin\theta$$

Example 8.44

The Cartesian co-ordinates of a point P are $(-4, -6)$. Convert these to polar co-ordinates.

The point P with these two co-ordinates is shown in Fig. 8.28.

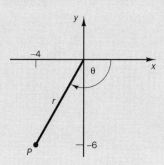

Figure 8.28

$$r = \sqrt{\left[(-4)^2 + (-6)^2\right]} = \sqrt{(16 + 36)} = 7.2$$
$$\theta = \tan^{-1}(y/x) = \tan^{-1}(-6/-4)$$
$$= \tan^{-1} 1.5 = 56.3°$$

The value of 56.3 is not in agreement with the position of P in the third quadrant. The smallest magnitude of θ is $180 - 56.3 = 123.7°$, and this is seen to be negative because it is in a clockwise direction.

The polar co-ordinates are $(7.2, -124°)$.

Example 8.45

Convert the polar co-ordinates $(18, 125°)$ to Cartesian co-ordinates.

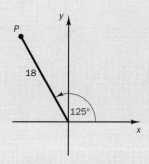

Figure 8.29

From Fig. 8.29:
$$x = r\cos\theta = 18\cos 125 = -10.3$$
$$y = r\sin\theta = 18\sin 125 = 14.7$$

The Cartesian co-ordinates are $(-10.3, 14.7)$.

The distance between two points with Cartesian co-ordinates

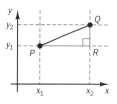

Figure 8.30

Figure 8.30 shows two points $P(x_1, y_1)$ and $Q(x_2, y_2)$.

Horizontal distance: $PR = x_2 - x_1$
Vertical distance: $QR = y_2 - y_1$
From Pythagoras: $PQ = \sqrt{(PR^2 + QR^2)}$
$$= \sqrt{\left[(x_2 - x_1)^2 + (y_2 - y_1)^2\right]}$$

Example 8.46

The location of two logic gates P and Q on a baseboard are given in polar co-ordinates as $(80\,\text{mm}, 60°)$ and $(100\,\text{mm}, 30°)$. Find the Cartesian co-ordinates and hence find the distance PQ.

At point P:
$$x_1 = r\cos\theta = 80\cos 60° = 40\,\text{mm}$$
$$y_1 = r\sin\theta = 80\sin 60° = 69.3\,\text{mm}$$

At point Q:
$$x_1 = r\cos\theta = 100\cos 30° = 86.6\,\text{mm}$$
$$y_1 = r\sin\theta = 100\sin 30° = 50\,\text{mm}$$

Distance $PQ = \sqrt{\left[(x_1 - x_1)^2 + (y_2 - y_1)^2\right]}$
$$= \sqrt{\left[(86.6 - 40)^2 + (50 - 69.3)^2\right]}$$
$$= 50\,\text{mm (correct to 2 significant figures)}$$

Self-assessment tasks 8.23

1. Convert the following Cartesian co-ordinates into polar co-ordinates:
   (a) $(16, -20)$, (b) $(-2, 5)$, (c) $(-3, -6)$
2. Convert the following polar co-ordinates into Cartesian co-ordinates:
   (a) $(12, 30°)$, (b) $(20, -130°)$, (c) $(15, 160°)$
3. The location, in mm, of two components is shown below. Express the locations in polar co-ordinates. Find also the distance AB.

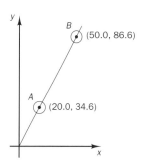

4. Five holes are equally spaced on a circle of diameter 250 mm as shown below. Find the Cartesian co-ordinates for each of the holes, and hence calculate the distances AB and BC.

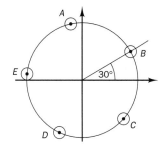

5. Three holes are to be drilled in a base-board as shown below. Calculate $\angle BAC$ and hence find the distances of the holes B and C from the horizontal and vertical datum lines through A.

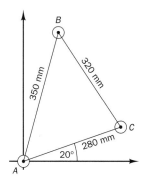

8.3 Use functions and graphs to model engineering situations and solve engineering problems

Representing engineering data

The relationship between two quantities in any engineering situation can be expressed graphically and usually by a formula or equation as well.

Representation of data with a formula or equation

A formula or equation is a convenient way of expressing the relationship between two quantities. In equations such as $y = 3x^2 + 2x - 1$, x is called the **independent** variable and y is the **dependent** variable, so called because its value depends upon the value of x. Such equations produce pairs of (x, y) values which can be used as Cartesian co-ordinates (see p. 264). Linear, trigonometric, exponential and logarithmic equations will be examined.

In equations such as $r = 2\cos\theta$, θ is the independent variable and r the dependent variable, the pairs of (r, θ) values produced are called polar co-ordinates (see p. 264). Polar graphs are examined on p. 275.

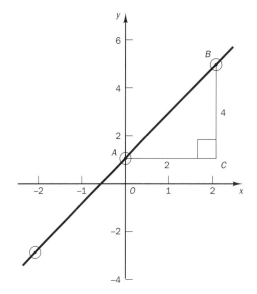

Figure 8.31

Graphical representation of two quantities (x, y) related by an equation

For each value of x a corresponding value of y is obtained from the equation. Each pair of values is used as the Cartesian co-ordinates of a point on a plane between two axes. The x values are plotted horizontally, and the y values (ordinates) vertically. These points trace out a curve or straight line which is the graph representing the equation. The shape of the graph is a good indication of how one quantity depends on the other.

Straight line graphs

Consider the equation $y = 2x + 1$.

If $x = 0$, $y = 2 \times 0 + 1 = 1$
If $x = -2$, $y = 2 \times -2 + 1 = -3$
If $x = 2$, $y = 2 \times 2 + 1 = 5$

These pairs of values provide the co-ordinates of three points $(-2, -3)$, $(0, 1)$, $(2, 5)$. The line through the three points is the graph representing the equation $y = 2x + 1$. It is a straight line as shown in Fig. 8.31.

Note: Two points are sufficient to produce a straight line graph; the third point acts as a useful check that the values of the co-ordinates are correct by ensuring that the three points all lie on a straight line.

It is now necessary to examine the significant features of a straight line. Moving along the line from left to right, that is, from A to B, we obtain

$$\text{Gradient of the line} = \frac{\text{Change in } y \, (CB)}{\text{Change in } x \, (AC)}$$
$$= \frac{5 - 1}{2 - 0}$$
$$= 2$$

The gradient is positive because an increase in x produces an increase in y. This calculation shows that y is changing twice as fast as x, or that the rate at which y is changing with respect to x is 2.

If we examine the equation $y = 2x + 1$ we see that

- The number in front of the x, called the **coefficient**, is 2, which is the **gradient** of the graph as shown in the calculation above.
- The constant term (the number on its own) is 1. This is the value of y where the line crosses the y axis, at $x = 0$. This value is called the **intercept**.

Using these definitions the general equation of a straight line is

$$y = mx + c$$

where m is the gradient of the line and c is the intercept on the y axis at $x = 0$.

Example 8.47

The effort $E(N)$ to lift a load W (kN) for a particular machine is given by the equation $E = 100W + 80$. Draw this graph for a load from 0 to 3 kN and determine the gradient and the intercept.

From the equation, when $W = 0$, $E = 80 N$:

$$W = 1 \text{ kN} \quad E = 100 \times 1 + 80 = 180 \text{ N}$$
$$W = 3 \text{ kN} \quad E = 100 \times 3 + 80 = 380 \text{ N}$$

This result produces three pairs of co-ordinates (0, 80), (1, 180), (3, 380) which are plotted between the axes as shown in Fig. 8.32

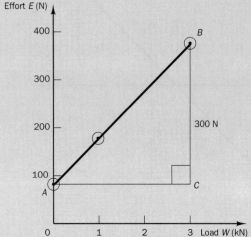

Figure 8.32

From the graph

$$\text{Gradient} = \frac{CB}{AC} = \frac{300 \text{ (N)}}{3 \text{ (kN)}} = 100 \text{ N/kN}$$

Intercept = 80 N

The values of a gradient and intercept can also be obtained directly from the equation.

Note: The unit of the gradient is N/kN, that is, increasing the load by 1 kN requires an increase of effort of 100 N. The intercept of 80 N shows that an effort of 80 N is required even with no load on the machine, which is required, mainly, to overcome friction forces.

Example 8.48

A straight line graph goes through the points $(-1, 6)$ $(2, 12)$. Sketch the graph. Find its intercept and gradient, and hence the equation which the graph represents.

The sketch is shown in Fig. 8.33. It is seen that the graph passes through the origin, that is, the point $(0, 0)$. Therefore the intercept is 0.

$$\text{Gradient } m = \frac{CB}{AC} = \frac{-12 - 6}{2 - (-1)} = -6$$

Note: The gradient is negative because when moving left to right x is increasing but y is decreasing. The equation is therefore

$$y = -6x + 0$$
$$y = -6x$$

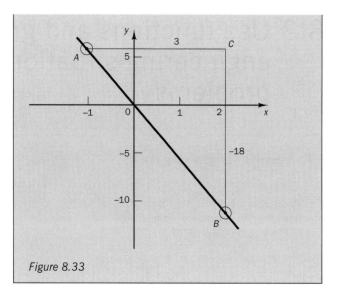

Figure 8.33

Self-assessment tasks 8.24

1. Find the gradient and intercept of each of the following straight lines and hence sketch the graphs.
   (a) $y = 2x - 1$ (b) $y = -x + 4$ (c) $y = x + 1$
   (d) $y = 3x$ (e) $y = -x$ (f) $y = 2$

2. Results from a test on a machine are shown as:

Load lifted W (kN)	2.0	4.0	6.0	8.0	9.0
Applied force F (N)	0.81	1.22	1.58	2.04	2.17

   Draw a graph of these results and hence find the law $F = aW + b$ of this machine.

3. A check carried out on a circuit showed that when the resistance was 5 Ω the current was 4, and when the resistance was 11 Ω the current was 2 A. The relationship is known to be of the form $R = a/I + b$. Find the values of a and b.

4. The resistance R Ω of a coil varies with temperature θ (°C) of the coil as shown:

Temperature θ (°C)	20	30	40	50	60	70
Resistance R (Ω)	60.4	65.4	69.2	75.6	79.8	84.8

   Draw the graph and verify the law $R = a\theta + b$, and find the constants a and b.

Quadratic graphs

Whereas in straight line graphs the two variables x and y have indices of 1, the quadratic graph is of the form

$$y = ax^2 + bx + c$$

where a, b, c are constants. For a particular graph, b and/or c may be 0. The quadratic graph has a parabolic shape, and its width and position will depend on the values of a, b and c. We shall plot a graph of

$$y = 2x^2 - 6x + 1$$

in the range $x = -1$ to $x = 4$. Table 8.5 shows values of y found for values of x between -1 and 4.

Table 8.5 Values: $y = 2x^2 - 6x + 1$

x	-1	0	1	2	3	4
$2x^2$	2	0	2	8	18	32
$-6x^2$	6	0	-6	-12	-18	-24
$+1$	$+1$	$+1$	$+1$	$+1$	$+1$	$+1$
y	9	1	-3	-3	$+1$	9

The x and y values from co-ordinate pairs which are plotted in Fig. 8.34.

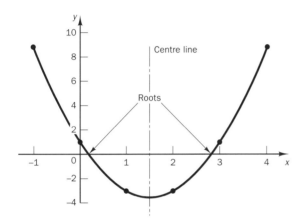

Figure 8.34

The quadratic equation has a number of significant features:

- The graph in Fig. 8.34 is symmetrical about the vertical line $x = 1.5$. The lowest point of the curve occurs on this line. The line of symmetry always occurs at $x = -b/2a$.

  From the graph $x = -(b/2a) = -[-6/(2 \times 2)] = 1.5$

  When $x = 1.5$,

  $$y = 2 \times (1.5)^2 - 6 \times (1.5) + 1 = -3.5$$

  Within the limit of accuracy of the drawing this agrees with the minimum value of y on the graph. If $b = 0$ the axis of symmetry is at $x = 0$; that is, the y axis.
- As with the equation of a straight line, the constant term $+1$ is the value of the intercept on the y axis.
- To solve the equation $2x^2 - 6x + 1 = 0$, we find the values of x at those points on the graph where $y = 0$; that is, where the graph cuts the x axis. These values, called the **roots** of the equation, are $x = 0.2$ and $x = 2.8$.

  Note: If the graph does not cut the x axis then the equation does not have a real solution. This is the case when the solution by formula (see p. 225) contains the square root of a negative number.
- When a is large the graph is narrow, and when a is small the graph is wide. If a is positive the graph has a $\cup$ shape; if a is negative the graph has a $\cap$ shape.

Example 8.49

A rectangular circuit board is to have a perimeter of 40 cm. Plot a graph showing how the area of the board varies with the length of the board. Find the dimensions of the board for the maximum area.

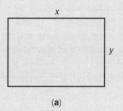

(a)

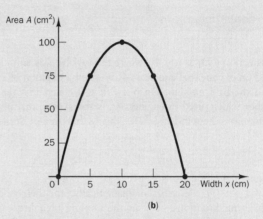

(b)

Figure 8.35

Let the length of the rectangle be x and the width be y, as shown in Fig. 8.35(b).

$$\text{Perimeter} = 2x + 2y = 40$$
$$y = 20 - x$$
$$\text{Area of the rectangle } A = xy = x(20 - x)$$
$$= -x^2 + 20x$$

The length x can only vary between 0 and 20 cm. We determine the values of A every 5 cm between 0 and 20 cm, as shown in Table 8.6.

Table 8.6 Values $y = -x^2 + 20x$

x	0	5	10	15	20
$-x^2$	0	-25	-100	-225	-400
$20x$	0	100	200	300	400
y	0	75	100	75	0

The graph of A against x is shown in Fig. 8.35(b).
From the graph the maximum value of the area $= 100 \text{ cm}^2$, which occurs at $x = 10$ cm, when the rectangle becomes a square.

Note: Comparing the equation with $y = ax^2 + bx + c$,

$$a = -1, \quad b = 20$$

The line of symmetry is at $x = -b/2a = 10$ so that the graph has a peak; that is, a maximum value at $x = 10$.

Self-assessment tasks 8.25

1. Draw a graph of $y = 2x^2 + 3x - 15$, by first of all calculating a table of values of y for values of x between -4 and $+3$. Use the graph to find the solution of the equation $2x^2 + 3x - 15 = 0$. Calculate the position of the line of symmetry and read from the graph the minimum value of y.
2. The stress (MN/m^2) in a plate varies with the distance x (cm) from the centre and is given by $S = -x^2 + 9x - 14$. Draw the graph for the range $x = 1$ to $x = 8$ cm. Find (a) the greatest positive stress in the plate and the position at which it occurs, (b) the position x at which there is no stress in the plate.

Exponential graphs

An equation such as $y = 4^x$, where the variable x is an index, is called an exponential equation. 4^x is called a function of x; 4 is called the base number. On p. 248 it was stated that the base number widely used in engineering is the exponential number e, where e is the number 2.718... being the value of the series

$$1 + \frac{1}{1} + \frac{1}{1.2} + \frac{1}{1.2.3} + \frac{1}{1.2.3.4} + \cdots$$

Exponential equations such as $y = Ae^{bx}$, $y = Ae^{-bx}$, $y = A(1 - e^{-bx})$, describe, mathematically, the behaviour of engineering situations, such as the cooling of a hot casting, the charging of a capacitor, etc.

Consider the exponential equations $y = e^x$ and $y = e^{-x}$. Values of the exponential functions are obtained for the range $x = -3$ to $x = 3$, using the calculator and the e^x key. These values are shown in Table 8.7.

Table 8.7 Values of the exponential function

x	-3	-2	-1	0	1	2	3
e^x	0.05	0.14	0.36	1	2.72	7.39	20.1
e^{-x}	20.1	7.39	2.72	1	0.36	0.14	0.05

The graphs of the two functions are shown in Fig. 8.36.

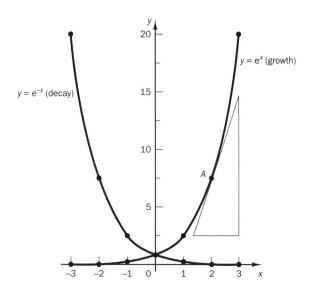

Figure 8.36

The graphs show that for the function e^x, y increases exponentially with x, and for the function e^{-x}, y decreases exponentially with x. It is also seen that the two graphs are mirror images of one another.

The graphs have one property which is unique; the gradient of the curve at any one point is always numerically equal to the height of the curve (the ordinate) at that point, that is, the rate at which the function is changing is always numerically equal to the value of the function at that point. This statement may be checked by finding the gradient at the point $A(x = 2)$, in Fig. 8.36. From the triangle, the gradient at A is 7.39, which is the same as the value of y at A.

It is this property that makes the exponential function so important in engineering. For example, it is reasonable to expect that the rate at which the electric charge leaves a discharging capacitor depends on the amount of charge left on the capacitor.

Example 8.50

The voltage v across the plates of a capacitor varies with the time $t(\mu s)$ according to the formula $v = 250\,e^{-0.4t}$. Draw a graph showing the voltage variation in the first $7\,\mu s$. From the graph find (a) the initial voltage when $t = 0$, (b) the rate at which the voltage is falling at $t = 3\,\mu s$.

The values of v are shown below and the graph is plotted in Fig. 8.37.

$t(\mu s)$	0	1	2	3	4	5	6	7
v (V)	250	168	112	75.3	50.4	33.9	22.6	15.0

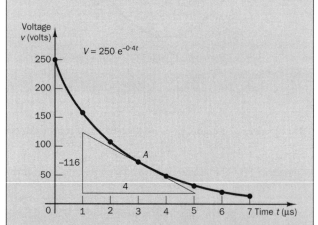

Figure 8.37

(a) Initial voltage $= 250\,V$
(b) Rate of change of $v =$ gradient of tangent at A

$$= \frac{16 - 132}{5 - 1} = -29\,V/\mu s$$

Note: The numerical rate at which the voltage is changing at A is $29\,V/\mu s$; this is *not* numerically equal to the voltage at A, which is $75\,V$. This will only be true for the equations $y = e^x$ and $y = e^{-x}$. In an equation such as the one above, the rate of change of the voltage is **proportional** to the voltage, that is, if the voltage is halved the rate of change will also be halved.

Exponential curves will have the following significant features:

- Referring to the equations $y = Ae^{bx}$ and Ae^{-bx}: A is the value of y where the curve crosses the y-axis. Positive b gives a growth curve, negative b gives a decay curve. Increasing b increases the gradient of the curve at all points.

- Referring to $y = A(1 - e^{-bx})$, the graph of which is shown in Fig. 8.38: A is the maximum value attained by y. The larger the value of b the quicker the curve will reach maximum value.

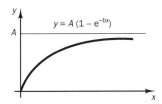

Figure 8.38

Self-assessment tasks 8.26

1. The voltage across the plates of a charging capacitor is given by $v = 30(1 - e^{-t/RC})$, where $R = 0.04\,M\Omega$, $C = 100\,\mu F$ and t is the time in microseconds. Plot a graph of v against t for the range $t = 0$ to $t = 15\,\mu s$. By drawing a tangent to the curve find the rate at which the voltage is increasing when $t = 5\,\mu s$. How does the rate of increase of voltage vary with time?
2. The formula for the force F in a drive belt is given by the formula $F = 800\,e^{-0.25\theta}$ where θ is the angle of contact in radians between the belt and the pulley. Draw the F–θ graph for the range $\theta = 0$ to $\theta = 3\pi/2$ rad, showing how the force in the belt depends upon the angle of contact with the pulley.

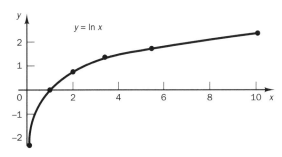

Figure 8.39

entering the value of x and pressing the $\boxed{\ln}$ key. These values are shown below:

x	0.1	0.5	1	2	3	5	10
$y = \ln x$	−2.3	−0.69	0	0.69	1.10	1.61	2.3

In the section on logarithms (p. 248) it was seen that $x = \log y$ is the logarithmic form of $y = e^x$. For this reason a logarithmic graph is a mirror image of the exponential graph. For example, with an exponential graph representing a practical problem, such as a hot casting which is cooling according to Newton's law of cooling, $\theta = \theta_0 e^{-kt}$, where k is a constant and θ is the temperature of the casting above the surroundings after a time t, and θ_0 is the initial temperature above the surroundings.

The graph of θ (vertically) against t (horizontally) will give an exponential decay graph. The graph of t (vertically) against θ (horizontally) will give a logarithmic graph.

Logarithmic graphs

The graph of $y = \ln x$ is plotted in Fig. 8.39 for a range of values of x from 0.1 to 10. As discussed on p. 248, the base of the logarithm written in this form is 'e'. For each of the values of x the value of y is obtained from a calculator by

Trigonometric graphs

Physical quantities such as a.c. currents and voltages, moving parts of pistons, etc., can be modelled with the trigonometric functions $\sin\theta$ and $\cos\theta$. The graph of both these functions are shown in Fig. 8.40.

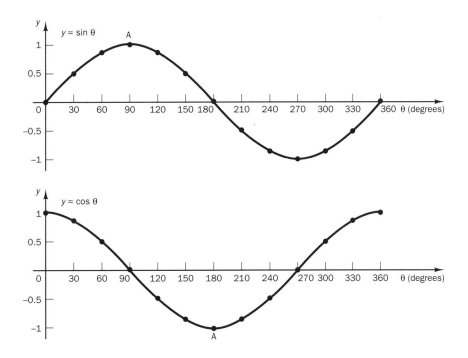

Figure 8.40

Table 8.8 Values of the sine and cosine

$\theta°$	0	30	60	90	120	150	180	210	240	270	300	330	360
$y = \sin\theta$	0	0.5	0.87	1.0	0.87	0.5	0	−0.5	−0.87	−1.0	−0.87	−0.5	0
$y = \cos\theta$	1.0	0.87	0.5	0	−0.5	−0.87	−1.0	−0.87	−0.5	0	0.5	0.87	1.0

The values of both functions are obtained in intervals of 30° over a range of 0–360° (see Table 8.8).

These trigonometric graphs have the following significant features:

- Both graphs are periodic and will repeat themselves after every 360°. The **period** of each is 360°.
- The maximum and minimum value of each function, is +1 and −1. This is called the **amplitude**. From the graph of $y = A\sin\theta$ the maximum value of y is A, that is, A is the amplitude.
- The graph of $\cos\theta$ is seen to have the same form as the graph of $\sin\theta$ but displaced to the left by 90°, that is, it is 90° ahead of $\sin\theta$, and hence can be written as $y = \sin(\theta + 90°)$, where 90° is the **phase angle**.

Radian measure of an angle

So far we have measured angles in degrees. Another measure used widely in engineering is **radians**. Referring to Fig. 8.41(a), 1 radian is defined as the angle AOB which cuts the circumference of a circle of radius 1 with an arc of length 1.

To convert radians to degrees we see that a complete revolution of this circle has a circumference of $2\pi 1$, that is, a complete revolution is 2π radians, which is the same as 360°. Therefore

Since this definition involves the division of two lengths, the radian measure has no units which is a major advantage in engineering calculations.

Conversions are carried out as follows:

- Degrees to radians:

$$40° = 40 \times \pi/180 = 0.698 \text{ radians}$$

$$60° = 60 \times \pi/180 = \frac{\pi}{3} \text{ radians}$$

- Radians to degrees:

$$1.6 \text{ radians} = 1.6 \times 180/\pi = 91.7°$$

$$\frac{\pi}{4} \text{ radians} = \frac{\pi}{4} \times 180/\pi = 45°$$

The general equation of a sine wave
$y = A\sin(\theta \pm \alpha)$

In the equation $y = A\sin(\theta + \alpha)$ the phase angle α is positive and the sine wave is said to *lead* $A\sin\theta$.

In the equation $y = A\sin(\theta - \alpha)$ the phase angle is negative and the sine wave is said to *lag* $A\sin\theta$. The two cases are shown in Fig. 8.42.

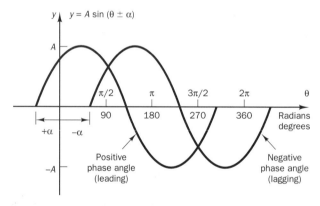

Figure 8.42

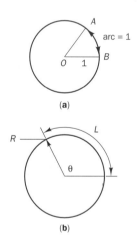

Figure 8.41

$$2\pi \text{ radians} = 360°$$
$$\pi \text{ radians} = 180°$$
$$1 \text{ radian} = 180/\pi = 57.30 \text{ radians}$$

(correct to 4 significant figures)

In general, as shown in Fig. 8.41(b):

Angle in radians = Length of arc/Radius

$$\theta = L/R$$

The rotating vector method of constructing a sine wave

A sine wave can be constructed by using a rotating line equal in length to the amplitude. This rotating vector is called a **phasor**. The construction is shown in Fig. 8.43.

The line OP is the phasor, which rotates in an anticlockwise direction, and is equal to the amplitude A of $y = A\sin(\theta + \alpha)$. OP_1 is drawn at an angle α to the horizontal in Fig. 8.43, indicating a positive value for the phase angle. If α were negative then OP would start below the horizontal axis, at an angle $-\alpha$. In the position OP_1 the angle $\theta = 0°$. P_1 is projected to Q_1 on the y axis, that is, the $\theta = 0°$ axis. The phasor OP is then rotated 30° in an anticlockwise direction to

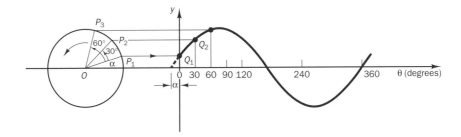

Figure 8.43

the position OP_2. P_2 is now projected to Q_2 on the $\theta = 30°$ ordinate, to give the second point on the sine wave. The process is repeated until the phasor returns to the position OP_1, when the sine wave will have completed one cycle. This rotating method illustrates the relationship between the sine wave and the process of generating a.c. voltages using a rotating coil.

Angular velocity ω

Let the phasor OP rotate at a constant angular velocity, ω rad/s. Every second the angle swept through is ω radians. Then in t seconds OP will have rotated through an angle

$$\theta = \omega t \qquad (11)$$

The general equation will then become

$$y = A \sin(\omega t \pm \alpha)$$

Note: In this form α must be expressed in radians.

Periodic time and frequency

Periodic time T

The periodic time is the time taken in seconds to complete one cycle of the sine wave, which is the same as the time for the phasor to complete one revolution. Therefore, when $t = T$, $\theta = 2\pi$ so that equation (11) becomes

$$2\pi = \omega T$$
$$T = 2\pi/\omega \qquad (12)$$

Frequency f

Frequency f is the number of sine waves completed in one second. It is measured in cycles per second or Hertz (Hz), where $1\,\text{Hz} = 1$ cycle/s. From the above we see that in T seconds the number of cycles generated $= 1$, and in $1\,\text{s}$ the number of cycles generated $= 1/T$, so that

$$f = 1/T$$

From equation (12), $f = \omega/2\pi$.

Example 8.51

An a.c. voltage has the formula $v = 340 \sin(314t - 0.7)$. Determine

(a) the maximum voltage
(b) the frequency of the voltage
(c) the phase angle in degrees

From the information make a rough sketch of the waveform by plotting the voltage against the angle of rotation in degrees, showing the main features.

Comparing the formula with the general equation $y = A \sin(\omega t - \alpha)$ we obtain

(a) Maximum voltage $= A = 340\,\text{V}$
(b) Frequency $f = \omega/2\pi = 314/2\pi = 50\,\text{Hz}$
(c) Phase angle $\alpha = -0.7\,\text{rad} = -0.7 \times (180/\pi) = -40°$ which shows that the wave is lagging

The sketch of the sine wave is shown in Fig. 8.44.

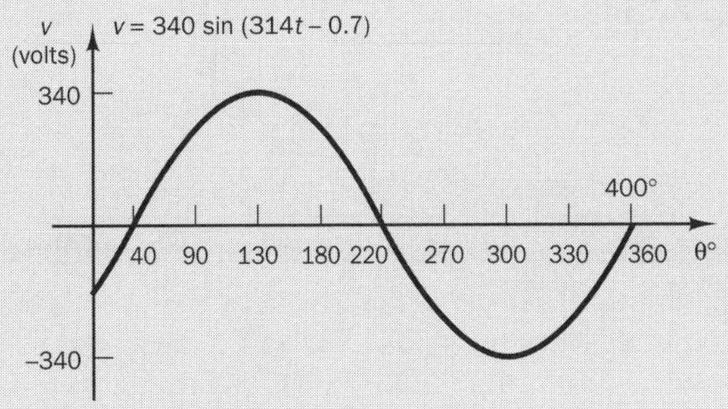

Figure 8.44

Addition of sine waves

In electrical engineering electric circuits are used to add a.c. voltages and currents. In mechanical engineering the sinusoidal movements of mechanical mechanisms may need to be combined. Adding together two sine waves with the same frequency will produce another sine wave of the same frequency. Adding together two sine waves of difference frequencies will produce a complicated waveform, which will not be in the form of a sine wave. Sine waves may be added together in two ways:

Method 1 Both sine waves are drawn and the ordinates (that is, the vertical values at each value of the angle) added together algebraically, to produce the combined waveform. This method may be used to combine sine waves with different frequencies.

Method 2 The rotating vector may be used after first obtaining the resultant of the two sine wave phasors. This method can only be used to combine sine waves with similar frequencies.

Both methods are described in the following examples.

Example 8.52

The movement of part of a packaging machine is made up of two movements given by $y_1 = 50 \sin \theta$ and $y_2 = 100 \sin(\theta + 50°)$. Both movements are measured in millimetres. Draw graphs of the two sine waves and find the resultant movement by adding the two together. Hence find the formula for the resultant movement.

Using method 1 the graphs of y_1 and y_2 are plotted for θ over one cycle of 360° as shown in Fig. 8.45, using the data from Table 8.9.

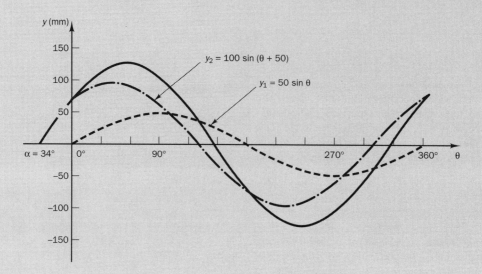

Figure 8.45

Table 8.9 Values: $y_1 = 50 \sin \theta$; $y_2 = 100 \sin(\theta + 50)$

$\theta°$	0	30	60	90	120	150	180	210	240	270	300	330	360
$\sin \theta$	0	0.5	0.87	1.0	0.87	0.5	0	−0.5	−0.87	−1.0	−0.87	−0.5	0
$y_1 = 50 \sin \theta$	0	25	43.5	50	43.5	25	0	−25	−43.5	−50	−43.5	−25	0
$(\theta + 50)$	50	80	110	140	170	200	230	260	290	320	350	380	410
$\sin(\theta + 50)$	0.766	0.985	0.940	0.643	0.174	−0.342	−0.766	−0.985	−0.940	−0.642	−0.174	0.342	0.766
$y_2 = 100 \sin(\theta + 50)$	76.6	98.5	94.0	64.3	17.4	−34.2	−76.6	−98.5	−94.0	−64.2	−17.4	34.2	76.6

The combined graph has ordinates $y_1 + y_2$ and these can be found either from the table or from the graphs, using algebraic addition, to take into account the signs. Because the two sine waves have the same frequency, the combined graph will be a sine wave with the same frequency. If we examine the combined graph we see that its phase angle is 34° and its peak value is 137 mm, so that its formula is

$$Y = 137 \sin(\theta + 34°)$$

Example 8.53

Using the rotating vector method draw a graph of the output current in amps from an electrical circuit which is given by

$$i = 5 \sin \omega t + 7.5 \sin(\omega t - \pi/3)$$

Referring to Fig. 8.46, we represent $5 \sin \omega t$ with a phasor $OA = 5$ units long drawn horizontally, since the phase angle is zero. The sine wave $7.5 \sin(\omega t - \pi/3)$ is represented by a phasor $OB = 7.5$ units long, drawn at $-\pi/3$ rad, that is, 60° below the axis because the phase angle is negative. OP_1, the resultant of these two phasors is now obtained by drawing a parallelogram. P_1 is projected across to give the first point Q_1 on the combined graph at $\theta = 0$. OP_1 is now rotated 30° to OP_2 and P_2 is projected across to fix the next point Q_2 on the graph at $\theta = 30°$. The process is continued until the whole cycle has been completed. Because we are adding two sine waves of equal frequency the output waveform is sinusoidal. From the graph it is seen that

> Amplitude A = 11.2 amps
> Phase angle α = −35° = −0.61 rad
> Output formula is $i = 11.2 \sin(\omega t - 0.61)$

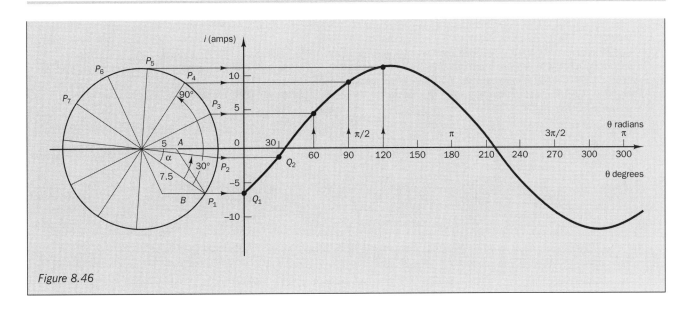

Figure 8.46

Self-assessment tasks 8.27

1. Draw the graphs of $y_1 = 20\sin\theta$ and $y_2 = 30\sin(\theta - 40)$ on the same axes. By adding together the ordinates (that is, the heights) of both graphs draw the combined graph $y = 20\sin\theta + 30\sin(\theta - 40)$. By examining the combined graph give its formula in the form $y = A\sin(\theta + \alpha)$.

2. Use the rotating vector method to draw the graph

$$v = 4\sin\omega t + 6\sin(\omega t - \pi/4)$$

What is the general formula for this graph?

3. The displacement x (m) of an oscillating mechanism is given by $x = 2.4\sin(5t - 0.1)$. Find
   (a) the maximum displacement
   (b) the time of one oscillation
   (c) the frequency of the oscillation
   (d) the phase angle in degrees, stating if it is leading or lagging.

   Sketch the displacement–angle graph showing the main points.

Example 8.54

The shape of the ground area illuminated by a motorway lamp is given by $r = 20(1 + \cos\theta)$, where r is the distance along the ground from the base of the lamp standard, and θ is the angle made by r from a fixed reference line. Plot the shape of the illuminated area.

The values of r in the following table are calculated from the equation.

$\theta°$	0	30	60	90	120	150	180
r (m)	40	37.3	30	20	10	2.68	0

$\theta°$	210	240	270	300	330	360
r (m)	2.68	10	20	30	37.3	40

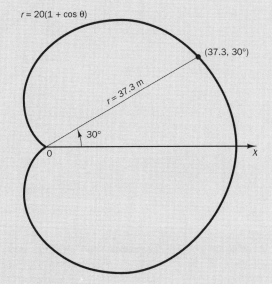

Figure 8.47

The illuminated area is shown in Fig. 8.47.

Polar graphs

In many engineering situations, such as drilling a set of holes in a circle, or measuring the intensity of light around a lamp, it is more convenient to use polar co-ordinates in place of Cartesian co-ordinates. Again, shapes such as cams have complicated Cartesian formulae but much simpler polar formulae. The application of polar graphs is shown in the following two examples.

Example 8.55

The output channel from a water turbine has a profile shape given by $r = 200\theta$ mm. Draw the profile for values of θ from 0 to 270°.

θ must be expressed in radians which have no units. If θ were expressed in degrees in the equation, then the distance r would be expressed in degrees which is impossible. The (mm) unit is contained within the coefficient 200.

Thus r is calculated using θ in radians, but for convenience the graph is plotted with θ in degrees. Table 8.10 is obtained from the formula and the profile plotted in Fig. 8.48.

Table 8.10

$\theta°$	0	10	20	30	90	120	180	240	270
$\sin\theta$	0	0.175	0.350	0.524	1.571	2.094	3.142	4.189	4.712
r (mm)	0	35.0	70.0	104.8	314.2	418.8	628.4	837.8	942.4

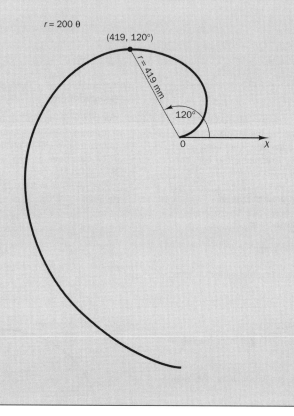

Figure 8.48

Self-assessment tasks 8.28

1. Draw the graphs of (a) $r = 2\cos\theta$, (b) $r = 2\sin\theta$ for values of θ from 0 to 360°.
2. A steel plate is to be marked out with the curve $r(\text{mm}) = 10 + 5\theta$. Plot this curve for θ from 0 to 3π radians.
3. The equation for the profile of a cam is $r(\text{mm}) = 50 + 40\cos\theta$. Draw this profile for the range of θ from 0 to 360°.
4. A display panel is to have the shape $r(\text{mm}) = 1 + \cos\theta$, for θ from 0 to 360°. Draw the panel to scale.

Solution of simultaneous equations graphically

On page 000 linear simultaneous equations were solved by calculation. Such equations may also be solved graphically, as shown in Example 8.56.

Each equation will have an infinite number of pairs of x, y values, which will satisfy the respective equation, but there will be only one pair of x, y values which will satisfy both equations. This pair of values is the solution of the equations.

Example 8.56

Use a graphical method to solve the simultaneous equations

$$y - 3x = 2$$
$$y + x = 12$$

The equations are rearranged to express y in terms of x as listed in Table 8.11, and values of y found for three values of x.

Table 8.11 Values: $y = 3x + 2$; $y = -x + 12$

x	y = 3x + 2	y = -x + 12
0	$y = (3 \times 0) + 2 = 2$	$y = 0 + 12 = 12$
2	$y = (3 \times 2) + 2 = 8$	$y = -2 + 12 = 10$
4	$y = (3 \times 4) + 2 = 14$	$y = -4 + 12 = 8$

The x, y pairs are plotted for each equation as in Fig. 8.49.

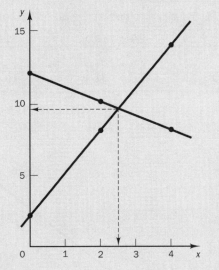

Figure 8.49

The solution is the x, y values at the point of intersection, that is $x = 2.5, y = 9.5$.

The graphical method can also be used when the equations are not linear, as shown in Example 8.57.

Example 8.57

The paths of two moving objects on a horizontal plane is given by the equations $y = -x^2 + 9x - 18$ and $y = -2.5x + 10$. If the distance x varies from 0 to 6 m, plot the graphs of the two equations and find where the two paths intersect.

Table 8.12 presents (x, y) values calculated for each equation.

Table 8.12 Values: $y = -x^2 + 9x - 18$; $y = -2.5x + 10$

x	y = -x² + 9x - 18	y = -2.5x + 10
0	$0 + 9 \times 0 - 18 = -18$	$-2.5 \times 0 + 10 = 10$
1	$-1^2 + 9 \times 1 - 18 = -10$	
2	$-2^2 + 9 \times 2 - 18 = -4$	
3	$-3^2 + 9 \times 3 - 18 = 0$	$-2.5 \times 3 + 10 = 2.5$
4	$-4^2 + 9 \times 4 - 18 = 2$	
5	$-5^2 + 9 \times 5 - 18 = 2$	
6	$-6^2 + 9 \times 6 - 18 = 0$	$-2.5 \times 6 + 10 = -5$

The graphs are shown in Fig. 8.50. From the graph the solution is

$$y = 1.2 \text{ m}, \quad x = 3.5 \text{ m}$$

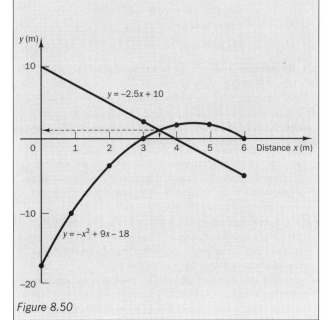

Figure 8.50

Self-assessment tasks 8.29

1. Solve, graphically, the following simultaneous equations:
   (a) $y = x$
       $y = 3x + 2$
   (b) $3x - 2y = 0$
       $x + 2y = -8$

2. Using Kirchhoff's laws has produced the following equations for currents in a circuit:

$$2i_1 - i_2 = 0$$
$$3i_2 + i_1 = 5$$

Plot the graphs of these equations and find i_1 and i_2.

3. Solve, graphically, the simultaneous equations

$$y = x - 1$$
$$y = x^2 - 3x + 2$$

4. Plot the graphs of the two equations and hence solve the equations to find F and x:

$$F = 4x^2 - 12x + 15$$
$$F = -2.5x + 25$$

Laws of experimental data

In experimental work it is often necessary to determine whether a set of values of x and y obeys a particular equation or formula. If the y values are plotted against x the points may, or may not, lie on or close to a straight line.

Linear law

If the points so plotted lie on a straight line the values obey a linear law $y = ax + b$. The values of the gradient a and the intercept b can then be obtained from the graph, so that the law for the data can be completely known.

Non-linear laws

If the values so plotted do not lie on a straight line it is possible that the values obey non-linear laws, such as,

$$y = ax^2 + b \qquad y = ax^2 + bx$$
$$y = a\sqrt{x} + b \qquad y = ax^b$$
$$y = a/x + b \qquad y = ab^x$$

It is not possible to decide from a curved graph to which law the data will apply. The technique is to modify the values plotted on each axis to produce a straight line. From the straight line so plotted it is possible to calculate the values of a and b. Each of the above equations can be reduced to give a linear graph $Y = mX + c$ as follows:

- $y = ax^2 + b$: Comparing with $Y = mX + c$, we see that plotting Y against X, that is, y against x^2, will produce a straight line if the values obey this law. From the graph the gradient is a and the intercept is b.
- $y = a\sqrt{x} + b$: Comparing with $Y = mX + c$, we see that plotting Y against X, that is, y against $x^{1/2}$ will produce a straight line if the values obey this law. From the graph the gradient is a and the intercept is b.
- $y = a/x + b$: Comparing with $Y = mX + c$, we see that plotting Y against X, that is, y against $1/x$, will produce a straight line if the values obey this law. From the graph the gradient is a and the intercept is b.
- $y = ax^2 + bx$: Divide throughout by x to give $y/x = ax + b$. Comparing with $Y = mX + c$, we see that plotting Y against X, that is y/x against x, will produce a straight line if the values obey this law. From the graph the gradient is a and the intercept is b.
- $y = ax^b$: Take logarithms to base 10 of both sides of the equation and apply the laws of logarithms

$$\log y = \log (ax^b)$$
$$= \log a + \log (x^b)$$
$$= b \log x + \log a$$

Comparing with $Y = mX + c$, we see that plotting Y against X, that is log y against log x will produce a straight line if the values obey this law. From the graph the gradient is b and the intercept is $\log a$, from which a is found from the reverse logarithm.

- $y = ab^x$: Take logarithms to base 10 on both sides of the equation and apply the laws of logarithms

$$\log y = \log (ab^x)$$
$$= \log a + \log (b^x)$$
$$= x \log b + \log a$$

Comparing with $Y = mX + c$, we see that plotting Y against X, that is log y against x, will produce a straight line if the values obey this law. From the graph the gradient is $\log b$ and the intercept is $\log a$, from which a and b are found.

This law is an exponential and will generally be found in the form $y = a e^{bx}$.

It should be noted that because experimental or actual data are subject to errors the various points are unlikely to lie exactly on a straight line. It is important, therefore, to choose the best fit straight line. The procedure is to calculate the gradient of the line from the graph to obtain the first constant. The intercept will not give the second constant if the x axis scale does not start at 0. In such a case the second constant is obtained by calculation from the original formula.

Example 8.58

The set of result shown below was obtained in a practical test to find how the stress y (MN/m^2) in a casting varied with the flange thickness x (cm). It is thought that the results obey approximately the law $y = ax^2 + b$. Verify that this is so by plotting a suitable graph and hence find the values of a and b. Calculate the maximum percentage error between any of the experimental result and the law obtained.

x (cm)	1.0	2.0	3.0	4.0	5.0
y (MN/m^2)	7.0	15	32	50	80

As described above, the law is linear if y is plotted against x^2. Therefore we produce a table of results of y against x^2 (Table 8.13), and plot these as in Fig. 8.51.

Table 8.13 Values: y against x^2

x (cm)	1.0	2.0	3.0	4.0	5.0
x^2	1.0	4.0	9.0	16.0	25.0
y (MN/m^2)	7.0	15	32	50	80

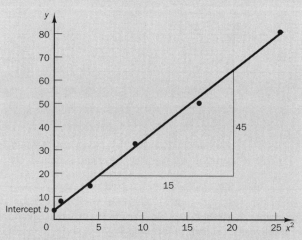

Figure 8.51

From Fig. 8.65 it is seen that the readings approximate closely to a straight line. Therefore the results obey the law. From the graph

$$\text{Gradient } a = 45/15 = 3.0$$
$$\text{Intercept } b = 4.0,$$

so that the law is

$$y = 3x^2 + 4$$

The largest deviation between a result and the straight line occurs when $x^2 = 16$. The true value from the graph is 52; the actual value is 50. Therefore,

Maximum percentage error $= (2/50) \times 100 = 4\%$

Example 8.59

In an electrical circuit the following results were obtained

Current I (A)	3.0	2.0	1.0	0.71	0.55
Resistance R (Ω)	2.0	4.0	10	15	20

Verify that these results obey the law $R = a/I + b$ and, if so, find the constants a and b.

As shown on page 278, this law reduces to a straight line when R is plotted against $1/I$. Therefore the table is recalculated to obtain values for $1/I$.

Resistance R (Ω)	2.0	4.0	10	15	20
$1/I$	0.33	0.5	1.0	1.41	1.82

The graph is shown in Fig. 8.52. It is a straight line which verifies that the results obey the law. From the graph

$$\text{Gradient } a = 12$$
$$\text{Intercept } b = -2$$

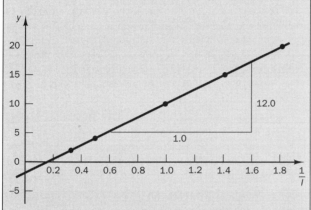

Figure 8.52

so that the law is

$$R = \frac{12}{I} - 2$$

Example 8.60

The following table shows the deflection x of a beam under given loads (N):

x (mm)	6.0	10.0	12.0	15.0	17.8	24.0
L (kN)	4.0	11.2	16.0	25.0	42.2	64.0

Verify that the results obey the law $L = ax^b$ and hence find a and b.

The linear form of the law is $\log L = b \log x + \log a$. This equation will produce a straight line graph when $\log L$ is plotted against $\log x$. The table is recalculated and the graph plotted as in Fig. 8.53

$\log x$	0.78	1.00	1.08	1.18	1.25	1.38
$\log L$	0.60	1.05	1.20	1.40	1.63	1.81

With the exception of one point, which is probably caused by experimental error, the results obey the law closely. From the graph

$$\text{Gradient } b = 0.80/0.4 = 2$$

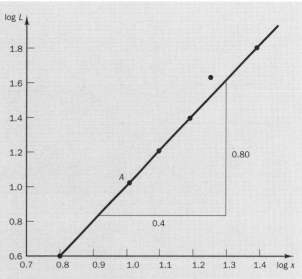

Figure 8.53

The intercept cannot be obtained from the graph because the scale on the horizontal axis does not start at the origin. We choose a known point, such as A, where $L = 11.2$ and $x = 10$ and substitute these readings, as well as $b = 2$, into the equation $L = ax^b$. Therefore

$$11.2 = a \cdot 10^2 = 100a$$
$$a = 0.112$$

The law is: $L = 0.11\, t^2$

Example 8.61

Verify that the law $L = ab^x$ holds for the maximum load L that can be carried by an alloy-steel bracket containing varing percentages x of a certain chemical.

x (%)	1	2	3	4	5
L (kN)	0.95	1.42	2.13	3.20	4.80

Hence, calculate the values of a and b.

The linear form of this law is $\log L = x \log b + \log a$. The straight line is obtained for this law if $\log L$ is plotted against x. Recalculating the table gives

x (%)	1	2	3	4	5
$\log L$	−0.02	0.15	0.33	0.51	0.68

The graph of $\log L$ against x is shown in Fig. 8.54 where the points lie on a straight line, thus verifying the law. From the graph

$$\text{Gradient} = \log b = 0.53/3 = 0.177$$
$$b = \boxed{\text{INV}}\boxed{\log}\; 0.177 = 1.50$$

The intercept cannot be obtained from the graph because the horizontal scale does not start at the origin. Using the values at the point A on the graph, where $L = 2.13$ and $x = 3$, we obtain

$$L = ab^x$$
$$2.13 = a \times 1.50^3$$
$$a = 0.63$$

The law is: $L = 0.63 \times 1.50^x$

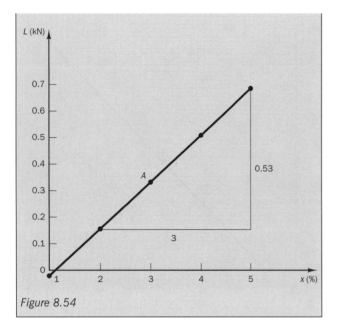

Figure 8.54

Self-assessment tasks 8.30

1. The stress σ in an alloy steel plate was recorded at different distances x and the following results were obtained:

   Distance x (cm)   4.0   7.0   8.0   10.0   11.0   12.0
   Stress σ (MN/m^2)   20.8   102   141   238   296   360

   Plot a suitable graph to show that the results follow the law $\sigma = ax^2 + bx$, and hence find the value of the constants a and b.

2. The readings of an ammeter were taken in an electrical circuit, as the resistance changed and the following table was generated:

   R (Ω)   51.4   53.8   56.0   58.0   61.0   64.0
   I (A)   7.00   4.99   3.96   3.31   2.70   2.26

   It is thought that R and I are related approximately by the law $R = a/I + b$. Plot a straight line to verify that this is so and find the constants a and b.

3. The following results, which were obtained experimentally, are thought to be related by the law $T = aL^2 + b$:

   Load L (kN)   1.0   1.5   2.0   2.3   2.5   2.7
   Torque T (kNm)   3.1   4.2   6.0   7.1   8.1   9.0

   Plot a suitable graph to check that this is so and find the constants a and b. Use the equation to estimate the torque with the load of 3.0 kN.

4. Show graphically that the values of the area of cross-section A of a beam, and the radius r of holes drilled in the beam, shown in the table, obey the law $A = ar^b$, where a and b are constants.

   Radius r (cm)   1.6   1.8   2.0   2.2   2.4   2.6
   Area A (cm^2)   2.48   2.06   1.78   1.55   1.36   1.20

   Find the values of a and b and use the formula to find the area when the radius of the hole required is 1.2 cm.

5. The temperature of a metal plate cooling in a wind tunnel varies with time according to the following table, where θ is the temperature above its surroundings and t is the time in minutes.

   Time t (min)   0   1.0   2.0   3.0   4.0   5.0
   Temperature θ (°C)   500   260   120   65   30   17

   Show that these results approximately obey the law $\theta = ab^R$, and find the constants a and b.

6. The table shows the current flowing in a circuit at different times:

   Time t (μs)   1.0   2.0   3.0   4.0   5.0   6.0   7.0
   Current i (A)   2.02   1.36   0.904   0.607   0.405   0.271   0.181

   Show that these values fit the formula $i = a e^{bt}$ and find the constants a and b.

Logarithmic graph paper

In Examples 8.60 and 8.61 graphs were plotted of the logarithms of experimental readings. To avoid calculating logarithms the results may be plotted directly on logarithmic paper. With normal graph paper the linear scaled axes have distances marked out directly proportional to the size of the numbers. For example, the length of axis between 1 and 2 is the same as the length of axis between 2 and 3. On a logarithm scale the length of axis between two numbers is proportional to the logarithm of those numbers, that is, logarithm scales are not linear. For example, the distances between 1 and 2 and between 2 and 3 are not the same. Furthermore, logarithm scales repeat themselves in cycles, since $\log 0.1 = -1$, $\log 1 = 0$, $\log 10 = 0$, $\log 100 = 2$.

Two types of logarithmic paper are available, and both are available in 1, 2 or 3 cycles:

- Log/log paper, in which both axis have logarithmic scales, which would be used to verify the law $y = ax^b$.
- Log/linear paper, in which one axis only has a logarithmic scale, the other scale being linear, which would be used to verify the law $y = ab^x$.

Examples 8.60 and 8.61 are now repeated using log/log and log/linear paper respectively.

In Example 8.60 the law is $L = ax^b$, the linear form being $\log L = b \log x + \log a$. In this case both axes need the logarithmic scale so that log/log paper is used. Also since both variables are within 1–10 and 10–100 bands, a 2-cycle paper will be required. The graph is shown in Fig. 8.55.

It is seen from the graph that the points lie on a straight line, thus verifying the law. The constants a and b are now determined by substituting values of V and t of two points *on the line*, into $\log L = b \log x + \log a$, and solving two simultaneous equations.

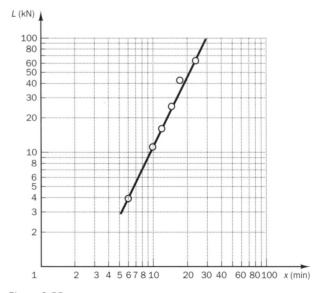

Figure 8.55

In Example 8.61 the law is $L = ab^x$, the linear form being $\log L = x \log b + \log a$. In this case $\log L$ is plotted against x so that log/linear paper is required. Since L is within 0.1–1 and 1–10 bands, a 2-cycle paper is required as shown in Fig. 8.56.

The graph is a straight line and thus the law is verified. Again to find the constants a and b two points are considered *on the line* and values of L and x used in a calculation with two simultaneous equations.

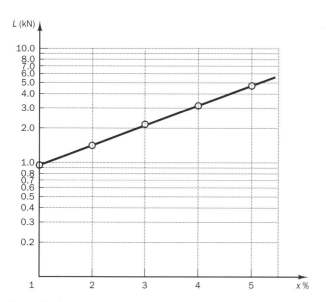

Figure 8.56

Areas under graphs

In engineering it is frequently necessary to find the areas under graphs. For example, the work done by a machine in overcoming a force is the *force × distance moved against it*. In a graph of force against distance the area under the graph is the work done. When graphs are linear it is easy to calculate the area, but when the graphs are curved calculating the area is more difficult.

One method of doing so is to divide the area under the curve into a number of vertical strips of equal width, as shown in Fig. 8.57(a).

Horizontal lines are drawn at the top of each strip as shown, creating rectangles from the strips. The area under the curve is approximately equal to the sum of the areas of the rectangles. An alternative method is to create the rectangles as in Fig. 8.57(b).

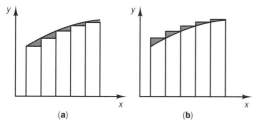

Figure 8.57

It can be seen that in Fig. 8.57(a) the sum of the areas of the rectangles is less than the area under the curve by an amount of the shaded areas. In Fig. 8.57(b) the sum is greater than the area under the curve by an amount of the shaded areas.

The actual area lies between these two values. A better approximation is to consider the strips as trapezia, by drawing straight lines such as BC across the top of strips, as shown in Fig. 8.58(a).

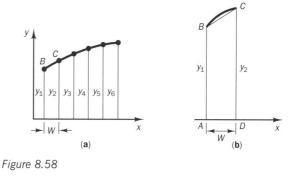

Figure 8.58

An enlarged view of the first strip is shown in Fig. 8.58(b). It is seen that the area under this portion of the curve BC approximates closely to the area of the trapezium ABCD. If w is the width of the trapezium

Area of the first trapezium $= \frac{1}{2}(y_1 + y_2)w$

Area of the second trapezium etc. $= \frac{1}{2}(y_2 + y_3)w$

Total area of the trapezia, on adding these expressions for the five strips, is approximately the area under the curve, that is

Area under the curve $= \left[\frac{1}{2}(y_1 + y_6) + (y_2 + y_3 + y_4 + y_5)\right]w$

This result is known as the trapezoidal rule, which extended to any number of strips is

Area under curve = $[\frac{1}{2}$(first + last ordinates) + sum of all other ordinates] × width of strip

The greater the number of strips the more accurate will be the result. As a compromise, 4 to 10 strips will avoid excessive calculation and produce a reasonably accurate result.

Example 8.62

The current–time formula for a circuit is given by $i = t^2 + 10$, where i is measured in amps. Draw the graph for the range $t = 0$ to $t = 4$ s. Calculate the total charge in coulombs transferred to the circuit, given by the area under the curve.

A table of values is first obtained in order to plot a graph, as shown in Fig. 8.59

t (s)	0	1	2	3	4
i (A)	10	11	14	19	26

Using four strips with $w = 1$ s in the trapezoidal rule we obtain

Area under the graph $= \left[\frac{1}{2}(10 + 26) + (11 + 14 + 19)\right] \times 1$

$= 62$ coulombs

Note: Using a method of integration, which is outside this course, will give an exact result for the area under the curve of 61.33. Thus, with four strips the error is only $0.67 \times 100/61.33 = 1.1$ per cent.

Figure 8.59

Self-assessment tasks 8.31

1. Use the table below to draw a force–distance graph:

Distance x (m)	1.0	1.3	1.5	1.75	2.0
Force F (kN)	7.0	8.7	10.0	11.9	13.9

   Use the trapezoidal rule to find the work done by the force (that is, the area under the curve.)
2. Find the area of the cam in question 3 on page 276 using the trapezoidal rule. If the cam has a thickness of 5 mm, and its density is 7.2×10^3 kg/m^3, find the mass of the cam.
3. Draw the graph $y = 40 \ln 2x$ for the range $x = 100$ mm to $x = 200$ mm. A template has the shape formed between the curve, the x axis and the ordinates $x = 100$ and $x = 200$. Find the area of the plate.
4. Draw the current–time graph $i = 50 \, e^{-2t}$ for $t = 0$ to $t = 1$ s. Hence find the charge transferred during this time (equal to the area under the curve).

Differentiation

In Example 8.50 it was seen how the gradient of a non-linear graph at any point could be found by drawing a tangent to the curve at that point.

The gradient gives a measure of the steepness of the graph at any point, which is a measure of how one quantity, say y, changes with the other quantity, say x. This rate of change has applications in current flow, velocity, acceleration, etc.

Differentiation is a mathematical method of finding the gradient of graphs without having to plot the curve and draw tangents.

Consider the problem of finding the gradient of a curve at a point A (Fig. 8.60). Another point B is taken close to A on the curve. The region A to B is enlarged in Fig. 8.60(b).

From Fig. 8.60(a) it is seen that the gradient of the straight line from A to B, that is, the chord AB, is very nearly equal to the gradient of the tangent at A. We can demonstrate this with a simple calculation for the curve $y = x^2$.

At A, $x = 3$; $y = AD = 3^2 = 9$
At B, $x = 3.001$; $y = BE = 3.001^2 = 9.006001$

Gradient of the chord AB

$$\text{Gradient} = \frac{BC}{AC} = \frac{BE - AD}{AC} = \frac{0.006001}{0.001} = 6.001$$

If we make B move closer to A and repeat the calculation, this gradient becomes closer and closer to 6, which will be its limiting value when B coincides with A. When this happens the chord AB becomes a tangent at A. This calculation with arithmetic numbers forms a basis of the method of finding an exact value for the gradient at any point on a graph. This process, which is a general one, is called *differentiation*.

One other point to note is the way that small quantities are expressed. In the above calculation AC is a very small addition to x, and is written as δx. Similarly, a small addition to a y value is written as δy. We now repeat the above calculation using δx and δy. Referring to Fig. 8.60(b), let the co-ordinates of point A be (x, y) and the co-ordinates of point B be $(x + \delta x, y + \delta y)$.

Substituting these values into the equation $y = x^2$ gives:

At A: $y = x^2$ (i)
At B: $y + \delta y = (x + \delta x)^2 = x^2 + 2x(\delta x) + (\delta x)^2$ (ii)

Subtracting equation (i) from (ii) gives

$$\delta y = BC = 2x(\delta x) + (\delta x)^2$$

Dividing throughout by δx

$$\delta y / \delta x = 2x + \delta x$$

But $\delta y / \delta x$ is the gradient of the chord AB. As B moves closer to A the chord AB becomes a tangent at A, that is, as $\delta x \to 0$. At this point $\delta y / \delta x$ is written as dy/dx and is called the *differential coefficient*. Therefore, the gradient of the tangent at A is

$$dy/dx = 2x$$

If $x = 3$ at the point A, $dy/dx = 2 \times 3 = 6$.

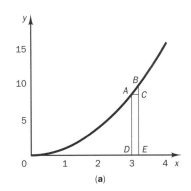
(a)

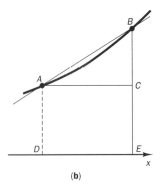
(b)

Figure 8.60

The formula for differentiating the general term $y = ax^n$ and a polynomial

The differential coefficient of higher powers of x can be obtained in the same way, as follows.

- When $y = x^3$: $dy/dx = 3x^2$
- When $y = x^4$: $dy/dx = 4x^3$
- When $y = x^5$: $dy/dx = 5x^4$

The general expression for any power can now be stated.

$$\text{When} \quad y = ax^n: \quad dy/dx = nax^{n-1} \qquad (13)$$

The following are examples of the use of this rule.

1. If $y = 5x^2$: $dy/dx = 2 \times 5x^1 = 10x$
2. If $y = 3x = 3x^1$: $dy/dx = 1 \times 3x^0 = 3$
3. If $y = 4 = 4x^0$: $dy/dx = 0 \times 4x^{-1} = 0$
4. If $y = 7x^{-2}$: $dy/dx = -2 \times 7x^{-3} = -14x^{-3}$

Example 8.63

Find the gradient of the curve and the angle of inclination of $y = x^2 + 2x - 4$ at the point $(1, -1)$.

Differentiate the polynomial by applying the rule to each term individually:

$$dy/dx = 2x + 2$$

At the point $(1, -1)$ $x = 1$, so that $dy/dx = 2 \times 1 + 2 = 4$.

The angle of inclination of the curve at this point is $\tan^{-1} 4$, s shown in Fig. 8.61 since gradient is (side opposite)/(side adjacent).

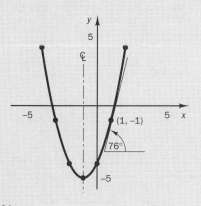

Figure 8.61

Differentiation of sin x and cos x

When trigonometric functions are differentiated the angle must always be in radians. In Fig. 8.62 OAB is a right-angled triangle, whose hypotenuse $OA = 1$, and the angle $AOB = x$ rad. OA is rotated to OC though a small angle δx. Figure 8.62(b) is enlarged to show the analysis.

Using the result from the radian measure of an angle (see p. 272),

$$\text{arc } AC = 1 \times \delta x = \delta x \qquad (i)$$

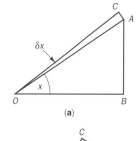

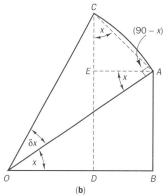

Figure 8.62

When $\delta x \to 0$:

- The angle $CAO \to 90°$ (as seen in Fig. 8.62(a))
- The arc AC is practically a straight line, becoming the hypotenuse of triangle AEC, in which case angle $ACE = x$ rad.

Differentiating sin x

$$y = \sin x = AB/OA = AB/1 = AB$$
$$y + \delta y = \sin(x + \delta x) = CD/OC = CD/1 = CD$$

Subtracting

$$\delta y = \sin(x + \delta x) - \sin x = CD - AB = CE$$

Therefore

$$\delta y / \delta x = CE/\delta x = CE/AC \quad \text{from result (i)}$$

When $\delta x \to 0$, arc AC becomes the hypotenuse, and $\delta y/\delta x = dy/dx$ and $CE/AC = \cos x$, that is:

$$y = \sin x, \quad dy/dx = \cos x \qquad (14)$$

Differentiating cos x

$$y = \cos x = OB/OA = OB/1 = OB$$
$$y + \delta y = \cos(x + \delta x) = OD/OC = OD/1 = OD$$

Subtracting

$$\delta y = \cos(x + \delta x) - \cos x = OD - OB = -DB$$

Therefore

$$\delta y/\delta x = -DB/\delta x = -EA/\delta x = -EA/AC \quad \text{from result (i)}$$

When $\delta x \to 0$, arc AC becomes the hypotenuse, and $\delta y/\delta x = dy/dx$ and therefore $CE/AC = -\sin x$, that is:

$$y = \cos x, \quad dy/dx = -\sin x \qquad (15)$$

Differentiation of the exponential and logarithm functions

Exponential function

In the section on logarithms (p. 249) the exponential function was written as

$$e^x = 1 + x + \frac{x^2}{1 \cdot 2} + \frac{x^3}{1 \cdot 2 \cdot 3} + \frac{x^4}{1 \cdot 2 \cdot 3 \cdot 4} + \cdots$$

Differentiating as a polynomial

$$\frac{dy}{dx} = 0 + 1 + \frac{2x}{1 \cdot 2} + \frac{3x^2}{1 \cdot 2 \cdot 3} + \frac{4x^3}{1 \cdot 2 \cdot 3 \cdot 4} + \cdots$$

$$= 1 + x + \frac{x^2}{1 \cdot 2} + \frac{x^3}{1 \cdot 2 \cdot 3} + \frac{x^4}{1 \cdot 2 \cdot 3 \cdot 4} + \cdots$$

Therefore

$$y = e^x, \quad dy/dx = e^x \quad (16)$$

Logarithm function

The logarithm function is $y = \ln x$. Changing into the exponential form

$$x = e^y \quad (ii)$$

Differentiating using equation (16),

$$dy/dx = e^y$$

But we require dy/dx.
Inverting

$$dy/dx = 1/e^y = 1/x \quad \text{from (ii)}$$

thus

$$y = \ln x, \quad dy/dx = 1/x \quad (17)$$

These results are summarised below, together with functions where the x variable has a coefficient a:

$$y = \sin x, \quad dy/dx = \cos x$$
$$y = \sin ax, \quad dy/dx = a \cos ax$$
$$y = \cos x, \quad dy/dx = -\sin x$$
$$y = \cos ax, \quad dy/dx = -a \sin ax$$
$$y = e^x, \quad dy/dx = e^x$$
$$v = e^{ax}, \quad dy/dx = a\,e^{ax}$$
$$y = \ln x, \quad dy/dx = 1/x$$
$$y = \ln ax, \quad dy/dx = 1/x$$

These results are used in the following examples:

(a) $y = \sin 7x, \quad dy/dx = 7 \cos 7x$
(b) $y = 4e^x, \quad dy/dx = 4e^x$
(c) $y = \ln 5x, \quad dy/dx = 1/x$
(d) $v = 5e^{-3x}, \quad dy/dx = 5 \times -3e^{-3x} = -15e^{-3x}$
(e) $y = 6 \cos 2x, \quad dy/dx = 6 \times -2 \sin 2x = -12 \sin 2x$

> **Example 8.64**
>
> Find the gradient of the curve $y = 2 \sin 3t$ at the point where $t = 1.6$.
>   Differentiating
>
> $$dy/dt = 2 \times 3 \cos 3t = 6 \cos 3t$$
>
> At $t = 1.6$:
>
> $$dy/dt = 6 \cos 3 \times 1.6 = 6 \cos 4.8 = 0.52$$

Repeated differentiation

An equation can be differentiated more than once, as shown by the repeated differentiation of $y = 5x^4 - 6x^3 + 3x - 2$:

$$dy/dx = 20x^3 - 18x^2 + 3$$

The second differentiation is written as

$$d^2y/dx^2 = 60x^2 - 36x$$

Third and fourth differentiation coefficients are written as d^3y/dx^3 and d^4y/dx^4.

> **Example 8.65**
>
> If $y = 5e^{-3t} + 2t$ find dy/dt and d^2y/dt^2.
>
> $$y = 5e^{-3t} + 2t$$
> $$dy/dx = 5 \times -3e^{-3t} + 2 = -15e^{-3t} + 2$$
> $$d^2y/dx^2 = -15 \times -3e^{-3t} + 0 = 45e^{-3t}$$

> **Self-assessment tasks 8.32**
>
> 1. Differentiate the following expressions
>   (a) $y = 5x^3 - 2x + 7$ (b) $y = (x + 2)(x - 1)$
>   (c) $y = 5e^{0.2x}$ (d) $y = 7 \sin x$
>   (e) $y = \ln 9x$ (f) $y = e^{-2x}$
>   (g) $y = \cos 3x$ (h) $y = 5 \ln x$
> 2. Use differentiation to find the gradient and angle of the curve $y = 2x^2 - 8x + 1$ to the x axis at the point (2, 3) (3, 10) and (1, 0). Make a sketch of the graph showing significant features, including the results you have obtained.
> 3. If $y = 7e^{0.1x}$ find dy/dx and d^2y/dx^2.
> 4. If $y = 2 \cos 5x$ find the value of dy/dx when $x = 1.2$ rad.
> 5. Find the gradient of the graph $y = 2 \ln 3x$ when $x = 1.5$.
> 6. What is the value of the gradient of the graph $y = 2e^{-3x}$ at the point $x = 0.2$.

Engineering applications of differentiation: rates of change

It was seen in the previous sections that differentiation was the rate of change of one variable y with another variable x. When applied to real situations in engineering these rates of change are actual physical quantities.

- When the rate of change is of distance s with time t, the differential coefficient is:

  ds/dt which is velocity (m/s)

- When the rate of change is of velocity v with time t, the differential coefficient is:

  dv/dt which is acceleration (m/s^2)

- When the rate of change is of electric charge q with time t, the differential coefficient is:

  dq/dt which is the current i (A)

Example 8.66

The equation for the distance s (mm) moved by a sliding mechanism in terms of the time t (s) is given by $s = 80 \sin 0.2t$. Find the velocity and acceleration when $t = 1$ s.

$$s = 80 \sin 0.2t$$

- Velocity:
$$v = ds/dt = 80 \times 0.2 \cos 0.2t = 16 \cos 0.2t$$

- Acceleration:
$$a = dv/dt = 16 \times -0.2 \sin 0.2t = -3.2 \sin 0.2t$$

- After 1 s:
$$v = 16 \cos 0.2 \times 1 = 15.7 \text{ mm/s}$$
$$a = -3.2 \sin 0.2 \times 1 = -0.636 \text{ mm/s}^2$$

The negative sign in the value of the acceleration indicates that at the time $t = 1$ s the mechanism is slowing down.

Example 8.67

The growth of a current i (A) in a coil is given by the formula $i = 5(1 - e^{-2t})$, where t is the time is seconds after closing the switch. Find the rate at which the current is changing at (a) $t = 0$, (b) $t = 1$s. Make a sketch of the current–time graph.

Converting the equation into a sum of terms

$$i = 5 - 5e^{-2t}$$

$$di/dt = 0 - 5 \times -2e^{-2t} = 10e^{-2t}$$

When $t = 0$: $\quad di/dt = 10e^0 = 10$ A/s

When $t = 1$: $\quad di/dt = 10e^{-2} = 1.35$ A/s

This shows that the rate of growth is decreasing with time, that is, the graph is getting flatter with time, and eventually reaches a final steady state, with no further change. The values of i for a range of time t is shown in Table 8.14, and the sketch is shown in Fig. 8.63.

Table 8.14 Values: $i = 5(1 - e^{-2t})$

t (s)	$i = 5(1 - e^{-2t})$
0	$i = 5(1 - e^0) = 0$
1	$i = 5(1 - e^{-2}) = 4.32$ A
2	$i = 5(1 - e^{-4}) = 4.91$ A

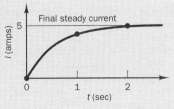

Figure 8.63

Self-assessment tasks 8.33

1. A missile is fired vertically and its height h (m) above the ground at a time t (s) is given by $h = 200t - 5t^2$. Find a formula for its velocity and acceleration. Find:
   (a) its height, velocity and acceleration after 4 s
   (b) its velocity after 20 s (what is the significance of this result?)
   (c) the greatest height reached.
2. The instantaneous charge q (C) in a capacitive circuit is $q = 1.3(1 - e^{-1.2t})$. Find the current dq/dt when $t = 1$ s.
3. The instantaneous current i (A) in an inductive circuit at a time t (s) is $i = 8e^{-5t}$. Find the rate at which the current is decreasing at $t = 0.02$ s.
4. A reciprocating machine slider is moving according to the equation $x = 200 \cos 15t + 120$, where x (mm) is the distance moved in a time t (s). Find:
   (a) the position of the slider after 0.1 s
   (b) the velocity of the slider after 0.1 s
   (c) the furthest distance moved by the slider
   (d) the maximum sliding velocity.
5. The distance s (m) moved by an object in a time t (s) is given by $s = 2t^3 - 13t^2 + 24t + 10$. Find the formulae for velocity and acceleration. Hence find:
   (a) the distance, velocity and acceleration after 2 s
   (b) the time t at which the velocity is zero
   (c) the time t at which the acceleration is zero.

Maxima and minima

It is often important in engineering to obtain maximum and minimum values of variables, such as the maximum stress that is possible in a bar, or the minimum impedance of an electric circuit. Provided that an equation describes the situation maximum or minimum values can be obtained by drawing a graph or by a process of differentiation. Consider the two graphs in Fig. 8.64. In Fig. 8.64(a) the graph has a maximum at the point A; in Fig. 8.64(b) the graph has a minimum at the point B.

It is seen that at both points the gradient $dy/dx = 0$ and this fact can be used to find the co-ordinates of either point, without the necessity of drawing a graph.

Because the condition $dy/dx = 0$ is the same condition for both maximum and minimum points, it is necessary to have a means of distinguishing between them. This is achieved by finding how the gradient changes about each of the points.

As seen in both graphs, moving from left to right:

- With a maximum point A, dy/dx changes from '+' to '−', that is d^2y/dx^2 will be negative.
- With a minimum point B, dy/dx changes from '−' to '+', that is d^2y/dx^2 will be positive.

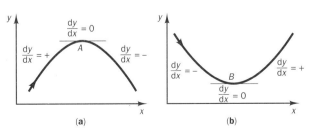

Figure 8.64

Therefore, to identify and find a maximum or minimum point the procedure is:

1. Set $dy/dx = 0$ and calculate the x value of the point.
2. Use this x value to find the y value; that is, the actual maximum or minimum value.
3. Find d^2y/dx^2.
4. Substitute the x value found in (1) into d^2y/dx^2:
   - If d^2y/dx^2 is positive the point is a minimum.
   - If d^2y/dx^2 is negative the point is a maximum.

Example 8.68

Find the maximum or minimum value of the curve

$$y = -3x^2 + 12x + 5$$

and sketch the curve.

Differentiating:
$$dy/dx = -6x + 12$$

Set $dy/dx = 0$:
$$-6x + 12 = 0$$
$$x = 2$$

Substitute $x = 2$ into $y = -3x^2 + 12x + 5$:
$$y = -3(2)^2 + 12(2) + 5$$
$$= -12 + 24 + 5$$
$$= 17$$

Hence there is a maximum or minimum point at (2, 17). Differentiating a second time:

$$d^2y/dx^2 = -6$$

In this case the second differential coefficient is a negative value without substituting $x = 2$. The point (2, 7) is therefore a maximum point.

To sketch the curve we know that it has a maximum at the point (2, 17) and that it cuts the y axis at $y = 5$ (that is, when $x = 0$). The sketch is shown in Fig. 8.65.

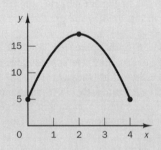

Figure 8.65

Example 8.69

The current i (A) in a circuit at a time t (s) is given by:

$$i = \tfrac{1}{3}t^3 - 3t^2 + 5t + 1$$

Find the maximum and minimum values of the current and the times when they occur. Sketch the graph for the range $t = 0$ and $t = 6$ s.

(a) For maximum or minimum points
$$di/dt = t^2 - 6t + 5 = 0$$
$$(t - 1)(t - 5) = 0, \text{ i.e. } t = 1\,\text{s or } t = 5\,\text{s}$$

If $t = 1$ $i = \tfrac{1}{3}1^3 - 3(1)^2 + 5(1) + 1 = 3.33\,\text{A}$

If $t = 5$: $i = \tfrac{1}{3}5^3 - 3(5)^2 + 5(5) + 1 = -7.33\,\text{A}$

(b) $d^2i/dt^2 = 2t - 6$.

At $t = 1$: $d^2i/dt^2 = 2(1) - 6 = -4$,
the (1, 3.33) point is a maximum

At $t = 5$: $d^2i/dt^2 = 2(5) - 6 = 4$,
the (5, −7.35) point is a minimum

The sketch is shown in Fig. 8.66

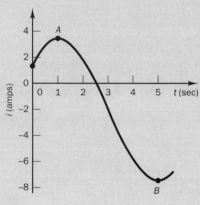

Figure 8.66

It is important to note that maximum and minimum points are not the largest or smallest values; rather they are the turning points where on the graph the values reverse, as shown at the points A and B in Fig. 8.66.

Self-assessment tasks 8.34

1. Use differentiation to find the co-ordinates of the maximum and minimum points on the curve $y = -x^2 + 2x + 3$. Sketch the curve showing the main points.
2. Use differentiation to find the maximum and minimum points on the curve $y = 2x^3 - 5x^2 - 4x$. Sketch the curve showing the main points.
3. The vertical height h (m) reached by a shell fired from a gun is given by $h = 50t - 5t^2$. Find the time and height of the shell when it is at its maximum height above the ground.
4. A solid rectangular block of metal with a square base of side x (mm) is to have a total surface area of $1{,}500\,\text{mm}^2$. Show that the volume of this block is given by $V = 375x - 0.5x^3$. Hence find the dimensions of the block to have maximum volume. Comment on the dimensions of the maximum volume.
5. A closed cylindrical tank is to have a volume of $2\,\text{m}^3$. Show that the total surface area A (m^2) of this tank is given by $A = 2\pi R^2 + 4/R$. Hence find the dimensions of the tank which has a minimum surface area. Calculate the value of this minimum area.
6. An open metal tray is to be made from a rectangular sheet of metal 1 m by 2 m. Four identical squares of side x (m) are to be cut out from the corners of the rectangle, and the sides folded to make a tray of depth x (m). Show that the volume V (m^3) of the tray is $V = 2x - 6x^2 + 4x^3$. Hence find the dimensions x for maximum volume.

8.4 Unit test

Test yourself on this unit with these sample multiple-choice questions.

1. The length of a bar is given as 3.7 m. State the shortest possible length of the bar. Select from
   (a) 3.6 m
   (b) 3.65 m
   (c) 3.69 m
   (d) 3.7 m

2. An engine casting has a mass of 600 kg, correct to 2 significant figures. The maximum mass of casting is:
   (a) 600 kg
   (b) 601 kg
   (c) 605 kg
   (d) 610 kg

3. The number 57 can be written as
   $32 + 16 + 8 + 0 + 0 + 1$.
   Which of the following is the number in binary?
   (a) 111111
   (b) 111001
   (c) 111010
   (d) 111100

4. The structure in Fig. 8.T1 is made of a cone on top of a cylinder. Which of the following is the expression for the total volume?
   (a) $\frac{1}{3}\pi x^2 h$
   (b) $\frac{4}{3}\pi x^3$
   (c) $\frac{1}{3}\pi x^3$
   (d) πx^3

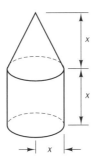

Figure 8.T1

5. Which of the following gives an answer of 1?
   (a) $4x - x - 3x$
   (b) $(3y)^0$
   (c) $3(y)^0$
   (d) y^{-1}

6. Which of the following is the correct factorisation of $x(a - b) + (-a + b)$?
   (a) $(a - b)(x + 1)$
   (b) $(a - b)(x - 1)$
   (c) $x(a - b)$
   (d) $(-a + b)(x - 1)$

7. In the right-angled triangle ABC in Fig. 8.T2, $\cos A = \frac{12}{13}$. Which of the following statements are correct?
   (a) $\sin C = \cos A$
   (b) $\cos B = \frac{5}{12}$
   (c) $\tan C = \frac{5}{12}$
   (d) $\tan A = \tan C$

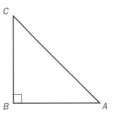

Figure 8.T2

8. $\cos 156°$ is equal to:
   (a) $\cos 24°$
   (b) $-\cos 24°$
   (c) $\cos -24°$
   (d) $\sin 66°$

9. If $\tan = -1.732$ which of the following are correct?
   (a) $\theta = 60°$
   (b) $\theta = -150°$
   (c) $\theta = 120°$
   (d) $\theta = 240°$

10. In the vector diagram (Fig. 8.T3) find the largest angle between two vectors.
    (a) 72°
    (b) 92°
    (c) 108°
    (d) 252°

Figure 8.T3

11. What is the unknown force F (newtons) in the force diagram (Fig. 8.T4)?
    (a) 0.45
    (b) 0.56
    (c) 0.90
    (d) 1.79

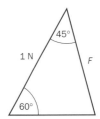

Figure 8.T4

12. A straight line passes through the points (−1, 0) and (0, −2). Which of the following is the equation of the line?

    (a) $y = -2x + 2$
    (b) $y = -2x - 2$
    (c) $y = -x - 1$
    (d) $y = -2x - 1$

13. The equation of the parabola in Fig. 8.T5 is:

    (a) $y = x^2$
    (b) $y = -x^2 + 4$
    (c) $y = x^2 - 4$
    (d) $y = -x^2 + 2x + 4$

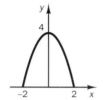

Figure 8.T5

14. The equation of the curve in Fig. 8.T6 is:

    (a) $y = 2 \ln x$
    (b) $y = 2(1 - e^{-0.5x})$
    (c) $v = 2e^{0.2x}$
    (d) $y = 2e^{-0.2x}$

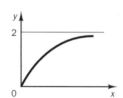

Figure 8.T6

15. The frequency and period of $x = 4 \sin(10\pi t - 1)$ is:

    (a) 4, 1
    (b) 10, 0.1
    (c) 10π, 10
    (d) 5, 0.2

16. The equation for the curve shown in Fig. 8.T7 is which of the following?

    (a) $y = 20 \sin(\theta - 30°)$
    (b) $y = 20 \sin(\theta + 30°)$
    (c) $y = 10 \sin(\theta - 30°)$
    (d) $y = 10 \sin(\theta + 30°)$

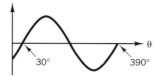

Figure 8.T7

17. Figure 8.T8 shows a linear graph when log y is plotted against log x. Which of the following equations matches the graph?

    (a) $y = 3.32x^{2.1}$
    (b) $y = 3.32x^{0.741}$
    (c) $y = 1.2x^{2.1}$
    (d) $y = 1.2x^{0.741}$

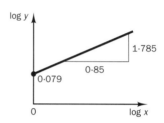

Figure 8.T8

18. The area under the curve in Fig. 8.T9 is:

    (a) 13
    (b) 14
    (c) 28
    (d) 56

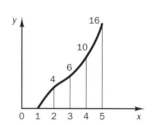

Figure 8.T9

19. The gradient of the tangent to the curve $y = x^2 - x + 2$ at the point (1, 2) is:

    (a) 1
    (b) 2
    (c) 3
    (d) 4

20. The current i flowing in a circuit is given by the equation $i = t^2 - 2t + 2$. What is the minimum current flowing?

    (a) 0
    (b) 1A
    (c) 2A
    (d) 4A

Answers to numerical self-assessment tasks

7.1
1. 1.25 kg
2. 40 cm

7.2 2. $r_A = 147.1\,\text{N}$; $r_B = 343.4\,\text{N}$

7.3 $r_A = 1{,}000\,\text{N}$; $r_B = 7{,}000\,\text{N}$

7.4 $-3{,}000\,\text{N}$

7.5 2.4 mm

7.6
1. 2.8 s; 17.7 m/s
2. 1.4 s; 18.4 m/s

7.7
1. $48{,}000\,\text{kg m s}^{-1}$; 4,000 N
2. $2.5\,\text{kg m s}^{-1}$
3. 225 m

7.8
1. 25×10^6 J
2. 6.25 J; 85 J
3. 135 kJ

7.9
1. 343 kJ
2. 31.3 m/s
3. 1,840 J; 70 m/s; 7.1 s; $52.5\,\text{kg m s}^{-1}$

7.10
1. 419 rad/s
2. 262 rad/s
3. 120 rev/min

7.11
1. 13.5 m/s
2. 4.7 m/s ($r = 30\,\text{cm}$); 9.4 m/s ($r = 60\,\text{cm}$)
3. 100 rad/s

7.12 3.72 mm

7.13
1. 1×10^6 Pa
2. 2.56×10^5 Pa

7.14
1. 720 kJ
2. 239 kJ; 26 °C

7.15
1. 2.95×10^6 J
2. 19.5 kJ

7.16
1. 650 V
2. 24 V

7.17
1. 4.3 A
2. 11.48 V
3. 26 Ω

7.18
1. 0.46 A
2. 16 Ω; 35 Ω
3. 0.7 Ω; 17 A

7.19
1. 0.75 A; 0.23 A; 0.5 A; 0.25 A; 0.25 A
2. $I_8 = 0.5\,\text{A}$; $I_9 = 0.75\,\text{A}$

7.20
1. $I_1 = 5.8\,\text{A}$; $I_2 = 4\,\text{A}$; $I_3 = 2\,\text{A}$
2. $I_1 = 6.13\,\text{A}$; $I_2 = 4\,\text{A}$; $I_3 = 2.13\,\text{A}$; $I_4 = 1.33\,\text{A}$; $I_5 = 0.8\,\text{A}$

7.21
1. 36×10^{-6} C
2. 60×10^{-6} C
3. 3×10^{-6} C

7.22 1,250 Hz; 6.5 V; 13 V; 9.19 V

7.23 1. 0.1 N; 1.5 N; 5 N

7.24 11.5 V

8.1
1. $+1, -12, -5, +11, +3$
2. $-1, -9, -2$
3. $+60, -24, -8$
4. $+3, -7, -8$

8.2
1. $8^4, 5^{11}, 3^4, 9^7, 6^3, 8^9$
2. $\frac{1}{8}, \frac{1}{7^4}, 5^3, 3 \times 2^4, (\frac{4}{3})^4$

8.3
1. $4.35 \times 10,\ 7.56 \times 10^2,\ 3.005 \times 10^2,\ 9.7 \times 10^{-3},\ 2 \times 10^{-3},\ 5.72 \times 10^{-4}$
2. 6.0×10^5
3. $2.455 \times 10^3,\ 4.092 \times 10$
4. $8.56 \times 10^{-5},\ 2.57 \times 10^{-4}$

8.4
1. 55, 33, 27
2. 1010, 11101, 101101, 10100111
3. 111, 11001, 1001101, 100000

8.5
1. (a) $9.8 \times 10^4\,\text{mm}^3$, $7.1 \times 10^2\,\text{cm}^3$, $2.93 \times 10^2\,\text{m}^3$
2. $3.1 \times 10^3\,\text{mm}^3$
3. $8.5 \times 10^4\,\text{mm}^3$
4. $1.9 \times 10^5\,\text{mm}^3$

8.6
1. $2.8 \times 10^2\,\text{g}$, $2.0 \times 10^3\,\text{g}$, $8.4 \times 10^5\,\text{kg}$
2. $7.82\,\text{g/cm}^3$, $7{,}820\,\text{kg/m}^3$
3. $110\,\text{cm}^3$

8.7
1. $12x, -5m, -15t, -2u + 2v$
2. $2x$
3. $d = 2A$

8.8
1. $-18t^6, 20y^4, -60m^6$
2. $-9f^3, -9$
3. $49w^6, 7w^6, 5^7$
4. $\frac{32}{3} \pi y^3$
5. $\frac{4}{25} \pi t^3$

8.9
1. (a) (i) 2.571, (ii) 3.751, (iii) 1.319, (iv) 0.4343
   (b) (i) 4.0×10^5, (ii) 75.2
2. 0.0122 s
3. (a) $2 \ln 3$, (b) $\log 5$, (c) $3/2$, (d) $\log a/b$
4. (a) 2.24, (b) 3.33, (c) 1.2

8.10
1. (a) $23a/20$,
   (b) $8t/21$,
   (c) $(76z + 100)/35$
2. $4r/3$, $16/3$

8.11
1. $-\frac{7}{3} y^2$, $-\frac{9}{5} s^{0.5}$, $\frac{11}{6} t^3$
2. $\frac{1}{3} y(7t - 3)^3$
3. $9c/abd$

8.12
1. $8t^2 + 10t - 7,\ 7a^2 - 19a - 6,\ s^2 + 2s - 8,\ r^2 - 9$
2. $u^3 + 9u^2 + 26u + 24,\ 2v^3 - v^2 - 8v + 4$

8.13
1. $3(2a + 3b - 4c)$, $2a(2p - 10q - 5r)$
2. $(u - v)(t - s)$, $5(y + 5z)(a - b)$

8.14
1. $(p - q)(p + q)$, $(m - 4n)(m + 4n)$, $12(5t - 1)(5t + 1)$
2. $\pi(R - r)(R + r)$

8.15
1. $(x - 6)(x + 4)$
2. $(x - 3)(x - 2)$
3. $(2x - 3)(3x + 5)$
4. $(3a - 1)(a - 2)$
5. $(2t - 5)(2t - 3)$
6. $(2m + 3)(4m - 3)$

8.16
1. $\alpha = (L - l)/lt$
2. $r = \left(R^2 - \dfrac{40}{\pi}\right)^{0.5}$
3. $r = R/(1 - 2R)$
4. $g = (4\pi^2 l)/T^2$

8.17
1. $y = 8$
2. $s = 7$
3. $x = -8$
4. $t = 5$
5. $z = -19/2$
6. $x = 1/5$
7. 60 cm^2
8. 8 A, 2 A

8.18
1. 1, 2
2. 2, −3
3. 4, −5
4. 75, 15

8.19
1. (a) 1, (b) 5/3, −2 (c) 1/2, 5, (d) 1/2, −7/2, (e) 1/3, 2
2. (a) −0.62, 1.62, (b) 1.77, 0.57, (c) 0.47, −1.27, (d) 1.60, −1.10, (e) 4.21, −0.71

8.20
1. 40°, 140° 2. 199°, 341° 3. 73°, 287°
4. 123°, 237° 5. 12°, 192° 6. 115°, 295°

8.21
1. 1.3 m, 77° 2. 44°, 4.3 m
3. 114 V, 35° 4. 30.62 mm, 9.518°
5. 1.9 m, 86° 6. 4,060 m, 3,440 m
7. 1.4 V, 134° 8. 230, 17°
9. 1.5 m, 1.2 m

8.22
1. 59.0 cm^2 2. 22°
3. 4.1 m^2 4. 34 cm^2

8.23
1. (a) (26, −51°), (b) (5.4, 112°), (c) (6.7, 243°)
2. (a) (10.4, 6), (b) (−12.9, −15.3), (c) (−14.1, 5.13)
3. (40.0, 60.0°), (120, 60.0°), 60 mm
4. $A(-26.0, 122)$, $B(108, 62.5)$, $C(92.9, -83.6)$ $D(-50.8, -114)$, $E(-124, 13.0)$ $AB = 147$ mm, $BC = 147$ mm
5. 59.8°, (62.0 mm, 344.4 mm), (263 mm, 95.8 mm)

8.24
1. (a) 2, −1, (b) −1, 4, (c) 1, 1, (d) 3, 0, (e) −1, 0, (f) 0. 2
2. $F = 0.2W + 0.4$
3. $V = 24/I - 1$
4. $R = 0.5\theta + 50$

8.25
1. −3.6, 2.1, min −16.1 at x −0.75
2. (a) 6.25 MN/m^2 at $x = 4.5$ cm
   (b) 2.0 cm, 7.0 cm

8.26
1. $2.15 \text{ V}/\mu\text{s}$

8.27
1. $47.1 \sin(\theta - 24.2°)$
2. $9.27 \sin(\theta - 27.2°)$
3. (a) 2.4 m, (b) $2\pi/5$, (c) $5/2\pi$ Hz, (d) 5.7° lagging

8.29
1. (a) (−1, −1), (b) (−2, −3)
2. 0.72, 1.43
3. (1, 0), (3, 2)
4. (3.2, 17), (−0.79, 27)

8.30
1. $\sigma = 3.1x^2 - 7.3x$
2. $R = 42/I + 45.4$
3. $T = 0.94L^2 + 2.12$, 10.58 kN m
4. $A = 5r^{-1.5}$, 3.80 cm^2
5. $\theta = 500 \times 2^{-t}$
6. $i = 3e^{-0.4t}$

8.31
1. 10.2 kJ 2. $10{,}400 \text{ mm}^2$, 0.373 kg
3. $22{,}700 \text{ mm}^2$ 4. 21.6 coulomb

8.32
1. (a) $10x^2 - 2$, (b) $2x + 1$, (c) $e^{0.2x}$, (d) $7\cos x$, (e) $1/x$, (f) $-2e^{-2x}$, (g) $-3\sin 3x$, (h) $5/x$
2. $0, 0°$; $4, 76°$; $-4, 104°$
3. $0.7e^{0.1x}$, $0.07e^{0.1x}$
4. 2.79
5. 1.33
6. −3.3

8.33
1. (a) 720 m, 160 m/s, -10 m/s^2
   (b) 0, highest point
   (c) 2000 m
2. 0.47 A
3. 36 A/s
4. (a) 134 mm, (b) −2,990 mm/s, (c) 320 mm, (d) 3,000 mm/s
5. (a) 22 m, −4 m/s, -2 m/s^2
   (b) 1.33 s and 3 s,
   (c) 2.17 s

8.34
1. (1, 4)
2. Maximum (−0.33, 0.70), minimum (2, −12)
3. 5 s, 125 m
4. $15.8 \times 15.8 \times 15.8$
5. radius 0.683 m, height 1.365 m, area 8.79 m^2
6. 0.21 m, maximum volume 0.192 m^3

Answers to unit tests

Question	Unit							
	1	2	3	4	5	6	7	8
1	(a)	(c)	(c)	(c)	(b)	(a)	(a)	(b)
2	(b)	(a)	(d)	(b)	(c)	(d)	(b)	(c)
3	(d)	(d)	(a)	(c)	(d)	(c)	(b)	(b)
4	(b)	(b)	(c)	(b)	(d)	(d)	(c)	(b)
5	(d)	(c)	(c)	(a)	(c)	(b)	(a)	(b)
6	(c)	(c)	(d)	(c)	(a)	(a)	(c)	(b)
7	(a)	(b)	(b)	(a)	(c)	(a)	(d)	(a)
8	(a)	(a)	(a)	(a)	(d)	(d)	(d)	(b)
9	(c)	(b)	(b)	(d)	(b)	(a)	(c)	(c)
10	(c)	(d)	(b)	(d)	(d)	(d)	(d)	(c)
11	(b)	(d)	(a)	(d)	(c)	(d)	(a)	(c)
12	(d)	(b)	(d)	(b)	(a)	(c)	(d)	(b)
13	(a)		(c)	(b)	(c)	(d)	(b)	(b)
14	(b)		(a)	(d)	(c)	(c)	(a)	(b)
15	(d)		(a)	(b)	(a)	(c)	(d)	(d)
16	(a)		(c)	(c)	(b)	(a)	(b)	(c)
17	(c)		(a)	(a)	(c)	(b)	(a)	(c)
18	(d)		(b)	(c)	(a)	(c)	(a)	(c)
19	(c)		(c)	(d)	(d)	(b)	(d)	(a)
20	(a)		(d)	(b)	(a)	(b)	(b)	(b)

Index

abrasives 76–7
acceleration 214, 215, 216
accelerator 115
accidents 187–94
accuracy 46–7, 218, 244
actuators 31–2
additives 101–2, 108
adhesives 71, 121
aesthetics of design 133, 141
algebraic operations 247, 250–51
alloys 94, 96, 97, 98, 99, 105–6, 119–20, 126, 128
American Iron and Steel Institute 143
ammeters 230
amplifiers 33
angles 257–66
 radian measures 272–3
 velocity 217, 273
annealing 72, 107, 116, 117–18
anodising 75, 97
arc welding 69–70, 120, 121
area *see* graphs; triangle
assemblies 81
assembly *see* joining and assembly
attributes 144
aviation industry 127–8, 168

balance, simple 209–10
beam 210–11
bending 63, 65, 102, 137, 211
bill of materials (BOM) 81, 158, 160
block diagram 23–4, 28–9
bond/bonding 76, 77, 121–2, 124
boring 56
boundaries (design) 145
Bourdon gauge 224
brackets, expansion of 251
brainstorming 146
brazing 120, 123
breakeven analysis 12–13
briefs, design 132–40, 141–5
buffing 75

capacitors 94, 229–30
carbon steels 106–7
carbo-nitriding/cyaniding 118, 119
carburising 118–19
cast iron 96–7, 115, 127
casting 61–2, 79, 114–15, 123
 see also die casting
catalyst 102, 115
centre-line average (CLA) 77
ceramics 102
cermets 103
change, rates of 284–5
charts 5–6, 151
chaser 55
checklists 139, 146–7
chemical engineering 167
chemical treatment 52, 74–5
circuitry 102, 150–51
 see also electrical systems
civil engineering 103, 126–7, 168
closed-loop 36–7, 41, 42–4
clothing *see* protective clothing
coating 76, 126
cold working 116–17
combinations, new 146
combustion engines 30
commerce *see* companies; trade
communication 5, 149
communications sector 173–4
companies 3–5, 7–10, 134, 142–3
composites 102, 128
compression 102, 122, 137
computer-aided design 155
concrete 103, 127
conductivity 78, 94–5, 112
conductors 233, 234
constraints
 design 137
 technical/operational 136–9, 143
consumer appliances 173–4
continuous control 43
contraction 115, 121, 128
control strategies 41–3
control systems 35–7, 40–41
converters, analogue to digital 34
conveyor systems 40
cooling 115–16
co-operative 4
co-ordinates, Cartesian/polar 264–6
corrosion 96, 97, 98, 112, 123, 125–6, 127
cosine *see* sine and cosine
costs 9, 10, 11, 12, 16, 19, 123, 133, 139, 142–3
 types of 16–18, 148

creep 125
crimping 67
current, electrical 225, 230
customer requirements 132–4, 144
cutting *see* screw-cutting threads

damping 45
'dead band' 43
degradation 113, 121
density and mass 108, 246–7
depreciation 16, 17, 18
design 9, 123, 125–6
 see also briefs; computer-aided design; constraints; environment and design
design portability 148
design solutions 145–8
diagnosis (service) 85
diagrams *see* communication; drawings/diagrams/sketches; graphics
die casting 61–2, 98, 115
differentiation 282–5
dimensionless quantity 213
direct labour 16, 18
dividing head 56
domestic appliances 98, 171
down time 9
drawing (process) 66
drawings/diagrams/sketches 80–82, 105–6, 144, 149–55
 see also flowcharts/diagrams
drills and drilling 59–60, 89
dynamic systems 214–20

economic sectors 2–3
economics, basic 164–5
economies of scale 12
economy and engineering 164–71
efficiency *see* workplace
elastic 109, 212
elastic limit 116, 212
electric motors 30
electrical engineering 168
electrical systems 225–33
electricity requirements 135
electrodes 120, 125
electromagnetic systems 232–7
electronics 138, 168
emergency systems 83, 87

emf 225, 235–6, 237
employment 16, 18, 166–7, 175–8
energy 110, 184–5, 200, 230
   chemical 25
   electrical 25, 138
   heat 221
   kinetic 23, 214, 216
   potential 214, 216
energy conversions 31–5
energy delivery 29–31
energy sourcing 197–9
energy utilisation 38
   see also work
engineering functions 8–10
engineering industry and technology 167–9, 170–71
   domestic sector 171–4
   and economy 164–71
   health sector 174–5
   see also environment
engineering subsystems 28, 29–35, 38–40
   functions 44–5
environment
   and design 139
   and engineering activities 195–6
   and manufacturing 199–200
environmental legislation 201–5
environmental stability 112–13
equations 250, 253–7, 276–7
equilibrium 117, 209
   diagrams 105–6
equipment see tools and equipment
ergonomics 132–3, 134, 141, 186
error 41, 42, 218, 219, 231, 244
etching 74, 105
European Union 171
evaluation 78, 86, 141, 147
expansion 95, 127, 128
experimental work, laws of 277–80
exploded view 82, 155
exponentials 249, 270–71, 284
expressions (mathematical) 251–2
extrusion 66, 96

factorisation 251
failure 123–6, 142
Failure Modes and Effect Analysis (FMEA) 142
Faraday's law 235, 237
fatigue 24, 124–5, 128
feedback from customers 8, 9, 12
feedback systems 39, 41
files 89
finance 7, 11, 12–15
finish see surface finish
fire precautions 83
first aid 83
fitness for purpose 141–2
fixed assets 18
Fleming's rules 234, 236
flowcharts/diagrams 85, 86, 151
fluxes 67, 120, 234–6
Fokker, Anthony 127–8
force 209, 217, 234
forging 63–5, 96
forming 54, 55, 65, 66
formulae, transposition of 252–3

fractions 250–51
fracture 124
franchise 4
frequency 225, 273
friction 100, 128, 209
function (subsystems) 44–5
functionality 23, 143–5

gain 33, 43, 44
galvanising 75, 98
Gantt chart 80
gas, inert 70
gas law 220–21
glass 102
glass reinforced plastic 102
grain structure 104, 114
graph paper 280–81
graphics 149–58
graphs 150, 151, 267–72, 275–7, 280–82
   solving equations 256–7, 276–7
   stress/strain 110
gravity die casting 62
grinding 76–8
gross national/domestic product 164, 165–6
guarding/guards 83–4, 135, 191

hacksaws 89
hammers 88
handling, mechanical 135–6
hardening 73, 116, 118–19
hardness tests 111–12
health and safety see safety; workplace
health sector appliances 174–5
heat 94–5
   see also measurement; temperature
heat affected zone 121
heat application 221
heat energy calculations 223–4
heat treatment 52, 72–4, 106, 123
honing 78
Hooke's law 93, 212
hot working 116, 117
hydraulic systems 25
hysteresis 43, 45, 94

impact test 110–11
indices 242, 247–8
induction 119, 235, 237
information 11–12, 143–5
inputs 23, 44
inspection (service) 85–6
investment appraisal 14–15
investment casting 62

joining/assembly 51, 67, 81, 119–22
joints 68, 71–2, 121–2

Kirchhoff's laws 228–9

lapping 78
lattice structure 116, 124
legislation see environmental; safety
length 209, 212
levers 89
lifting and carrying 135–6, 191
linear motion 214, 215–16, 217

liquid state 104, 105, 114–15
logarithms 248–9, 250, 271, 280–81, 284
lost wax process 62

machinery, safety 135
magnetic fields 232–7
magnetic materials 94, 106
maintenance 9, 84, 87–9, 148
make-or-buy decisions 13–14
management information systems 39
manometer 224
manufacturing see environment; processes
market research 8, 9, 12
marketing 7–8
mask 74
mass see density and mass
material removal 51, 58
materials (non-metallic) 99–102, 107–9, 115, 117, 121
   characteristics of 114
   properties of 78, 93–5, 123
   selection of 123, 126–8
   shaping of 51
   structure of 103–7, 107–9
   types of 95–103, 143
   see also metals
materials testing 109–13
matrix system 6
maxima and minima 285–6
measuring 218, 245
   current 230–31
   distance 212
   instruments 89, 218, 223–4, 230–31
   pressure 224
   temperature 223–4
   voltage 231
   weight 219
mechanical engineering 168
mechanisms 137–8
metallurgy 117, 122
metals 94, 114–15, 115–17, 119, 121
   structure of 103–4, 107
   types/properties of 94–9
micrometer 218
microscopy 104–5
milling 56–9
modulus of elasticity 99, 110, 213
moment 209
momentum 215
monitoring systems 35
Morse tapers 59
motion 214–18
motor industry 128, 168
motors, electric 30
moulding 65–7, 117
multimeter 231
multiplication 241, 247–8

net present value 15
Newton's laws of motion 215, 217
nitriding 118, 119
normalising 72–3, 107, 118
nuclear engineering 199
nugget 70
numbers 241–4, 247, 250–51

objectives tree 147
Ohm's law 226
on/off control 42–3
op-amp 33, 37
open-loop systems 36, 39–40, 40–41, 41–2
organisational structures 5–6
oscilloscope 231
output 23, 24, 148, 180
ownership patterns 3–5
oxidation 127
oxides 96, 97, 102

paints and painting 76
parts list 158–60
pattern 61, 115
payback 14
peak-to-peak value 225
performance 141, 147
performance variables 44
phase change 221, 222
planning controls 202
planning, production 80
plastics 100, 107–8, 112, 115, 121
plating 74–5, 96
pliers 88
point load 211
polishing 75–6, 104
pollution 196–200
power 214, 215
power supply unit (PSU) 29–30
preparation/pretreatment 121–2
pressure 224
pressure gauge 224
processes
    evaluating 78–9
    identifying 9–10, 23, 51–2
    production 9–10, 80–82
processing capability 145
processing methods 114–17, 123
projections 153
properties *see* materials
protective clothing 83, 189–90
prototype 9, 11
pulse counters 34
Pythagoras, theorem of 251

quality assurance 85
quality control 10, 11–13, 14, 133, 141
quantity (production) 133, 213
quenching 73, 118

radian *see* angles
radiation 38
ratios *see* trigonometric ratios
reaming 56
reference signal 42
regulations (services) 87

reliability 8, 9, 10, 14, 133, 142
repeatability 47
research and development 8–9, 11
resistance 94, 226
resistors in combination 226
resonance 46
resources (for design) 138–9
response time 46
rms value 225
rotary table 56
rubber 102, 108

safety 29, 123, 124, 148
    legislation 83, 134, 135, 186
    procedures 834, 867, 135, 136
    see also workplace
schedules (production) 80
schematics 82, 144, 158
screw-cutting threads 54–5
screwdrivers 88
sectioning 153
semiconductors 94, 102, 109
sensitivity 38, 44
service, performance of 85–9
service sector 2–3
serviceability 143
set points 46
shaping of materials 51
shareholder 3
shear force 122, 137, 211–12
signal modulation 34–5
signal-processing units 32–5
significant figure 244
sine and cosine 260–63, 283
sine waves 272, 273–5
sinter(ing) 103, 115, 122, 128
size decisions 133–4
sketches *see* drawings/diagrams/sketches
'slip' 99, 125
soldering 67–8, 119–20, 123
solenoid 233
solid state 104, 105–6, 115–17
span of control 5
spanners 88
specific heat capacity 221
specification 140, 147
spring balance 219
stability 29, 46, 95, 112–13
standards 126, 134, 142, 149
static systems 209
stock 10, 12
storage 10, 84
strain 127, 212
stress 93, 102, 114, 115, 123, 124, 125, 127, 128, 212
    see also graphs
surface finish 52, 75–8
symbols 150–51

tangent 218
tank cutter (trepanning) 60
techniques, specific 5278, 82
temperature 42–3, 73, 106, 220, 221, 222, 223–4
tempering 73–4, 118
tensile strength 97
tensile test 109–10
tension 122, 137, 212–13
testing, destructive 9, 109
tests, materials 109–13
thermal systems 220
thermocouple 31, 73, 224
thermostat 43
time 148, 214, 216, 273
time lag 45
Tinman's Solder 67
tolerance 128
tools and equipment, selection/maintenance of 56, 84, 87–9
torque 30
torsion 137
trade 3–5, 167, 169
transducers 31, 35, 42
transformers 236–7
trepanning 60
triangle, area of 263–4
trigonometric graphs 271–2
trigonometric ratios 257–60
turning 52–3, 53–4
turns ratio 237

vacuum forming 66, 117
variable costs 16, 18, 19
VDU workstations 134
velocity 214, 216, 217–18, 273
vibration 24
vice, machine 56
voltage 225, 231
volumes 245–6

waste discharge systems 37–8
waste management 202–3
waste products and disposal 185–6, 195, 196–200, 203–4
welding 68–71, 99, 120–21, 123
Wheatstone bridge 32
Whiteheart process 97
wiring 71–2
Wohler, Friedrich 125
work 25, 214, 215–16
workforce 16, 18, 175–7
workplace 134, 136, 177–8
    efficiency/productivity 180–81, 183–6
    safety/comfort/health 179, 181–3, 186–94
wrought materials 97, 123, 127